BOSTON STUDIES IN THE PHILOSOPHY OF SCIENCE

VOLUME XIII

LOGICAL AND EPISTEMOLOGICAL STUDIES IN CONTEMPORARY PHYSICS

SYNTHESE LIBRARY

MONOGRAPHS ON EPISTEMOLOGY,

LOGIC, METHODOLOGY, PHILOSOPHY OF SCIENCE,

SOCIOLOGY OF SCIENCE AND OF KNOWLEDGE,

AND ON THE MATHEMATICAL METHODS OF

SOCIAL AND BEHAVIORAL SCIENCES

BOSTON STUDIES IN THE PHILOSOPHY OF SCIENCE

EDITED BY ROBERT S. COHEN AND MARX W. WARTOFSKY

VOLUME XIII

LOGICAL AND EPISTEMOLOGICAL STUDIES IN CONTEMPORARY PHYSICS

Edited by

ROBERT S. COHEN AND MARX W. WARTOFSKY

D. REIDEL PUBLISHING COMPANY

DORDRECHT-HOLLAND / BOSTON-U.S.A.

Library of Congress Catalog Card Number 73–83557

Cloth edition: ISBN 90 277 0391 4
Paperback edition: ISBN 90 277 0377 9

Published by D. Reidel Publishing Company,
P.O. Box 17, Dordrecht, Holland

Sold and distributed in the U.S.A., Canada and Mexico
by D. Reidel Publishing Company, Inc.
306 Dartmouth Street, Boston,
Mass. 02116, U.S.A.

Printed in The Netherlands by D. Reidel, Dordrecht

TABLE OF CONTENTS

PREFACE

In this volume of the *Boston Studies in the Philosophy of Science*, we present a collection of articles on philosophical issues in contemporary physics. The principal domain of these investigations is quantum physics. There are also articles on questions in classical mechanics (Hooker), and relativity theory (Papapetrou and Stachel), as well as a monographic essay in evolutionary epistemology (Yilmaz), applying the conceptual and mathematical understanding of special relativistic quantum field theory to set forth a theory of the evolution and adaptation of perceptual structures. Finally, in addition, there are two essays on classical issues in the philosophy of nature – one, on types of continuity (Čapek), which suggests an analogy between the perceptual and the quantum domains; the other, on causality, the first translation into English of a minor classic in the philosophical understanding of modern physics, H. Bergman's well-known but little-read *Der Kampf um das Kausalgesetz in der jüngsten Physik* (Vieweg, Braunschweig, 1929). On the occasion of this publication, Professor Bergman has kindly contributed an introductory essay, 'Personal Remembrances of Albert Einstein'.

Of the seven essays on quantum mechanics, four are on quantum logic (Marlow, Heelan, Bub and Demopoulos, Van Fraassen), the last being a critical survey of various current proposals for quantum logics; the remaining three (MacKinnon, Stachel and Van Fraassen) are concerned with both the formal issues and the ontological commitments of quantum physics.

All but two of these contributions were originally presented as working papers to the Boston Colloquium for the Philosophy of Science during 1969–72. The exceptions are Van Fraassen's 'Labyrinth of Quantum Logics', originally read at the first biennial meeting of the Philosophy of Science Association (Pittsburgh, 1968) and, of course, Bergman's monograph.

A companion volume from the same period (*Boston Studies*, Vol. XIV) is devoted to methodological and historical essays on the natural and social sciences.

ROBERT S. COHEN
MARX W. WARTOFSKY

Boston University Center
for the Philosophy and History of Science
August 1973

HÜSEYİN YILMAZ

PERCEPTION AND PHILOSOPHY OF SCIENCE

ABSTRACT. It is argued that an evolutionary theory of perception and knowledge, and a perceptual philosophy of science is, at present, a realistic and useful way of looking at the whole cognitive endeavor of living beings including man. This theory conceives perceptions, concepts and theories as ways of adapting to the useful regularities of environmental niches, and further extending these niches by exploration. This may lead, at times, to readaptation to less complex niches, etc., but in the case of man the scientific means of observation and exploration seem to lead to ever greater expansion of his environment until, at least mentally, his niche becomes the whole observable universe. Man's problem of knowledge is thus immensely more complex and also unitary compared to organisms which fill special niches. The task is nevertheless made conceivable because of the few comprehensive, and overriding regularities encountered throughout the universe. In this essay the evolutionary epistemology is assumed, defended and restated and eventually is made the foundation of a partially deductive system from which, with the help of other information of a physical, environmental nature the properties of sensory, perceptual and logico-conceptual organizations are, in essence, inferrable. We hope that this reversal provides further insights into the nature of knowledge and proves to be useful in helping us to create new and better concepts and theories consistent with our own status as products of biological and psycho-social evolution.

Some years ago as a physicist working on quantum mechanics and general relativity, I started to think seriously on the nature of physical science and its theories. My purpose was, of course, to find some general and meaningful concepts through which I might understand the process of science and theory construction better, and hopefully develop some methods and strategies for the construction of new theories in the future. Such an activity, I presume, belongs to the dangerous province of the philosophy of science.

I am sure everyone considers such matters from time to time and my impressions of how to proceed were really quite nebulous. At first I seemed to feel that a thorough study of modern physical theories and the latest developments in particle physics and field theory might provide valuable clues. Gradually, however, I got discouraged from this view, mainly because these fields were still quite unsettled and rapidly evolving. If we are looking for clues of permanent value, I felt, we should be studying structures which have already evolved permanent features. By this criterion, however, we might just as well give up the whole of physical science as a topic to study since hardly anything in it is more than three hundred years old.

Apparently, then, we must study cognitive structures much older than modern science, yet exhibiting features somehow meaningfully resembling modern science. Possibly, a study of the elementary perceptual phenomena could be the thing to do. I felt that the phenomenon of color perception would be a convenient starting point in this respect because in its long evolution the human eye must have assimilated some permanent features of light and radiation in the environment. The sun is there for more than a billion years, practically unchanged. The structure of the earth and its atmosphere, the biological environment, and natural objects were essentially the same or very nearly so during the whole evolutionary history of color vision. It should then be reasonable to assume that the dire necessity of survival must have instilled into the structure of color perception of living beings some permanently relevant features of this environment. These features, whether conscious or not, should manifest in our daily lives as cognitive acts because such features would be part of an overall inner model we, as organisms, build regarding the environmental niches in which we live and survive.

Thus, if we hold color perception as a kind of unconscious and ancient science, its study might be revealing with regard to the philosophy of science. With intent to learn something about how to do science from elementary perception, I therefore started to study the method by which the human eye perceives colors and colored objects.

I soon discovered, however, that there were serious difficulties in the way. First of all, although a great deal was known about color vision in an experimental and physiological sense, no theory of color perception in the sense of well-formulated modern theories of physics existed. What passed as a theory was actually a description of some mechanism hypothesized to exist in the eye of some particular species. No overall functional or organizational concept, no environmental or evolutionary principle was utilized. True enough, a color space was constructed on the basis of psychophysical experiments performed since the time of Newton and adjusted to the mechanismic theories of Young, Helmholtz, and Hering, but no one seriously questioned why this particular model of color space had to be the one realized and no other. The response functions, transformation properties, signatures, etc. of this color space were not discussed in close relation to evolution and survival, hence lacked the qualities of optimality, economy and efficiency which, in our opinion, were to be among the desirable features of a physical theory. Furthermore, there seemed to exist some old, and as yet, unexplained phenomena in color perception, as evidenced by the excitement generated through the two-color projections of E. H. Land. In other words, to learn something from color vision in regard to the methodology of science was probably just as difficult as to learn it from any other field.

Somehow, this pessimistic conclusion did not discourage me. I felt intuitively that the difficulties were not insurmountable. Given enough time and some luck, I could perhaps produce a theory of color vision of the desired kind myself. Since we know a great deal about the physical nature of light, the spectral composition of the sun, diurnal and atmospheric variations, etc., we can, at least in principle, specify the physical environment of light. Again, at least in principle, we can identify the information carrying parameters of light distributions for the purpose of using them in a perception device. Now if we imagine an organism in this environment and assume that through long struggle for survival the organism is to evolve a fairly efficient perception device for its needs, then we can turn the problem into a mathematical one. This is because, given enough time for evolutionary forces to assert themselves fully, and given sufficiently versatile materials and biophysical processes of a statistical nature, any survival device should tend to become optimum. If we use this statement of optimality in the sense of a variational principle we should obtain as many equations as there are variables to be optimized. In other words, we have, at least in principle, a well-defined mathematical problem which can be solved (Appendix 1). The solution of this problem is expected to result in a functional theory of color perception, largely independent of mechanism, and, essentially valid for all species living in the same environment and survival conditions.

It turns out that a program of this kind can actually be carried out (Appendix 3). We shall only mention here that in the first stage it provides all the relevant features of what is known as the classical color theory. In a second step it provides transformation formulae which explain experiments of the kind E. H. Land popularized. Most interesting from a philosophical point of view is the fact that the transformation formulae carrying the perception of colors from one illuminant to another are formally the same as the Lorentz transformations of special relativity. In other words, color perception may be said to satisfy a relativity principle of essentially the Einsteinian structure.

In this new perspective, the theory of color perception can indeed teach us something with regard to the philosophy of science. First of all, elementary perception may not be so elementary, namely, there may be in its concepts and laws which are, in principle, as advanced as those of modern physical theories. Secondly, elementary perception may be considered as an unconscious, well-tested science of the same, perhaps even greater, importance than modern science. In fact, we may argue that color perception as a field of science is so essential to our species that the human eye had evolved the color space and its transformation formulae millions of years before Albert Einstein advanced a similar theory for the physical space. Conversely it becomes reasonable, perhaps even emperative, to regard advanced science and theory as an elaborate and extended perception device. With this we imply the simple, and, I am sure very old, idea of considering scientific instruments as an extension of sensory powers and the formation of concepts as something like the extension of percepts and so on. If such a possibility were seriously entertained, it would lead to a general view of contemplating the whole cognitive development, from the simple differential reactions of lower organisms to the most advanced theories of man, as a continuous flow of evolutionary adaptations. In this process man's natural environmental niche would gradually expand, at least in the sense of aided and extended explorations, until it becomes the whole observable universe and the progress of science would be similar to the continual evolution of perception devices from their primitive less complex forms to their improved more comprehensive stages of proficiency. This idea, in turn, would raise an interesting philosophical question as to how far one can descend into the realm of sensations without losing perceptual ingredients and how far one can ascend into the realm of modern physical theories and still find in them perceptual elements. It would further be possible to inquire whether and in what way this general conception of the cognitive process can help us to formulate general ideas and theories in the whole perceptuo-cognitive domain.

With regard to the first part of the question, it is interesting to note that the psychophysical power law of S. S. Stevens has recently been derived from two premises of perceptual origin (Appendix 2). In other words, at the level of psychophysics the sensory phenomena are still determined by perceptual principles. Consequently, a clear distinction between sensory and perceptual phenomena is not possible even at this level and the more reasonable attitude is, perhaps, to include psychophysics as part of the overall cognitive proposition. We may further point out that I applied the same method by which the color theory was formulated to the perceptions of human speech, to the perceptions of voice and music (Appendix 4), to the categorization processes, distinctive features, and to the reversible figures and patterns. The theories so obtained seem to stand up when confronted with experiments in their respective domains. Work in progress on the jnd and Weber's law, and also on the projection, completion, prediction and gestalt processes including some higher order processes appearing in linguistics are encouraging enough to suggest that in essence the method might be applicable to the whole sensory-perceptual field, including the mathematics and logic. In this connection we may perhaps note that apart from an inherent difficulty of parametrizing the physical stimulus, the treatment of taste, olfaction and tactile sensations leads to very similar structural and transformational properties. In fact, from the work of Pluchik on clinical psychology, and of Pauling on orthomolecular psychotherapy, it appears that even the subjects of emotion and personality might be similarly structured as a response to the chemical stimulus of the cerebral environment.

With regard to the second part of the question, it seems possible to derive advanced physical theories like special relativity through essentially the same methods, provided some basic regularities of the phenomena under study are known. For example, given the fact that motion is observed in terms of clock and ruler readings and that every physical system in motion exhibits both wave and corpuscular manifestations in every observation, we may ask for a minimally redundant kinematical description. We have come to the conclusion that (Appendix 5) this would lead to the special theory of relativity. We call attention to the essentially perceptual character of this statement. The clock and ruler readings are similar to psychophysical magnitudes of color theory and wave and particle concepts are similar to elementary percepts. They are manifestations, without necessarily being intrinsic properties of the physical systems. Finally, the persistence of these percepts in every observation is similar to the invariance of object-colors in color theory. In this connection we would like to mention also that David Bohm and the present speaker have independently argued the perceptual character of the special relativity theory on even more elementary grounds. Similarly, the problem of boundary conditions in general relativity and cosmology led Plebanski and others to formulate attitudes essentially perceptual in nature. The present speaker tried to overcome some difficulties in non-inertial frames and in general relativity by introducing new concepts of observation reminiscent of a perceptual process (Appendix 7), leading to a new approach to general relativity.

In quantum theory the Copenhagen interpretation regards science, including mathematics, as man's way of seeing reality and Niels Bohr puts great emphasis on mental and linguistic processes. In our opinion especially the statistical and probabilistic aspects of the quantum field theory contain perceptual ingredients (Appendix 6). In this connection we should perhaps point out that the whole concept of a field in physics might be a purely perceptual construct. Remember that no one has ever observed a field directly. In the last analysis all of our observations regarding elementary processes are of the nature of clicks in the Geiger counters, pointer readings on some measurement devices or spots on photographic emulsions. The fact that mentally a concept of field is interposed between every two observations, without requiring detailed observability of this field, seems to make quantum field theory a perceptual construct from the beginning. Similarly the concept of a point particle as used in physics, turns out to be more of a perceptual construct than something demonstrably real. No one has been able to ascertain that those clocks in the Geiger counters and those spots on the photographic plates imply point particles of the sort we would like to imagine. For such reasons the equations we write down on paper do not seem to describe nature but our information about nature. This is just what is expected from a perceptual construct, for we do not expect the inner model we build about the environment to be made out of the same materials as the objects of that environment. So to speak, both our perceptions and scientific theories seem to be of the nature of interpretations and projections but when these interpretations and projections become sufficiently consistent with nature to help us in our daily life and our scientific endeavor we seem to call them perceptions and scientific theories. In other words, it seems that advanced physical theories contain perceptual constructs conditioned by economy, efficiency, and consistency of description of information, so that it is not unreasonable to consider them as extensions of man's natural perceptual endowment (Appendix 8).

This general conception of the unity of our cognitive faculties ranging from elementary sensory phenomena to the most advanced physical theories is, in our opinion, quite satisfying, and gives credence to the movement created by Ernst Mach and others on the nature and the unity of science. The philosophy of science thereby implied is also very close to Mach's viewpoint of mental economy and convenience in retention, exposition, and application of knowledge although there are differences both in our firmer insistence on the evolutionary nature of epistemology as a selection retention process and in our greater freedom beyond the purely pragmatic, utilitarian conception of science. It is, to a great extent, operationally oriented but is not as stringent as the orthodox operational philosophy with regard to the requirement of a one-to-one correspondence between concept and operation. It appears that, if consistently beneficial, a concept of great perceptual utility and economical value would be incorporated as a theoretical construct without requiring the direct observability of that construct. This would happen even when the concept offers no other benefit than aesthetic reward, say, in the form of elegance and symmetry and in fact might take precedence over or lead to compromise with economy. There seems to be considerable methodological and epistemological value in such a conception of man's cognitive endeavor because, in the first place, it offers a range of relationships and analogies among various branches of knowledge for cross-breeding and interdiciplinary activity, and in the second place, it allows sufficient freedom to invent, organize and perfect new and imaginative approaches to conceptualization until the phenomena under consideration begin to make consistent overall sense within the human criteria that we are to set for ourselves. This author's endeavor has been, and still is, to first formulate an evolutionary theory of perception and knowledge, and a perceptual philosophy of science along these lines and then reverse the process to test the validity of such a contention. With the reversal we here mean the process of taking the evolutionary theory as a principle and deriving from it, in the light of environmental regularities, perceptual and scientific theories. We believe this activity ought to be one of the vital goals of the evolutionary epistimology.

In summary, there seems to be no doubt that perceptual powers of higher organisms, including man, are evolved for the purpose of survival and match rather closely to some of the more significant regularities of the external world. In a sense, it may be said that the living beings tend to build an internal model of their environmental niches so that their responses to the necessities of life may be sufficiently veridical to maintain their continued existence. As evolution proceeds to higher forms of social and psycho-social adaptation, new and more elaborate inner models are expected to develop, and, as it happened in the human case, the process might eventually transcend its original beginnings, leading to the formation of science and theory (Appendix 9).

Man's ability to communicate allows a comparison of the experiences which are independent of the individual inner feelings and emotions, and are, to some degree, objective. Furthermore, man's active, manipulative powers lead to the construction of tools and aids, and eventually to an enormous extension of sensory powers in the form of detectors, measuring devices, and experimental arrangements which explore the normally undetected and normally unseen aspects of the world. When the information uncovered by these new means of knowledge and extended environment becomes, through the process of verbal and cultural communication, the common intellectual experience and interest of so many human beings, higher and social forms of criteria for simplicity, objectivity, acceptability coherence and completeness become necessary. The fact that our mental powers are limited, and our lives brief, imposes on us further conditions of economy in description, retention, and application of knowledge, and eventually leads to formal axiomatic theories in which one tries to cover as large a portion as possible of the total experience with the aid of as few as possible of descriptive statements.

APPENDICES

In going over the manuscript the moderator pointed out that some of the background work is not readily available to readers who might be interested in examining them. He therefore suggested condensed expositions of some of the essential ideas in the form of appendices. We hope that the following short outlines from various articles and from a monograph to be published by the MIT Press serve this purpose.

1. Nature of a Perceptual Theory

It appears that some of the basic overall features of a perceptual faculty can be derived from the consideration of three main items; (a) The physical properties of the stimulus energy and its environmental distribution. (b) The needs of the organism in terms of individual and group survival. (c) Constraints arising from the availability, usability and compatibility of materials and processes. In other words, it seems that the environmental regularities, survival needs and engineering constraints ought to determine, to a large extent, the overall perceptual function.[1] One can formulate these ideas in a reasonably well defined mathematical form. First, from the knowledge of the carrier (stimulus energy) and its environmental distribution we can identify information bearing parameters, ξ, to be used. Then we can imagine at least in principle, a certain function or functional of these parameters which corresponds to a cost function for the survival of an organism. The integral of this function over the space of these parameters may be taken as the overall cost function. This would be an ideal cost which does not include the constraints coming from the material, engineering and metabolic considerations. If the materials are not fully available, usable, or compatible in a bio-engineering sense there will be some constraints which will have to be included. We may write the cost functional as an integral

$$A = \int L(u, \bar{u}, \xi)\, \mathrm{d}\xi + \sum_i \lambda_i A_i \tag{1.1}$$

where u's are functions of the information bearing parameters and $\bar{u}$'s are some control functionals. Now, according to the theory of evolution, one may argue that those organisms that were able to survive for a very long time, namely, the ones we still see around, had either optimized such a function or came close to (optimizing) it by evolving an efficient device.[2] The simplest form corresponds to omitting the constraints and this yields an ideal, normative theory free of material constraints. This ideal form is hardly ever realized in nature because either the materials and processes cannot be found with ideal characteristics or the infinite time necessary for the evolution of an ideal device cannot be assumed in practice. Nevertheless, the ideal case is theoretically very important because it allows us to have an idea of the influence of the material restrictions and the stage of evolution of the device that is actually found in nature. The principle of optimality assumes that the quantity A tends to a stationary value, preferably a minimum

$$(1.2) \qquad \delta A = 0, \qquad \delta^2 A \geqslant 0$$

where the material constraints are, in principle, treated through a method of Lagrangian multipliers or some equivalents thereof.

The assumption of optimality raises subtle questions of stability or absoluteness of the minima. This question cannot be settled easily but we can at least assure the 'relative hierarchy' of the minima by using the Darwinian idea of mutation. These mutations can be considered as large statistical fluctuations which explore rather efficiently nearby minima leaving only those minima which are very remote;[3] so remote that either the necessary mutations to carry the system there cannot occur with appreciable probability over the evolutionary history of the species, or else both devices will be developed and used by the organism in a complementary or alternative sense. In fact the latter might happen even in the context of the same minimum as a realization of parallel devices based on more or less different constructions. A similar hierarchy exists in passing from biology to perception and to science in which analogous and parallel devices of selection coexist.

A number of very general remarks may be made about such a formalism: (a) if the environment and the survival needs of two organisms are sufficiently close, and if the biochemical processes are sufficiently versatile, then the perception devices of two completely different organisms will have similar functional properties. A case in point is the lensed eye of man and the composite eye of the bee. Man and bee are very different organisms and their eyes are very different constructions, yet their color visions are functionally the same with slight accountable variations. (b) if certain variables ξ do not appear explicitly in the integrand of Equation (1.1) but only through a set of functions $u_i(\xi)$, then the integral is automatically invariant under the substitution of such variables. Such variables make up a geometrical background in which perceptions are represented in a relational manner. (This amounts to saying that perceptually we tend to adapt to overall regularities so that they no longer carry information. Attention is then directed completely to the uncommon and unexpected.) In essence, the substitutions in question introduce transformations under which the percepts are invariant (c) due to the optimal nature of the representation, perceptions tend to model external relational properties in a homological fashion. (d) In the case of multidimensional perceptual phenomena like color and speech the functions $u_i(\xi)$ tend to form a linearly independent truncated, orthogonal set which afford an expansion into a perceptual function space in the form of a series

$$P(\xi) = \alpha u_0(\xi) + \beta u_1(\xi) + \gamma u_2(\xi) \ldots \tag{1.3}$$

where $u_i(\xi)$ and ξ_i are interpretable as basis vectors and coordinates respectively. Since the integrand of Equation (1.1) is independent of ξ for such perceptions, the variation of A due to the variations of ξ can only transform $u_i'(\xi)$ into another set $u_i(\xi)$ such that A is stationary. This systematic transformation then becomes the expression of the perceptual adaptation property of the device of perception in question. We may write

$$u_i'(\xi) = \sum_k L_{ik} u_k(\xi) \tag{1.4}$$

where L_{ik} represents a transformation. Depending on the nature of phenomena this transformation can be unitary, orthogonal, unimodular, conformal, isometric, and so on. (e) In the case of short or ambiguous stimuli (for example the stop consonants or the Necker cube) decision pressure resulting from the urgency of life situations leads to categorical responses where there are only a small number of alternatives. (f) In this case $u_i(\xi)$ may now be considered as distinct percepts rather than the basis vectors of a continuum, and, α_i take on the meaning of probability coefficients and eventually lead to statistical equilibrium and stability considerations as a consequence of the process of optimization. (g) If the environment changes or radiation to new niches occurs new cost functions are in question so that some old structures may become deficient and new ones have to be evolved in their stead, (h) the material, engineering and metabolic constraints lead to many subsidiary conditions and compromises so as to mold the idealized theory into a more realistic one which will be a closer representation of the device that is actually found in nature.

One may point out that Equation (1.1) may be considered as a summary of what we already know from other sources and in itself not as very useful. We partly agree on this point since the cost function or functional is usually not easily determinable. On the other hand, we are not prepared to overlook its value because, it presents, at least in principle, all that one is trying to do in a condensed and unified form. The situation is essentially the same in higher physical theories where the principle of stationary action is a condensed formulation of what we otherwise know. Yet a theoretical physicist could hardly be persuaded to abandon it because it offers distinct advantages of mathematical and inferential economy, consistency, and convenience and suits his taste in the pursuit of comprehensive and frugal presentations of his topic.

The above principle as applied to perception may be conceived as a special case of a more general principle which applies in a succession of hierarchies to the field of biology as a whole. Examples of application of the general principle are many and range from the shape of erythrocytes (red blood cells) to the narrow size inside the bone of a bird. The so-called 'principle of optimal design' expresses these and similar examples in a convenient form, although its rigorous proof is not yet available.[4] The ultimate physical basis of such principles seems, however, missing from the arguments presented so far. For example, if sunlight were not received by the Earth it is easy to see that the eye would not evolve, but it is not easy to see why erythrocytes would not evolve. Yet we know that nothing will evolve on Earth without the reception of sunlight. On the other hand, the common impression that we are using sunlight energy to keep things evolving is not true. The Earth radiates away as much energy as it receives from the sun and on the average no energy is used up or stored. The driving force behind all happenings of a meaningful nature is not energy but a quality called negentropy. It seems conceivable to us that both the principle of optimality and its corollary, the principle of optimum design can eventually be derived from first principles by taking into consideration the ultimate cost, negentropy, in a proper way.[5] For the present purposes, however, there is no necessity to go to that level since perceptual phenomena appear only in the relatively higher organisms where more basic consequences of the general principle are already implemented. Thus, our optimization process represents a subtler, but nonetheless hierarchially second order statement which makes reference to more basic biological perceptual facts that are previously established. We may therefore hold Equation (1.1) as a principle in its own right although we know that it rests on things which themselves are to be optimized.

This presentation, in a sense, implies a hierarchy of cognitive and perceptual structures ranging from ordinary genetic evolution of the *DNA* alphabet to the evolution of scientific theories by a succession of natural selection-retention mechanisms. It also makes clear, however, that such hierarchies are not always clearly separable from each other and often they are not even discernible. If one pushes the hypothesis testing to perception it makes sense. It might still be defensible in psychophysics and detection theory but loses its meaning in cellular and virus levels and becomes thermal fluctuation and radiative transition. It really also loses its meaning in the furthest forefront of science where most of the fresh knowledge comes as a surprise without anybody making a hypothesis about it.

2. The Psychophysical Law

In order to be aware of the external world the higher organisms developed sensory elements which respond when suitable stimulus energies impinge upon them. These sensors are sufficiently structured and contain great many atoms and molecules in order to rise above the thermal fluctuations. A higher organism is a colony of lower organisms we call cells. In a sense the cells are more basic than the colony which was evolved to serve their survival. In their cooperative effort to survive together the cells underwent division of labor and some of them ended up becoming sensors. The sensors are then finite both in number and dimensions and the sensory representation of the physical variables must, of necessity, be coarser-grained than the physical variables. The latter might even be continuous. In the case of sensing a frequency ‘continuum’ by a series of resonators, e.g., it can be seen that if ω and ω' are close to each other to the extent of falling on the same sensor, their effect will be mixed additively, whereas, if they are separated by more than one sensor apart they can be sensed independently of each other (Figure 2.1). Furthermore, if we build the array so that they cover the frequency extension in certain overlapping ways a sensory space with a certain implication of continuity can be obtained. By neural interconnections the neighboring elements may then be made to

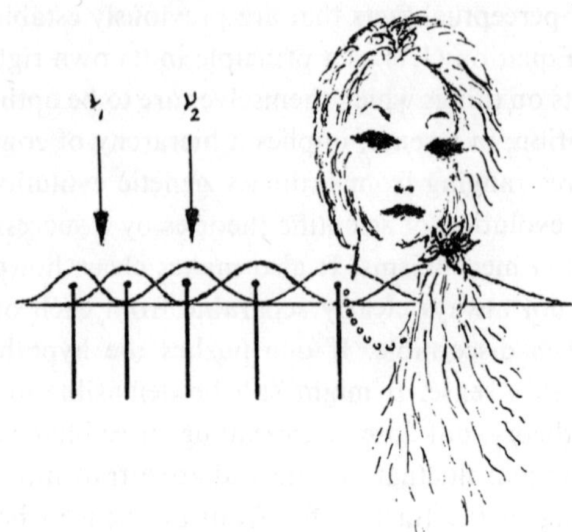

Fig. 2.1. Creation of a sensory space.

interact in a reinforcive or inhibitive fashion, for purposes of integration and discrimination, which further shape the behavior of such a space. In other words, a sensory variable, *s*, corresponding to a physical variable *i* can be constructed out of the cooperative outputs of sensors of finite range and dimensions. Our next problem would then be the form of *s* as a function of *i*.

In order to investigate this problem we now remember that as a consequence of the previous appendix the organism will be modeling some significant features of the external world in some sense of consistency and usefulness, and the model so obtained is required to be stable against the variations induced by other features and regularities in the environment. We may express such ideas in a more precise manner by two separate statements: (a) Perceptual organizations tend to model the environment in a homological manner, (b) Percepts tend to remain invariant under environmental changes.[1]

The first statement implies that, to a physical stimulus variable *i*, there corresponds a sensory variable *s*. In other words, variable *i* is imaged or represented by the variable *s*. The second statement implies that sensory variables exhibit a relativity property such that, when the values of a physical variable undergo a systematic change, the values of the sensory variable change in such a way as to leave the perceptual content of a situation invariant. This principle attributes perceptual significance to the relations among stimuli (i_1, i_2), (i_2, i_3),..., as well as to the magnitudes of the stimuli.

Many variables undergo systematic changes in typical perceiving situations. For example, light intensity may vary from high to low, thereby affecting all the intensities reflected from objects in the visual field by a constant positive factor K. Similarly, the intensity of speech may be high or low, depending on the speaker or the distance of the speaker. Again, the temperature of the room, the color of the illumination, the spectrum of an ambient noise, and many other factors may each add systematic variations to the physical variables to which the organism is trying to respond.

Let us consider, then, a relational physical property $f(i_1, i_2)$ that the organism finds it useful and survival-promoting to model in its perceptual response. The internal relational property $P(s_1, s_2)$, which may be called a percept, will be a representation of the external relational property. Omitting a scale parameter that depends only on units, we may write

(2.1) $$P(s_1, s_2) = f(i_1, i_2).$$

We next consider the second of the two statements. It implies in effect that environmental stimulus changes tend to make no difference to relational percepts. Accordingly, in the aforementioned type of multiplicative variation, $P(s_1, s_2)$ will be invariant under the substitution $i \to Ki$, that is,

(2.2) $$P(s_1, s_2) = f(i_1, i_2) = f(Ki_1, Ki_2).$$

The desired invariance will be guaranteed if, as one possibility, f is any function of the ratio i_2/i_1

(2.3) $$f(i_1, i_2) = f(i_2/i_1) = f(i_1^{-1} \cdot i_2).$$

Finally, in accordance with the first postulate, we note that $P(s_1, s_2)$ is essentially a homological model, or analogue, of the external relational property $f(i_1, i_2)$. Therefore, in order to maintain close correspondence between the external world and the internal model, the same invariance property under magnitude substitution $s \to \lambda s$ will tend to be realized. In other words, P is a function of the ratio s_2/s_1 of the internal perceptual variables. From Equation (3.1) we then have

(2.4) $$P(s_2/s_1) = f(i_2/i_1).$$

Let $i_1 = i$, $i_2 = i + \Delta i$, and similarly $s_1 = s$, $s_2 = s + \Delta s$, and write Equation (2.4) as

$$(2.5) \qquad P\left(1 + \frac{\Delta s}{s}\right) = f\left(1 + \frac{\Delta i}{i}\right)$$

with boundary condition $P(1) = f(1)$. Expanding by Taylor's series and keeping only the first-order terms (which is appropriate if Δi and Δs are small compared to i and s),

$$(2.6) \qquad P(1) + P'(1)\frac{\mathrm{d}s}{s} = f(1) + f'(1)\frac{\mathrm{d}i}{i}$$

where $P'(1)$ and $f'(1)$ are derivatives evaluated at $s_2 = s_1$ and $i_2 = i_1$ and therefore constants. By using the boundary condition and integrating, we obtain

$$(2.7) \qquad s = Ai^{\alpha}, \qquad \alpha = f'(1)/P'(1)$$

which is the power law of Stevens.[2]

Note that the derivation of his first-order law makes no use of the concept of the just noticeable difference (jnd) or resolving power. Nor is Weber's law required. The Δi in Equation (2.5) differs from a jnd in the same sense that a differential differs from a measure of dispersion. In other words, Δi is a difference, whereas a jnd is a measure of variability, scatter, or noise.

What would happen to the psychophysical law if the relational process concerned differences rather than ratios? Under the postulate of homological modeling, $P(s_1, s_2)$ cannot satisfy an additive invariance unless $f(i_1, i_2)$ also satisfies such an invariance. An additive invariance leads, by a parallel series of arguments, to the formula

$$(2.8) \qquad P(s_2 - s_1) = f(i_2 - i_1).$$

Then, by a similar procedure of integration, we arrive at a linear law

$$(2.9) \qquad s = Ai + B, \qquad A = f'(0)/P'(0)$$

which, interestingly, may be looked upon as a power law of unit exponent. Unit exponents are found in the perception of such continua as length, repetition rate, time duration, and the like.

Some general remarks may be made about the psychophysical law:

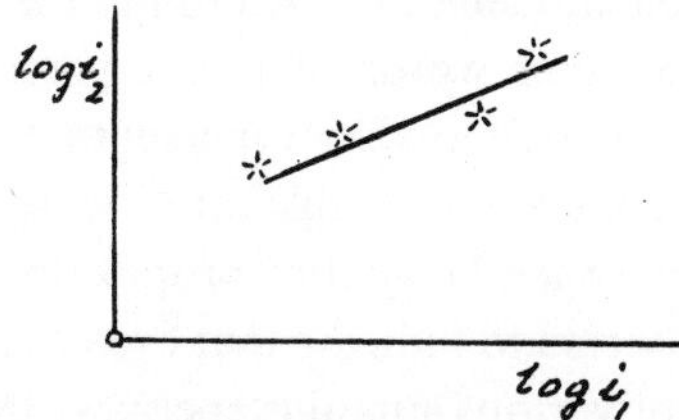

Fig. 2.2. Cross-modality under Steven's law.

(a) Two different sensory modalities can be compared directly (Figure 2.2) through the cross-modality formula

$$(2.10) \qquad \frac{\log i_1}{\log i_2} = \frac{\alpha_1}{\alpha_2}$$

extending the applicability of power law to unrelated continua. (b) The power law can be extended to multidimensional continua like color vision and speech perception

$$(2.11) \qquad |s| = A\,|i|^{\alpha}, \qquad W = e^{j\beta}$$

where s and i are vectors and $e^{j\beta}$ is an orthogonal transformation, j an operator and β an ordinary real or complex number. These forms are the basis of some of Edwin H. Land's two-color projections where a coordinate system defined on a log-log scale exhibits linear invariances[3]. (c) It also appears that whenever the underlying organization has the nature of a modeling and has relational invariances such as independence under large variations of environmental conditions a power law is realized. Thus, in relational biology a growing organism seems to obey a power law (the so-called allometric law of Huxley) in each of the variables along which the organism develops. Similarly, if emotions are functions of brain chemistry as sometimes hypothesized their magnitudes should be power functions of concentrations at least to a first order approximation, so on. Recent advances in clinical psychology, especially the work of Plutchik on emotions[4] and of Pauling on orthomolecular psychotherapy seem to be consistent with these considerations.

It is sometimes claimed that the sensory phenomena are the raw materials of higher perceptual organizations. The content of this appendix shows that this is not quite so. The perceptual principles apply even at the level of sensory phenomena and determine the behavior of psychophysical functions. In other words, it is not possible to draw a sharp distinction between psychophysical and perceptual events at the level of psychophysics and the more reasonable attitude is, perhaps, to conceive them as part of a more general cognitive proposition embodying both. In fact modern research seems to indicate that even at a level below psychophysics the detectors of various stimulus energies already come optimally structured through principles of a perceptual nature as well as the more basic biological and metabolic requirements.

3. Perception of Light and Color

As it became quite clear at the end of the previous section, it is possible to derive a theory of color vision by directly applying the formulae (2.16). All one has to do is to conceive color vision as a multidimensional psychophysic and to assume further that (a) Color space is three dimensional. (b) Brightness is positive definite. Then it can be shown in a simple way that there exists a color cone a cross-section of which is the usual color wheel, colors possess psychophysically the attributes of brightness, hue and saturation. Furthermore, it can be shown that there exist certain transformations under which object colors tend to be invariant and these transformations imply a kind of relativity in color perception. This strictly mathematical approach, however, has the deficiency that it obscures the nature of the elementary necessities that are involved and removes the contact with evolution to a secondary plane. For this reason we shall present the main features of the theory in a different way.[1]

In order to simplify the problem without losing its evolutionary flavor, we may first idealize the natural evironment by assuming that the normal illumination covers only a finite band of frequencies extending from ν_1 to ν_2 and contains constant energy in this interval. Then the sensitivity curve to brightness will have these same properties, namely, the sensitivity is constant in this band and zero outside. This is quite easy to understand because in a statistical sense the sum-total of all objects send to the eye essentially the same distribution as the illuminant and to optimize the response (see Equation (3.6)) the sensitivity will have to imitate the statistical distribution. Furthermore, we know from elementary color experience that the corresponding sensation is achromatic.

Next we consider, in a statistical sense, objects which have reflection properties that favor, within the illuminant band, higher, lower, middle areas, etc. To distinguish these objects from the achromatic ones and from each other further response functions will have to be considered. Assuming that these responses proceed evolutionarily from simpler to more complex functions and tend to avoid sensory overlap in their outputs, one concludes the new functions to be sine and cosine functions. This is because under very general conditions of continuity and of boundary in the range of $v_1 - v_2$ the simplest functions satisfying the above requirements are sine and cosine functions. In other words, the sensory representation of light distributions will be under such circumstances, a Fourier expansion

$$f(v) = \gamma \cdot 1 + \alpha \sin\phi + \beta \cos\phi + \cdots \tag{3.1}$$

where 1, $\sin\phi$, $\cos\phi$, ... are response functions in the interval v_1 to v_2, which is now represented by the angle variable $\phi \to 0$ to π and γ, α, β ... are the intensities (coordinates) of color experience. In this way color becomes a vector f in a multi-dimensional sensory space. Note that this sensory space, which is closed on a circle, is completely different from the physical space of frequencies which is linear. Thus the sensory variables may, at certain circumstances, form spaces which are even topologically quite different compared to the underlying physical variables.

There are various arguments indicating that the number of dimensions involved cannot be more than a few.[2] We shall only mention here that it cannot be infinite since this would require infinitely sharp detection mechanisms which a biological system cannot implement. Thus the series must cut-off somewhere. Consistent with the observed behavior of the normal human eye we take the cut-off at the third term. We can then formally represent colors as vectors in a three-dimensional, orthogonal system of coordinates. Furthermore, we must recognize that the space represented by these coordinates has an unfamiliar property compared to ordinary three-dimensional space, namely, the γ-coordinate has to be always positive owing to the positive definiteness of light intensity. This restricts the possible color experiences to those vectors lying in the upper half of the coordinate system. Actually the restrictions are even severer than this because if $\gamma = 0$ then α and β must also vanish as these latter arise from variations of intensity itself and cannot differ from zero unless there is some energy in the first place. This means that the vector f cannot lean all the way to become parallel to (α, β)-plane. There must exist a limiting angle beyond which no color exists. Since this would be true for every direction in this space, the human color experience exists only in-

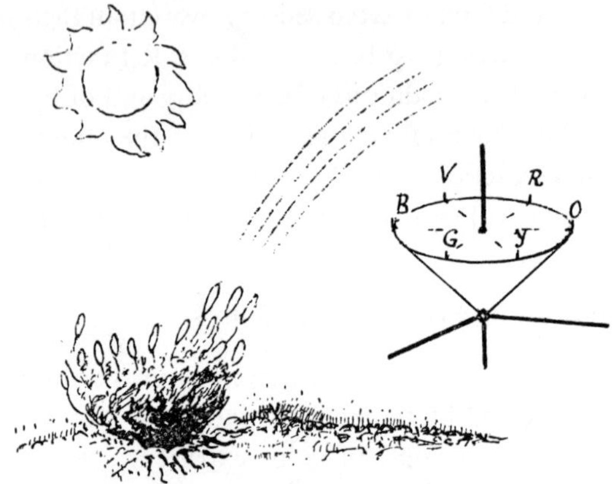

Fig. 3.1. Color cone and the color wheel.

side a limiting cone (Figure 3.1). Then it becomes evident that the psychological attributes of color roughly correspond to the conical coordinates

$$\begin{aligned} &\phi \rightarrow \text{hue}, \\ (3.2) \qquad &\sigma = \rho/\gamma \rightarrow \text{saturation}, \\ &\gamma \rightarrow \text{brightness, lightness} \end{aligned}$$

where $\rho = \sqrt{(\alpha^2 + \beta^2)}$. The cross-section at, say $\gamma = 1$ can be considered as a Newtonian color wheel. The maximum possible saturations $\sigma \rightarrow \Sigma$ are found at the periphery and satisfy the equation $\alpha^2 + \beta^2 = \Sigma^2$. In this manner we have obtained a simple theory which gives correctly the classical psychophysical patterning of our color experience.

Further thinking along the evolutionary lines leads to a very interesting concept of transformations which is consistent with the requirement (1.4) of the previous section. Consider that the illuminant light is changed into a composition which is colored relative to the original system while still containing all the frequencies in a fairly smooth fashion (so smooth that it does not require any higher function than the first three Fourier terms to describe it). In nature such changes of illuminants exists, as, for example the blocking of the sun by a small cloud which turns the illuminant from yellow sunlight to blue sky quality. Under such a change the object colors must remain invariant because if they didn't color denominations of the objects would lead to errors. To implement such an invariance property into the human color space, we assume that the eye performs a transformation of coordinates. Hence to each illumination we associate a reference frame (Figure 3.2) and require that all these frames

Fig. 3.2. Transformation in color space.

are related via the invariance of object colors. Let us consider two such frames and, omitting for the present the small shift of the origin from 0 to 0′, try to see what kind of relations are implied by such an invariance.

If the relations are at least approximately linear we can write

$$\alpha' = \alpha - \sigma\gamma \tag{3.3}$$
$$\gamma' = \gamma - \frac{\sigma}{\Sigma^2}\alpha$$

where σ is the saturation of the new illuminant as seen from the old reference frame. The first of these equations simply states that white will remain white in both frames. To see this, notice that $\sigma = \alpha/\gamma$ is the definition of saturation which here refers to the new illuminant as seen from the old frame and this particular case leads to $\alpha'=0$, hence achromatic. The second equation implies that in both frames pure colors correspond to pure colors. To show this let $\sigma \to \Sigma$ in the old frame and compute $\sigma'=\alpha'/\gamma'$ in the new one. We shall then obtain $\sigma' \to \Sigma$, namely, Σ is invariant. The linearity of the equations then guarantees the invariance of object colors in general in $\pm\,\alpha$ directions.

All these arguments depend so far on the ratios of these quantities and do not say anything about the intensities. How do we deal with different illuminant intensities? From what is said, we may write

$$\begin{aligned} \alpha' &= A(\alpha - \sigma\gamma) & \alpha &= B(\alpha' + \sigma\gamma') \\ \gamma' &= A\left(\gamma - \frac{\sigma}{\Sigma^2}\alpha\right) & \gamma &= B\left(\gamma' + \frac{\sigma}{\Sigma^2}\alpha'\right) \end{aligned} \tag{3.4}$$

where the second set is written down by analogy and for obvious reasons with a change of sign in σ. This analogy or symmetry of equations is the expression of equivalance of all natural illuminants, namely, any natural illuminant is as good as any other for purposes of color classification. This statement is independent of scales A and B. Assuming that sensory space has mathematical consistency we may then require these two sets of equations to be the solutions of each other. By substituting α and γ from the second set of equations into the first set and requiring $\alpha'=\alpha'$, $\gamma'=\gamma'$ we find

$$AB(1 - \sigma^2/\Sigma^2) = 1, \quad A(\alpha^2 - \Sigma^2\gamma^2) = B(\alpha'^2 - \Sigma^2\gamma'^2). \tag{3.5}$$

For $\sigma=0$ the meaning of this equation is a brightness (scale) transformation. A particularly interesting case occurs when both illuminants are equal in their respective brightnesses. Then $A=B=1/\sqrt{1-\sigma^2/\Sigma^2}$ and one sees that Equation (3.4) reduces to a form identical to Lorentz transformations. From analogy we may then conclude that, if the theory holds in second order approximations as well as in the first order, each observer tends to see the other room brighter by the factor $1/\sqrt{1-\sigma^2/\Sigma^2}$ (the grass on the other side of the fence is not only greener but also brighter). It is evident from the similarity of the transformations formulae that many of the concepts of special relativity are applicable to color theory. For example under the smooth changes of illuminant described above any two colors matching under one illuminant will still match under a different illuminant (invariance of metamers), etc. These and other consequences of the transformation formulae can be demonstrated by simple and inexpensive experiments anyone can afford to perform at home (Figure 3.3).

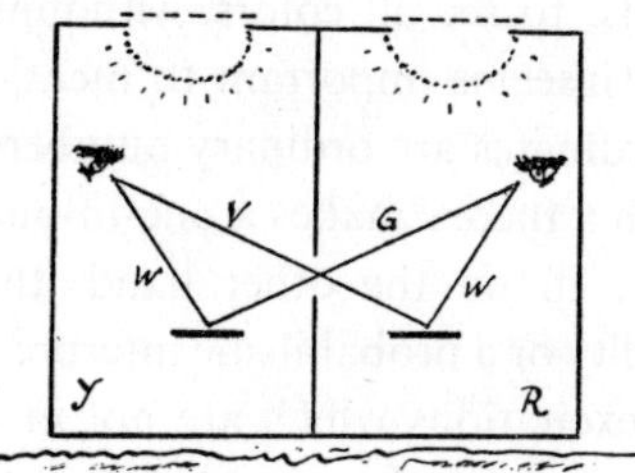

Fig. 3.3. Relativity in color vision.

In our opinion these formulae represent a modern relativistic viewpoint on the nature of color perception.[3]

The inhomogeneous terms are small in magnitude and seem to arise from the fact that from one illumination to another the length of the color vector changes in physical terms. To compensate this, an overall scale plus a homogeneous transformation is not sufficient. A small, properly chosen, translation helps the situation. Although small in magnitude the inhomogeneous transformation can be shown to lead to large qualitative consequences. For example, in some two-color projections publicized by Dr. Edwin H. Land one perceives all hues although on the screen only two hues say, a yellow and a red, exist. This phenomenon can be explained by saying that the plane defined by the two hues l_1 and l_2 is intersected by the new γ' axis which is not parallel to it due to the inhomogeneous shift. Then it is evident that the l_1 or l_2 plane will exhibit all hues as judged from the new frame. Of course, in such an arrangement some hues will be brighter than others and, due to the psychophysical power law, certain linear relationships will exist when these are drawn on a log-log paper, etc. Furthermore, if the natural counterpart of such projections is thought of as the illumination of a scene by the two colors of these projections it is possible to argue from evolution that the power function of α and β ought to saturate before that of γ 's. The result is that any sufficiently intense spot should tend to be perceived as achromatic in the normal eye. Thus in these projections, whites as well as grays tend to be represented because the color-cone tends to become narrower at high intensities. We shall not elaborate here on the averaging process defining the achromatic axis.

There are, of course, many other effects in color vision this simple, overall theory does not cover. One of these is the tendency, in the above-mentioned experiments, to see all colors. Although we feel this claim is somewhat exaggerated it seems important to mention its origin. A linear theory where the coordinates are ordinary numbers seems not to explain it. This is because such a theory makes a one-to-one correspondence with the physical stimulus. If, on the other hand, the coordinates have a property of multistability or a probabilistic interpretation, a given physical stimulus can lead to excitations which are not in one-to-one correspondance with the stimulus. In such a case the logical preliminaries of the theory would begin to diverge from the previous situation where the numbers are simply intensity coefficients and not probability indicators. In Appendix 4 we shall see an example of such a perception. In the case of colors, the desired explanation seems to be in the boundary conditions satisfied by a perceptual field at object or pattern boundaries. Hence, one must contemplate also a non-classical, field theory-like reinterpretation of color experience which we shall not go into here. We only mention that the work by Ratliff on Mach bands and figural interactions is indicative of the same direction of conceptualization. The idealized, normative theory presented above may further be modified into a more realistic structure through additional requirements of materials and engineering kind without altering its overall conceptual framework and topological properties. For example the natural illuminant $I(v)$ is more like a black-body distribution having a maximum at about the wavelength 550 Å instead of being a constant between some v_1 and v_2. With a normalization condition, $\int u_0(v)^2 \, dv = 1$, the response $\int I(v) \, u_0(v) \, dv$ would be optimum if $u_0(v)$ is the same as $I(v)$ apart from a multiplicative factor. This says that the sensitivity function $u_0(v)$ has also a maximum at about 555 Å, so on. But for a lensed eye like ours there is a further condition on $u_0(v)$ because of dispersion. Since shorter and longer wavelengths form images at different focal planes we would lose pattern resolution unless we narrowed the above $u_0(v)$ further. In this way we can see that $u_0(v)$ will be a Gaussian type function narrower than the illuminant $I(v)$ but peaked at approximately the same wave-length as $I(v)$.

$$(3.6) \qquad u_0(\nu) = e^{-\kappa^2(\nu-\nu_0)^2/2} = e^{-x^2/2}$$

where ν_0 is the frequency corresponding to 550 Å and κ is a numerical constant defining the width of the sensitivity curve. The next two functions $u_1(\nu)$ and $u_2(\nu)$ may now be inferred from the Gramm-Schmidt method to be

$$(3.7) \qquad u_1(\nu) = x\, e^{-x^2/2}, \qquad u_2(\nu) = (2x^2 - 1)\, e^{-x^2/2}$$

where $x = \kappa(\nu - \nu_0)$. When one plots these curves one finds that they are very similar, structurally, to sine and cosine curves. Furthermore it is easy biologically to construct $u_1(\nu)$ and $u_2(\nu)$ when $u_0(\nu)$ is given. Consider the slightly shifted forms $u_+ = u_0(x+\varepsilon)$, $u_- = u_0(x-\varepsilon)$. By linearly combining these and $u_0(x)$ we obtain

$$(3.8) \qquad u_1 = u_+ - u_-, \qquad u_2 = u_+ + u_- - 2u_0 .$$

In other words once $u_0(\nu)$ is evolved the organism can, in principle, construct u_1 and u_2, by neural summation from the response function $u_0(\nu)$ and its slightly differentiated forms, $u_0(\nu \pm \Delta\nu)$.

These three functions u_0, u_1 and u_2 are as they stand, very similar to Hering's curves of color sensitivity. Furthermore the Young-Helmholtz type of curves can be obtained from these functions merely by a linear substitution[4]. It is, of course, also evident that the transformation formulae like (3.4) can be incorporated into the more realistic form above without in any way altering the conceptual framework of the previous, idealized form.

4. Perception of Voice and Music

The most prominent feature of the voice producing organs, the vocal chords, is its reed-like operation. Sound produced by this action consists of a pulse series. The Fourier analysis of such a sound is always composed of a fundamental, equal to pulse repetition frequency, plus harmonics. The envelope of the Fourier spectrum is indicative of the shape of the vocal pulse and determines it up to phases. Such sounds with a fundamental plus its harmonics also come out of plucked strings, tubes and many musical instruments. Furthermore, many animals, including mammals, frogs, and birds, produce their sounds in a way similar to man and possess in their voices the same general structure, namely, a fundamental plus its harmonics. Since, evolutionarily, this is an important class of sounds, the human ear is expected to deal with it in a meaningfully special way by taking advantage of its common property (fundamental plus harmonics). In other words, the human ear will tend to perceive voice as a single percept and make something simple out of it, whereas the pure tone, which rarely occurs in nature, may be, to the ear, something more complicated in comparison.

A way to implement this requirement may be to assign a single pitch to the complex of harmonics and interpret the energy distribution over the harmonics (envelope) as a quality (timbre) of the source. The envelope properties may be interpreted as or similar to vowel or phoneme qualities, leaving us with the problem of the pitch of a voice pattern in general. This kind of pitch is called variously as the fundamental pitch, periodicity pitch or the residue pitch of the voice pattern[1].

We may now ask two different but related questions: (a) What is the pitch quality that is most useful to associate with a voice pattern as a cognitive determinant? (b) What is the sensitivity function (existence region) that is most useful to evolve for the perception of a voice pattern? The first question is fairly easy to answer because the differences between consecutive harmonics are always equal to the fundamental. This means that even if the distribution underwent severe distortion due to (vocal track) filtering, noise, masking, reverberation, etc., the periodicity (fundamental) pitch is still the same. Hence, the periodicity pitch is a most reliable determinant to seek for. Next question is how to extract this information. Here one's first impulse would be to postulate that in the ear some mechanical non-linearity has to be involved. However, such a nonlinearity would give the percept different relative magnitudes at low and high amplitudes, disappearing altogether at very low, overall intensities. For example a very weak voice of a lost child would lose this valuable determinant and mother could not differentiate it from other environmental sounds. Since this sort of construction defeats its own purpose, one should look for other ways of doing the job.

In an abstract sense, one can imagine a great many sensors covering the whole audible range and then neurally connecting consecutively all those which correspond to multiples of a given frequency, ν. Then the responses derived from each of these series may be conveyed to some common neurons which will report to the brain. Depending on which neuron reports, one would have a sensation of the periodicity pitch and the whole process would be linear. Thus, we may consider the harmonic series

$$(4.1) \qquad 0, \nu, 2\nu, 3\nu, 4\nu, \ldots$$

as a single percept, the fundamental, ν, being a determinant. Now consider for example three consecutive harmonics $f_+ = (n+1)\nu, f, f_- = (n-1)\nu$ (such a triad is easy to produce by amplitude modulation techniques). We shall have a periodicity pitch, p

$$(4.2) \qquad p = f/n.$$

In other words, even if the fundamental and all other harmonics except those three are missing, we shall still hear the fundamental pitch (Figure 4.1). Likewise, two of these consecutive multiples will tend also to excite

Fig. 4.1. The case of the missing fundamental.

the fundamental pitch. Such phenomena are in fact observed and are referred to, in the literature, as "the perception of the missing fundamental." What would happen if the above triplet is displaced upward in frequency without changing the difference? Since the detector neurons will have an overlapping set of resonance curves, the complex would now excite the intermediate neurons belonging to slightly higher frequencies such that the above relations would still be valid, but the pitch linearly shifted until the displacement of the complex is so large that the next consecutive set starts becoming realized. Then there would be a pitch-jump back to ν. In other words, there would be a set of discrete pitches p_n associated to a given complex (categorization). Which pitch will be perceived would then be only a statistical matter. Since this would manifest as a fluctuation of pitch (if rapid enough) the sound of such a complex would be heard somewhat wobbly even when it is physically quite stable.

Such phenomena are also observed to exist in the perception of voice pitch[1].

It is interesting to point out that these considerations are based entirely on a frequency domain and therefore preserve the place-theory concepts. Somewhat similar constructions can also be advanced on a time-domain basis. For example, when the phases are arranged properly, the complex will exhibit pronounced maxima corresponding to the periodicity pitch (envelope) and this could be extracted by a peak picking mechanism[2]. The difference between the two mechanisms would then be a dependence on phase. Indeed, there exists a phase dependence on the perception of pitch, namely, when in phase the complex is heard somewhat harsher and slightly lower in pitch and differently colored in quality but the main formula $p=f/n$ still holds. Since natural voice possesses similar characteristics one would be tempted to assume the existence of a volleying mechanism in the ear. However, there are secondary phenomena which are not explainable by either of the two constructions. For this reason, we shall concentrate below on a more abstract method which is reminiscent of a quantum theory for fields. Obviously, some peripheral or neural mechanism or both must be responsible for what we perceive, but presently we are interested more in the requisite function than in making an identification[3].

Let us consider the series (4.1) as analogous to a problem of field-quantization and write an operator, F_ν, for the frequency, f

$$(4.3) \qquad F_\nu = a_\nu^+ a_\nu + \Lambda (a_\nu e^{2\pi i\nu t} + a_\nu^+ e^{-2\pi i\nu t})$$

where Λ is a coupling parameter and a^+, a_ν are operators satisfying the relations

$$(4.4) \qquad a_\nu a_{\nu'}^+ - a_{\nu'}^+ a_\nu = \delta_{\nu\nu'} .$$

We can construct a state vector $|n\rangle$ such that the operator $N_\nu = a^+ a_\nu$, will produce as its eigenvalues, the positive integer values of Equation (4.1)

$$(4.5) \qquad N_\nu |n_\nu\rangle = n_\nu |n_\nu\rangle , \qquad n_\nu = 0, 1, 2, 3, \ldots .$$

Then for $\Lambda=0$ the result of Equation (4.2) is recovered. For $\Lambda\neq 0$ the relation $p=\nu=f/n$ will be modified slightly but, as we have seen, there is a condition on Λ, namely, for integer values of $\delta f/\delta\nu$ the influence of Λ must disappear. Furthermore it is expected from evolutionary arguments that the effect will not depend on f. An elementary way to satisfy these conditions is to assume the form

$$(4.6) \qquad \Lambda = A(\delta f - n\,\delta v)^{\alpha} f^{\beta}.$$

A straightforward perturbation treatment will then give, for the operator F_v, the expectation value

$$(4.7) \qquad \bar{F}_v = \langle n | F_v | n \rangle = f + \frac{A^2 n^2}{2f^2} (\delta f - n\,\delta v)^{2\alpha} f^{2\beta}.$$

From the arguments given previously there are good evolutionary reasons for the linear dependence of pitch sensation on the frequency variables. This means that we must set $\alpha = \frac{1}{2}$, $\beta = 1$. Hence the expected pitch sensation becomes

$$(4.8) \qquad p = \frac{f}{n} + \frac{A^2 n}{2} n(\delta f - n\,\delta v)$$

where experimentally, $A \simeq 0.04$–0.05. In other words, there will be two additional phenomena quite strange in behavior: (a) the pitch will appear slightly higher than f/n when f is shifted upward by an amount δf, (b) the pitch will appear to drop slightly compared to f/n when the difference frequency v is increased by an amount δv. Such phenomena were indeed observed some years ago by Ritsma, Cardozo and Schouten. These two effects, called collectively as the 'second effect' of pitch, appear to be virtually impossible to explain through assumptions of a mechanical nonlinearity or neural vollying. To our knowledge the abstract approach sketched above seems, so far, the only theoretical structure which offers a possible theoretical explanation. Furthermore, the appearance of natural integer numbers, fluctuations and equivocations implicit in such a theory, seem to provide a reasonable foundation for a statistical theory of music or at least the western music, with a better understanding of harmony, melody, dissonance and rhythm. Of course here we do not mean that the whole aesthetic scope of music which is largely a cultural phenomenon is bound up with this elementary subject of residue perception. We do mean, however, that the tonal preliminaries of music seems to have evolved under the influence of voice mechanism and therefore bear the marks of the residue theory. Recent studies, for example, of Ann A. Pierce (Brandeis University thesis, 1958) bear out this point.

If such a quantum field theory analogy in the perception of pitch were ever taken seriously, then the second question we asked previously might be answered as follows: consider the statistical distribution of pitches over the populace as occurs in the everyday life. For males this will be a black-body type distribution peaked around 100 Hz.

$$(4.9) \qquad \bar{\nu} \simeq \frac{\nu}{Be^{\beta\nu} - 1}.$$

For example if one takes $B=2$ one finds $\beta \simeq \frac{1}{190}$ Hz^{-1} in order to have a peak around 100 Hz. This distribution practically vanishes for $\nu \simeq$ $\simeq 700$ Hz so it is a reasonable estimate. Now the number of harmonics implemented for residue perception at a given value of ν will tend to be approximately proportional to this distribution so that

$$(4.10) \qquad \bar{n} \simeq CM \frac{\nu}{Be^{\beta\nu} - 1}$$

where $C \simeq 0.55$ will determine the existence region of the residue[5] for a modulation depth $0 < M < 1$ (Figure 4.2). Comparison with the results

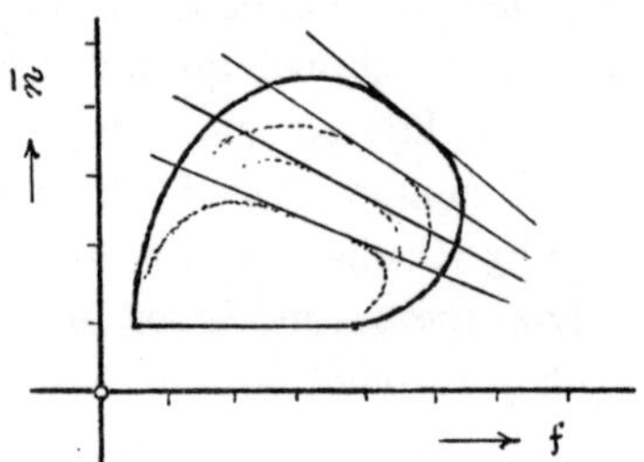

Fig. 4.2. Existence region of the residue.

of experiment show that this is indeed so, although of course somewhat variable from one individual to another. Variations, however, are not large and for some individuals, the formula is quite accurately realized.

We must confess, however, that this theory of residue pitch has always been criticized as too abstract, and possibly a wrong interpretation of something which might be the result of a very simple mechanism. I have no doubt that this is a valid criticism in the field of sensory phenomena where much of our traditional thinking is tied to mechanism. But I think I will never regret having formulated it because it was essentially in the course of pondering how the human ear could possibly overcome the problems of divergence in such a theory that the statistical generalization of the Lorentz frames and the idea of the statistical field theory presented in Appendix 6 were conceived. It does not make too much difference if those theories themselves prove to be not quite correct (is there any such thing as a correct theory?) because it was a most interesting instance where a purely perceptual phenomenon suggested something reasonable to try in the realm of the most advanced theories of physics.

5. Theory of Space and Time

It is a common observation that when viewed from a speeding car the trajectories of rain drops appear tilted. Conversely if the trajectories are slanted, there exists a suitable motion of the car for which the trajectories become vertical as observed from the car. In classical mechanics such observations are explained by the Galilean transformation

$$(5.1)\qquad \begin{aligned} x' &= x + Vt \qquad & x &= x' - Vt' \\ t' &= t & t &= t'. \end{aligned}$$

Now imagine that we were dealing with 'light waves' instead of rain drops. Intuitively we would hold the rays of light as analogous to the trajectories of rain drops and expect a similar effect. In 1728, only one year after Newton's death, the expected effect (called aberration of light) was indeed observed by Bradley. To derive the amount of the tilt as a function of the velocity of the vehicle we may proceed as follows: Let the light rays fall at an angle α to the ground. If, in the moving vehicle, the rays are perpendicular to the floor, then applying the Galilean transformations the relation ought to be

$$(5.2)\qquad \sin\alpha = \frac{V}{c}$$

and this is, experimentally, the correct result. However, this extremely simple, almost self-evident, derivation contains a serious contradiction which we now consider.

The contradiction arises from the wave conception of light. To see the problem remember that rays of light are defined mathematically as perpendicular trajectories to wave fronts. Let in one frame of reference the rays be perpendicular to the floor. Then in this frame every part of a wave-front is falling to the floor simultaneously. But, according to the Galilean transformations, this statement would still hold if one views the wave from a moving vehicle. The reason for this is that in the Galilean transformations (5.1) time is not affected by the motion. It follows that, according to the Galilean formulae, the rays, defined as perpendicular trajectories to wave fronts, should still remain vertical to the floor in every frame of reference and no tilt ought to be observed. It seems therefore that either we must reject the wave nature of light and assume light to be composed entirely of mechanical particles, or else revise the Galilean transformations of mechanics in such a way that the above tilting of the rays as well as the particle trajectories becomes derivable[1].

Confronted with such a choice a particle physicist of our era, especially the post-1955 breed, would be at a loss. For after the advent of axiomatic *S*-matrix theory, it became increasingly clear that the field conception might not at all be necessary for the description of fundamental processes. In fact, no one has ever observed a field directly. In the last analysis, all of our physical information comes from particle and particle-like manifestations of elementary systems. On the other hand the wave nature of light must be considered fully established. Starting with Huygens' (1678) concept of waves as a basis, an impressive array of experiments ranging from Newton's rings (1668) and Young's double slit experiments (1800) to the production of electromagnetic waves and the diffraction of X-rays supporting the wave theory, were explained. To a pre-*S*-matrix physicist, these experiments established the wave nature of light beyond any reasonable doubt. If we were to think and act like a classical physicist, we should therefore be inclined to accept the reality of the waves and proceed to modify the Galilean transformation.

The modification in question demands that if the rays, like the trajectories of some imagined particles, are, in any way changed in direction, then there must exist a corresponding slant of the wave front in the moving frame such that rays remain perpendicular to wave fronts in every frame of reference. This condition amounts to saying that in the frame where the rays are tilted the corresponding wave fronts do not fall everywhere simultaneously to the ground, but that there is a time difference $\delta t = d/c = Vx/c^2$ between the origin and the point x (Figure 5.1). Hence we must modify the Galilean formulae, at least into

$$(5.3)\qquad \begin{aligned} x' &\to x + Vt \qquad & x &\to x' - Vt' \\ t' &\to t + \frac{V}{c^2}x \qquad & t &\to t' - \frac{V}{c^2}x' \end{aligned}$$

in order for the experimentally observed relation, Equation (5.2), to come out correctly. Here the reverse transformations are simply written down from symmetry. This symmetry is the expression of the equivalence of the two reference frames, which is the basis of the mechanical relativity.

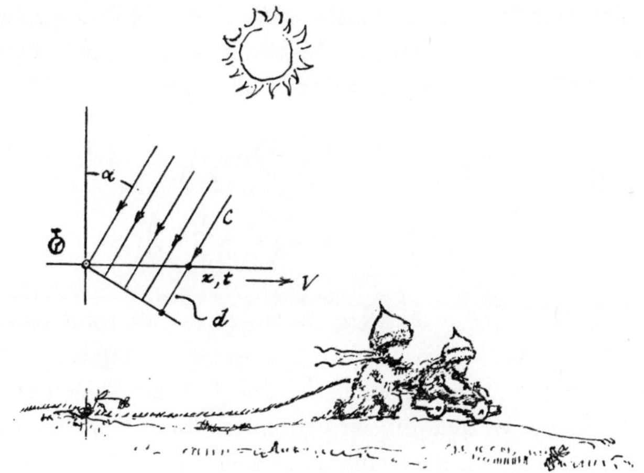

Fig. 5.1. Interpretation of Bradley's discovery.

It simply means that any reference frame is as good as any other and there is no way of singling out one over another. In modifying the transformations to accommodate the tilting of rays, we certainly do not want to lose this equivalence which is one of the most important principles of classical physics.

We note that the modified transformations (5.3) already contain the principle of constancy of the velocity of light. To see it, let $x/t=c$, which means that we are considering a point moving with velocity of light in the unprimed frame. Then by calculating $x'/t'=c'$ in the primed frame, one obtains $c'=c$. This shows that Einstein's 1905 statement of the constancy of the velocity of light could be replaced by the statement:

"light rays and their wave fronts are orthogonal in all inertial frames of reference." Thus through the identification of a light ray as a kind of particle trajectory this apparently innocent reinterpretation of Einstein's second postulate makes direct contact with Bradley's 1728 discovery and implies a wave-particle duality which is essential for the quantum theory of light[2].

Despite these interesting features, the formulae (5.3) still contain a serious deficiency, because, unlike the Galilean transformations, the second set is not the solution of the first set and vice versa. This is of course a mathematical inconsistency and must be overcome before we can ever take these formulae seriously. The solution of this problem turns out to be quite simple although the result is somewhat unexpected. Suppose in Equation (5.3) we multiply the right hand side of each equation by a factor A

$$\begin{aligned} x' &= A(x + Vt) & x &= A(x' - Vt') \\ t' &= A\left(t + \frac{V}{c^2}x\right) & t &= A\left(t' - \frac{V}{c^2}x'\right). \end{aligned} \tag{5.4}$$

The factor is the same everywhere because: (a) we want to keep the orthogonality of rays and wave fronts within each frame. (b) We want to preserve the symmetry between the two frames. Then treating this factor as an unknown quantity we can determine it from the condition that the two sets will be the solutions of each other. This mathematical consistency requirement gives $x = A(x' - vt') = A^2(1 - V^2/c^2)\,x$, hence

$$A = \frac{1}{\sqrt{1 - V^2/c^2}} \tag{5.5}$$

leading to Lorentz transformations. We note the fact that these transformations, considered from a purely geometrical point of view, could have omitted the requirement (b) above, namely, the symmetry between the two frames. If we denote a Lorentz transformation by the symbol $L_{\alpha\beta}$ this symmetry implies the orthogonality of the transformations

$$L^T_{\alpha\beta} = L^{-1}_{\alpha\beta}\,, \qquad \det L_{\alpha\beta} = 1\,. \tag{5.6}$$

The equation (5.6) is the expression of the existence of clocks and rigid rulers which can be used in both frames without recalibration. If we did not have this property the primed and unprimed frames could have different scales A and A' such that $AA'(1-V^2/c^2)=1$. The assumption of different scales of this sort goes, in a sense, beyond the usual special relativity and requires experimental evidence for its justification. Notice, however, that the wave-particle duality and the invariance of the velocity of light would not, in any way, be destroyed. Furthermore it is possible to argue that Equation (5.6) could always be satisfied, at least in a local sense, by utilizing measuring devices – clocks and rulers – which operate in terms of the same phenomena that are to be measured. As a consequence recalibration would not be necessary and the validity of (5.6) would be automatic. With these considerations in mind one could claim, with considerable justification, that the usual special relativity is the least complicated theory consistent with the idea of waves and particles and a requirement of their duality.

Further thinking along these lines, however, leads to great difficulties of interpretation. First of all the wave and particle concepts appear to a classically oriented mind, mutually exclusive, so that the assumption of their simultaneous presence for one and the same phenomenon is quite unintelligible. Secondly, as we have stated earlier, no experiment has ever yielded a direct observation of a wave. All of our information seems to come from particle-like manifestations on photographic plates, in geiger counters etc. On the other hand point-particles of a classically imaginable kind do not make consistent sense because even in a simple double-slit experiment the observed diffraction pattern is not explainable unless the particle has some volatile extension and a fundamental lack of identity to be able to pass through both holes simultaneously. Furthermore, a classically imaginable point particle of finite mass could not possess intrinsic angular momentum whereas experiment shows that many of the so-called elementary particles possess intrinsic angular momentum we call spin. Under such circumstances it is hard to attach to the useful concepts of waves and particles any tangible significance other than to say that they are useful. It is as though the human mind has abstracted these two concepts from gross macroscopic experience and later projected them into microscopic domain where, strictly speaking, they do not hold. Then it becomes all the more puzzling that special relativity should be derivable from these two concepts – waves and particles – and their unintelligible duality[3].

Our way of resolving this dilemma is to argue that special relativity itself refers to a collective macroscopic image of matter and represents a mental structure very similar to our perceptual faculties. Thus, for example, the concepts of space and time may arise just because of the presence and the motions of matter, and the properties of matter may be such that we can extract information only in a macroscopic statistical sense[4]. Then, in its endeavour to adapt to the information acquired, the human mind would create certain images of matter similar in nature to elementary percepts, and try to comprehend the situation in terms of these images. Wave and particle concepts of physics might correspond to such perceptual constructs. Within such a point of view the question of the actual existence of waves and particles would lose much of its import. Furthermore, the derivability of special relativity from these concepts and their assumed duality would no longer be surprising although the seeming incompatibility of the two concepts would continue to cause logical strains. The latter difficulty, however, could be attributed to the incomplete stage of our adaptation and be interpreted to imply further revisions, perhaps even to the extent of questioning the logic itself.

The special theory of relativity, as it is customarily understood defines a four dimensional metric geometry through the invariance of

$$s^2 = -x^2 - y^2 - z^2 + c^2t^2 = x_\mu x^\mu \tag{5.7}$$

where $x_\mu x^\mu \geqslant 0$ is a positive definite quantity. Although immensely appealing from a kinematical point of view this way of expressing the content of the theory creates a serious problem. It makes it logically acceptable that waves and particles can exist side by side and independently of each other in the same general space defined by (5.7). Furthermore it implies that this space can have a meaning all by itself even when no material objects are assumed to exist anywhere in the universe. We find these implications contrary to our simple derivation of Lorentz transformations from the inseparable presence of the wave and particle attributes and to one of our perceptual premises that the concept of space-time ought to arise on account of the presence of material particles. We shall therefore write the content of the theory in a different way, namely[5]

$$(x_\mu x^\mu - s^2)\, P = 0\,, \qquad x_\mu x^\mu \geqslant 0 \tag{5.8}$$

where P is some quantity representing the existence of a particle at x. The essential point here is that if $P=0$, that is, if no particle is present, (5.7) does not have to hold and the concept of a space-time geometry becomes logically meaningless in the absence of matter. In other words the space-time framework which is so ingrained even in our every day activities as so real and imposing is closer in nature to our perceptual constructs and does not have independent reality apart from the presence and the properties of physical objects. So to speak space, like color, exists only in the mind of the beholder.

We may put the idea into a sharper focus through the familiar definition of momentum $p_\mu = m\Delta x_\mu / \Delta s$, $p_0 > 0$, and rewrite Equation (5.8) as

$$(5.9) \qquad (p_\mu p^\mu - m^2)\, P = 0\,.$$

This equation shows that such a definition of momentum has no meaning when $P=0$. On the other hand when $P \neq 0$ we are in liberty to apply a Fourier transformation on the ensemble of solutions, $P(p)$, so that

$$(5.10) \qquad \phi(x) = \int P(p)\, e^{-ipx}\, \mathrm{d}^4 p$$

where p, and x are shorthand for p_μ and $x^\mu = (ct, x, y, z)$. This transformation converts (5.9) into a wave equation

$$(5.11) \qquad (\partial_\mu \partial^\mu + m^2)\, \phi(x) = 0$$

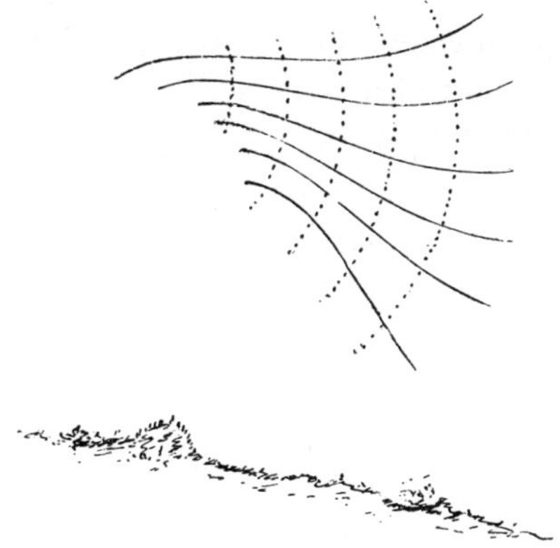

Fig. 5.2. Orthogonality of rays and wave-fronts.

as it may be checked by differentiation. Furthermore, it can be shown that the solutions of this equation satisfy the ray and wave-front relation postulated earlier (Figure 5.2) and by a correspondence implicit in (5.10), namely[6],

$$p_\mu = i\partial_\mu \tag{5.12}$$

our rays can now be thought of as families of particle trajectories. In other words, under the new definition (5.8) of space-time the wave and particle aspects may be thought inseparably present and their much confused duality becomes clearly explained by their essential relatedness and mathematical unity[7]. One may object that (5.9) might be introducing a space-time structure for each particle (configuration space) and complicating further the situation. This in a sense it does. But through the definition of momentum with a scale m we actually have (5.8) which defines, at least for non-interacting particles, a kinematical space-time background valid for all particles. Thus although in a weaker sense than (5.7) we can still maintain the idea of a general space-time framework. Figuratively speaking the space-time framework as represented by (5.8) may be thought of as a conceptual response of the human mind to the more physical energy-momentum behavior implicit in (5.9). In a sense Equation (5.10) defines (in terms of clock readings and rigid rulers) and explores what such a space can and cannot do. For example, for any physically detectable change in a system some transfer of energy and momentum is necessary and this induces, due to the definition of x_μ by (5.10), a change Δx_μ at the transfer region such that $\Delta p_\mu \Delta x_\mu \sim 1$. In other words Equation (5.8) already contains the uncertainty relations provided the definition of x_μ by (5.10) is interpreted to imply a wave-particle duality, (5.9), also for the objects that are the carriers of information (photons, mesons, etc.).

These equations also provide a particularly simple interpretation of the much confused quantum mechanical interference. Consider the case of the double-slit experiment with electrons (Figure 5.3). We argue as follows: According to Equation (5.9), which now replaces (5.8), the electrons with momenta p are coming from left. This momentum is sharply defined because the solutions of (5.9) are of the form $P \to A\delta(p)$. Likewise the solutions of (5.8) are of the form $P \to B\delta(x)$. Thus an electron will behave as a particle, $p^2 - m^2 = 0$, whenever observed, in the

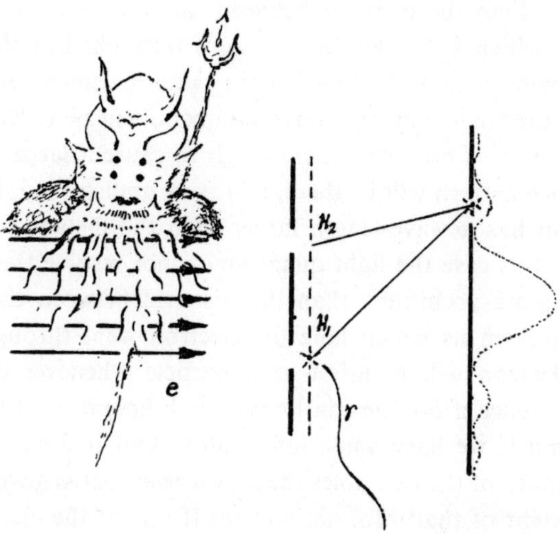

Fig. 5.3. Electronic interference at a double-slit.

sense that it will transfer a definite momentum if absorbed and this will happen at a definite point. However, the exact position this will manifest is not determined because (5.9) implies (5.8) only up to an integration constant, $x \to x+K$. We say that space-time is homogeneous and therefore for fixed momentum any position is as good as any other. Converse is also true, namely, if position is fixed then there exists a set of momenta which by virtue of the Lorentz transformation have equivalent members each of which is as good as any other. Now according to (5.11), which is a consequence of (5.9) and homogeneity (the factor e^{ipx} in the Fourier transformation is a representation of homogeneity) the electron satisfies a wave equation which states that as long as the sharply defined energy-momentum (frequency) is not disturbed the final observation will display interference pattern, albeit only in a statistical sense. However, if momentum is disturbed by an intervening event, say, the emission or scattering of a light quantum at H_1 then a transfer of energy and momentum Δp_μ to this light quantum has occurred. Equation (5.10) tells us that such a change must accompany a corresponding change Δx_μ in our space-time assignment so as to satisfy $\Delta p_\mu \Delta x_\mu > 1$. Three special cases are of interest: (a) If the light quantum is energetic enough to have a wavelength smaller than the distance between the two holes it will be able to inform us which hole the electron went through. But then Δp_μ will be large enough so that the interference pattern, which requires sharp definition of the frequency and wave numbers, will be destroyed. (b) If the light quantum is not energetic enough to cause a large transfer Δp_μ the interference pattern will be there, relatively undisturbed, but then the light quantum has a wave-length larger than the distance between the two holes. In this case the light quantum cannot resolve the position of the electron more accurately than the distance between the two holes and cannot inform us which hole the electron went through. In other words the electron will manifest as a particle whenever detected but display interference if no one can know which hole it went through. On the other hand if we have some information that it went through one and not the other of the two holes the interference pattern will be washed out to the extent of that information.[8] (c) If one of the electrons is disturbed in a systematic manner $x \to x + K$ where K is constant (Aharanov-Bohm effect) as in $\int (pu + A \cdot j)\,\mathrm{d}s \to p(x + \int (e/m)\,A\,\mathrm{d}s)$, then the interference pattern will only be displaced without being diluted or washed out.

The above viewpoint, apart from leading us straight into the heart of modern quantum mechanics, sheds strong light on a problem that has always been somewhat mysterious. We have in mind the so-called principle of causality. To present the argument with sufficient clarity we first point out that the Equation (5.11) reveals the need for a source term $\eta(x)$ such that

$$(5.13) \qquad (\partial_\mu \partial^\mu + m^2)\,\phi(x) = \eta(x).$$

The reason is simply that without such a term the field $\phi(x)$ itself can never be produced or removed since it would not interact, and would not exchange energy-momentum, with anything. Thus such an object could not be detected. The fact that we can detect the presence of fields, albeit indirectly, implies the necessity of such a term as part of the field equation (by Fourier transform, a similar term is necessary for the particle equation). The general solution of (5.13) on the other hand is of the form

$$(5.14) \qquad \phi(x) = \int \Delta(x, x')\,\eta(x')\,\mathrm{d}^4x' + \int \Delta(x, x')\,\overleftrightarrow{\partial'_\mu}\phi(x')\,\mathrm{d}\sigma'^\mu$$

where $\Delta(x, x')$ satisfies the fundamental inhomogeneous equation

(5.15) $(\partial_\mu \partial^\mu + m^2)\, \Delta(x) = \delta(x)$.

Equation (5.14) implies that $\phi(x)$ is a superposition of waves $\Delta(x, x')$ at x emanating from the source or boundary points x'. Now according to (5.8) the quantity P and by (5.10) ϕ and Δ can exist only if $x_\mu x^\mu \geqslant 0$. In other words we cannot accept all possible solutions of (5.15) as physical. The physically relevant solutions $\Delta(x)$ must satisfy

(5.16) $\Delta(x) = 0$ if $x_\mu x^\mu \leq 0$

which is an expression of the principle of causality. By multiplying with $m^2 > 0$ and using $p_0 = m x_0 / s > 0$ the condition can also be written $\omega_0 x^0 > 0$, $\omega_0 = m|x_0|/s$. In this form the causality condition appears as a consequence of the positive definiteness of energy and simply says that $\Delta(x)$ will contain positive frequencies for positive time differences and negative frequencies for negative time differences (Figure 6.1). It is equivalent to the 'outgoing-wave' or 'radiation' condition used in antenna theory.[9] Two essential points now emerge: First, the positive energy imposition may be thought to imply a substratum (call it vacuum) which is stable so that all waves and particles correspond to excitations over the lowest energy state of that substratum. Second, the propagator $\Delta(x, x')$ being independent of any particular field or particle, summarizes our knowledge of the basic space-time framework but so far there is no way of fixing the fundamental $\delta(x, x')$-function except that it is in some sense relativistically invariant. We shall come back to it in the next section. For the time being it is quite gratifying to observe how the perceptual approach has provided a connection between the positive energy requirement and causality and thereby reduced by one the number of fundamental principles of field theory.

We can see at once that a reformulation of physical theory along the lines of the present discussions could offer distinct advantages of simplicity, economy and scope. In the next appendix we shall carry the arguments one step further and try to show that some of the contemporary problems troubling quantum field theories could be prevented without violating any of the known principles of physics.

6. Statistical Theory of Fields

The considerations of the previous appendix seem to imply that despite the unobservability of the waves and particles in a direct and detailed fashion we must take these two concepts very seriously. In the following we shall regard waves and particles as certain macroscopic and statistically conditioned percepts of a mathematically meaningful kind and assume that they can help us in writing down possibly relevant equations on paper. These equations may then be solved and certain macroscopic or statistical inferences amenable to experimental inquiry may be extracted. Furthermore, it is quite possible that in manipulating these equations some general and physically meaningful statements independent of particular waves and particles may be identified, which then become the expression of a common kinematical framework on which individual waves and particles may be projected. Starting with Equations (5.9) and (5.11) as our bases we now notice that a Lorentz transformation carries a particle with momentum p into another particle of the same kind but with a different momentum p'. Likewise a plane-wave solution $\phi_k(x)$ of the same wave equation with propagation vector k is carried into another plane-wave solution of the same wave equation but with a different propagation vector k'. Thus if the wave and particle concepts are useful and reflect the logical completeness of some higher level theory the momentum vector p could legitimately be represented by the propagation vector k. Consider, then, a plane-wave solution, $\phi_{k_0}(x)$, of the wave equation belonging to a momentum eigenvector k_0. If from this we should reach a new momentum k by a Lorentz transformation, we must obtain a new plane-wave solution, $\phi_k(x)$. Suppressing tensor indices, etc., for simplicity, we may write

$$\phi_k(x) = L_{kk_0}\phi_{k_0}(x) \tag{6.1}$$

where the Lorentz transformation is written symbolically and acts on the eigenvalue k_0. Since such solutions can be superposed we may also write

$$\phi(x) = \sum_k \alpha_k \phi_k(x) = \sum_k \alpha_k L_{kk_0}\phi_{k_0}(x)\,. \tag{6.2}$$

We shall assume that in Equation (6.2) all the permissible frames of reference appearing in the conventional theory are included, and the solution $\phi(x)$ is in this sense complete. It is clear that the set of Ω-transformations defined by

$$\Omega = \sum_k \alpha_k L_{kk_0} \tag{6.3}$$

acts as a certain generalization of Lorentz transformations. However, such a generalization can be legitimate only if it can be reduced to ordinary Lorentz transformations in the limiting circumstances. For example, if all the α_k coefficients except a particular $\alpha_{k'}$ become zero, this particular $\alpha_{k'}$ must have unit magnitude[1]. We shall write the general condition of reducibility as

$$(6.4) \qquad \Omega^*\Omega = \sum_{kk'} \alpha_k^* \alpha_{k'} L_{kk_0}^T L_{k'k_0} = 1$$

which, by virtue of the orthogonality, Equation (5.6), of the Lorentz frames implies that

$$(6.5) \qquad \sum_k |\alpha_k{}^2| = 1 .$$

With this condition Equation (6.2) admits two equivalent interpretations: (a) The generalized frame $\Omega = \sum_k \alpha_k L_{kk0}$ is made up of all permissible Lorentz frames L_{kk_0} with a probability amplitude α_k being assigned to each L_{kk_0}. (b) The generalized wave $\phi(x)$ is made up of all permissible waves $\phi_k(x)$ a probability amplitude α_k being assigned to each $\phi_k(x)$. In this manner the simultaneous consideration of wave and particle aspects leads to a statistical interpretation of a frame of reference, which further revises our usual understanding of special relativity. Note that Equation (6.5) expresses the normalizability of every permissible field $\phi(x)$ although it does not contain information as to how the normalization itself shall be achieved. In other words, the normalizability condition (6.5) has to do with some kinematic background properties of all fields as is evident from Equation (6.3) rather than the dynamics of the individual waves corresponding to individual particles.[2] The origin of this statement goes back all the way to Equation (5.8). We emphasize strongly that (5.8) is essentially an eigenvalue equation and therefore its consequences are independent of the numerical magnitude of $P(p)$. Likewise its Fourier transform, Equation (5.11), is insensitive to the numerical magnitude of $\phi(x)$. This means that P or ϕ can be normalized, the true significance of the equation being in the eigenvalue structure of the equations rather than the magnitudes of the fields themselves.

To explore further the nature of the restrictions imposed by the above considerations, we may write the condition (6.5) in the familiar, infinite volume limit

$$\text{(6.6)} \qquad \frac{V}{(2\pi)^3}\int \mathrm{d}^3k\,|\alpha_k|^2 = 1$$

which translates the condition (6.5) into

$$\text{(6.7)} \qquad |\alpha_k|^2 \leq k^{-3-|\varepsilon|} \quad \text{for} \quad k \to \infty$$

$$\text{(6.8)} \qquad |\alpha_k|^2 \leq k^{-3+|\varepsilon|} \quad \text{for} \quad k \to 0$$

where ε is a small nonzero quantity. We shall see in Appendix 2 that α_k-coefficients appear explicitly in the propagators and affect the numerical values of the integrals over intermediate states. Since the ultraviolet divergences occurring in quantum electrodynamics are at most quadratic we can see that the condition (6.7) is already sufficient to make high energy limit convergent.

Another, perhaps less obvious consequence of the normalizability condition is that it imposes certain restrictions on the variation $\delta\phi(x)$ of the field in a Lagrangian treatment. To see this remember that the Ω-transformations are generalized Lorentz transformations and as such the condition (6.5) is kinematical in nature rather than dynamical. The essence of a variational treatment, on the other hand, is that one ignores the dynamical constraints and considers all possible variations of the field in their absence while obeying the kinematical constraints.[3] Therefore if $\phi(x)$ satisfies any such conditions, then $\phi'(x)=\phi(x)+\delta\phi(x)$ must also. But we shall find a little later that the fields are, in nature, operators and their normalization is to be implemented through their commutators (or anti-commutators). By integrating the commutator (anti-commutator) $[\phi'(x), \phi'(x')]_\pm$ over $\mathrm{d}V \to \mathrm{d}\sigma_\mu$ and cancelling $[\phi(x), \phi(x')]_\pm$ we find

$$\text{(6.9)} \qquad \int_\sigma f^\mu(x')\,[\phi(x), \delta\phi(x')]_\pm\,\mathrm{d}\sigma_\mu = 0$$

where $f^\mu(x)$ is a nonzero vector and '$-$' and '$+$' indicate commutator and anticommutator, respectively. As a consequence of normalizability this equation implies that $\phi(x)$ and $\delta\phi(x)$ are orthogonal in the operator space of $\phi(x)$. Note carefully, however, that Ω-transformations do not determine the vector $f^\mu(x)$ nor do they contain any information as to why the normalization is to be achieved through commutators. We shall see that this information comes from the action principle.

Next, we consider the problem determining the function $w_k = |\alpha_k|^2$ satisfying the boundary conditions. To do this we need a new statement. We shall use the following entropy-information principle: Let any description by aggregates of Lorentz frames satisfy a regularity condition such that $w_k = w(k)$ is a positive-definite function. Then we can define an entropy

$$(6.10) \quad S = -\sum_k |\alpha_k|^2 \ln |\alpha_{k\varepsilon}|^2$$

which measures, in some sense, the information content of the aggregate. Since entropy can be interpreted logically as a lack of information we maximize the entropy subject to the condition that the kinematical characteristics of elementary particles are expressible as averages over the eigenvalues of fields. As explained in Note 2 this procedure yields

$$(6.11) \quad \alpha_k = A\, e^{-(\zeta_\mu + i\chi_\mu)\, k^\mu - 1/2(\lambda_{\mu\nu} + i\theta_{\mu\nu})\, s^{\mu\nu} - \cdots}$$

where A is a normalization factor, and ζ_μ, $\lambda_{\mu\nu}$ are Lagrangian multipliers. In principle this distribution also covers observables which are functions of the kinematic observables. For example p^2 is an invariant function of p, hence the form $e^{-\kappa^2 p^2/2}$ is a possible distribution. The essential point is that the kinematical description is limited by these formulae although we are not totally ignorant about the behavior of the distributions.

The expression (6.11) seems to have a connection with the fundamental principle of quantum statistics which states:[4] "In calculating average values of properties of a quantized system with energy between W_1 and W_2 the same weight is to be assigned to every accessible state with energy in this range, in default of other information." If Equation (6.11) is taken seriously the phrase "the same weight" should now be replaced by "the weight $w = Ae^{-2\zeta k - \lambda s \ldots}$." We shall see shortly that the problem of field quantization is independent of this statement. In fact, if the fields are quantized with this *a priori* probability in mind, ultraviolet divergences would not occur in quantum electrodynamics.

We may now give a careful look into the conventional action principle. For the purpose of economy we shall consider here only the salient points and for detail the reader is referred to the excellent exposition by Jauch and Rohrlich.[5] Consider an action integral

$$(6.12) \quad W = \int L(\phi, \partial_\mu \phi)\, \mathrm{d}^4 x$$

and a unitary transformation $U^{-1}=U^*$ such that

$$(6.13)\qquad \phi(x)=U\phi(x_0)\,U^{-1}.$$

Latter implies that we are considering the field as an operator. If U is an infinitesimal unitary transformation, we may write

$$(6.14)\qquad U=1+iF\,,\quad F=F^*$$

where F is the generator of the infinitesimal transformation U. Equation (6.13) now translates into

$$(6.15)\qquad \delta\phi(x)=\phi(x_0)-\phi(x)=i\,[F,\phi(x)]\,.$$

The action principle states that the generator F and the action W are related as[6]

$$(6.16)\qquad \delta W=F(\sigma)-F(\sigma_0)=W_{\sigma\sigma_0}=\int_{\sigma_0}^{\sigma} L\,\mathrm{d}^4x$$

where σ and σ_0 are two space-like surfaces. Proceeding as in Note 4, one has

$$(6.17)\qquad F(\sigma)=\int_{\sigma}[(\pi^{\mu}\partial^{\nu}\phi-g^{\mu\nu}L)\,\delta x_{\nu}-\pi^{\mu}\delta\phi]\,\mathrm{d}\sigma_{\mu}$$

with $\pi^{\mu}=\partial/\partial_{\mu}\phi$. From Equation (6.17) it follows that the variation in F owing to $\delta\phi$ arises from the $\pi_{\mu}\delta\phi$ term alone. Introducing this variation of F (owing to $\delta\phi$) back into Equation (6.15), we find

$$(6.18)\qquad \delta\phi(x)=i\,[\phi(x),\int_{\sigma}\pi^{\mu}\delta\phi]\,\mathrm{d}\sigma_{\mu}$$

which corresponds to Equation (1.104) of Note 4 with tensor indices suppressed. By using the mathematical identity

$$(6.19)\qquad [A,BC]=[A,B]_{\pm}C\mp B\,[A,C]_{\pm}$$

where '$-$' and '$+$' denote commutator and anticommutator, respectively, we obtain

$$(6.20)\qquad \delta\phi(x)=i\int_{\sigma}[\phi(x),\pi^{\mu}(x')]_{\pm}\delta\phi(x')\,\mathrm{d}\sigma_{\mu}\mp i\int_{\sigma}\pi^{\mu}(x')\times$$
$$\times\,[\phi(x),\delta\phi(x')]_{\pm}\,\mathrm{d}\sigma_{\mu}.$$

At this point the conventional action principle leads to a difficulty, because although it implies that the normalization itself will have to be achieved through the commutation properties (the first integral of Equation (6.21) below is in the form of a normalization), no information regarding the second integral is contained in it. As mentioned in the previous section our Ω-transformations provide this information. In fact, the second integral vanishes. We obtain the commutator relation

$$(6.21) \qquad \delta\phi(x) = i \int_{\sigma} [\phi(x), \pi^{\mu}(x')]_{\pm} \delta\phi(x') \, d\sigma_{\mu} .$$

Since in the conventional theory the second integral of Equation (6.20) is 'assumed' to vanish anyway, one might conclude that from here on our theory becomes identical to the conventional theory. This, however, is not the case. The conventional assumption does not provide any further information, whereas the invariance of the Lagrangian under the generalized transformation implies further the existence and the normalizability of the α_k-coefficients. This is very essential, because without a clear separation between the kinematical constraints and the dynamically allowable variations, there is no meaning to the idea of a variational principle, especially for operators.[7] Thus although the two theories shall have identical forms in terms of $\phi(x)$ the differences will become apparent as soon as we expand $\phi(x)$ into some eigenoperators or try to construct propagators from the normalization conditions, Equation (6.21). The essential difference lies in the fact that the conventional theory equates the commutator (anticommutator) in Equation (6.21) to a $\delta^{\mu}(x-x')$ function, whereas we are prevented from doing this because of the α_k coefficients. However, if we let the Ω-transformation tend to a particular Lorentz transformation in Equation (6.21), we obtain the familiar

$$(6.22) \qquad [\pi_k^{\mu}(x), \phi_{k'}(x')]_{\pm} = i \, \delta_{kk'}^{\mu}(x - x') .$$

Consequently, the present theory provides the commutation relations

$$(6.23) \qquad [\pi^{\mu}(x), \phi(x')]_{\pm} = i \, \delta^{\mu}(x, x') = i \sum_{k} |\alpha_k|^2 \, \delta_{kk'}^{\mu}(x - x') .$$

With the help of this new $\delta(x, x')$ function, we can now construct the propagators. Thus, in the absence of derivative coupling (e.g., electrodynamics),

$$(6.24)\qquad \Delta(x, x') = \frac{-i}{(2\pi)^4} \int_c \mathrm{d}^4 k \, |\alpha_{k\varepsilon}|^2 \, \frac{e^{-ik(x-x')}}{-k^2 + m^2}$$

$$(6.25)\qquad S(x, x') = \frac{-i}{(2\pi)^4} \int_c \mathrm{d}^4 p \, |\alpha_p|^2 \, \frac{e^{-ip(x-x')}}{-i\gamma p + m}$$

where, as a consequence of the principle of causality, the contours c in the complex k_0 plane are to be taken the same way as in the conventional theory (Figure 6.1). Of course the definition of the vacuum state $|0\rangle$, and

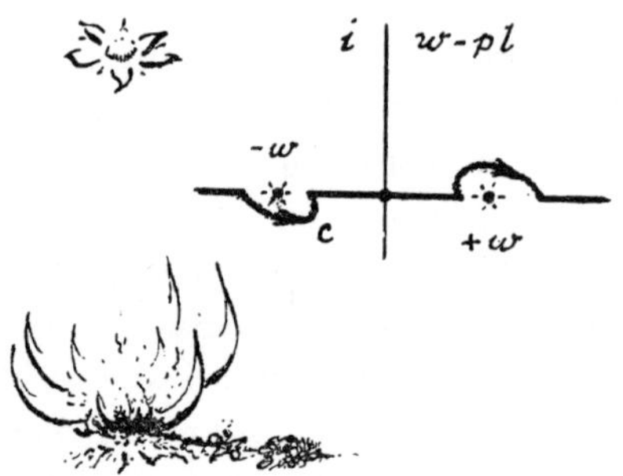

Fig. 6.1. Contour for a causal propagator.

all other preliminaries of the usual field theory are likewise assumed valid the only difference being the modification introduced into the propagators. Equations (6.24) and (6.25) are now the functions to be used in the evaluation of Feynman graphs. For electrodynamics $\Delta(x, x')|_{m\to 0} = D(x, x')$ and as we can see $w_k = |\alpha_k|^2$ explicitly appears in them. It is evident that under the negative exponential dependence of w_k (on energy-momentum) given by Equation (6.11) all the ultraviolet divergent integrals of quantum field theory will become absolutely convergent[8] although the infrared divergences will still survive.

It can be verified that Δ's satisfy equations similar to the conventional ones, namely,

$$(6.26)\qquad \Delta(x, x') = [\phi(x), \phi(x')]_-, \quad (\partial_\mu \partial^\mu + m^2)\, \Delta(x, x') = \delta(x, x'),$$

$$(6.27)\qquad S(x, x') = [\psi(x), \psi(x')]_+, \quad S(x, x') = (\gamma^\mu \partial_\mu - m)\, \Delta(x, x')$$

and so on. We note that although the conventional Feynman propagators are recoverable from these by setting $|\alpha_k|^2 = |\alpha_p|^2 = 1$, the conventional theory as a whole is not recoverable because of the normalizability conditions.[9] In this connection we should perhaps emphasize again that the action principle and the resulting field quantization is independent of the entropy principle. It is in general possible to consider non-equilibrium situations where entropy is not maximum or use semi-empirical procedures to determine the function $w(k)$ although the entropy principle appears to be eminently proper. It is, for example, possible to consider forms like $e^{-\zeta|k|}$, $e^{-\varkappa^2|k|^2}$ etc., but it seems necessary, in complicated graphs, that exponents refer to the total kinetic energy of the intermediate state.

Comparing Equations (6.26) and (5.15) we now see that the causal function $\Delta(x, x')$, which characterizes our kinematical description is indeed determinable (due to Equation (6.27) $S(x, x')$ is not necessary to consider separately). It is, however, different from what would have been expected through a classical appreciation of relativistic ideas. The entropy-information principle implies in fact that there exists an upper bound to the amount of detailed information $\Delta(x, x')$ can provide. This is reflected in the circumstance that $\delta(x, x')$ cannot equal a sharply defined Dirac δ-function. As to the stability of the vacuum it is guaranteed by the positiveness of energy. This can be explicitly formulated by splitting $\phi_\kappa^+(x)$ and $\phi_\kappa^-(x)$ which act as creation and annihilation operators. The fact that there is no state lower, in energy, then the vacuum can be translated into $\phi^-(x)|0\rangle=0$. With this condition we may refer back to Equation (6.16) which gives the equations of motion under the extremality, $\delta W=0$, alone. The requirement that this extremum will be a minimum, $\delta^2 W>0$, is related to the fact that as applied to the vacuum, only the positive frequency components of $\delta\phi(x)$ contribute to action. Despite these remarks we should keep in mind, however, that quantum mechanical action is an operator and therefore its minimality should be discussed in reference to its expectation values, and, strictly speaking, in the classical limit. Finally it should perhaps be pointed out that the commutation relations in the usual form of (6.22) has little to do with the problem of quantization. This is because (6.22) is a statement about the orthonormality of unit vectors of a function space and this could be conceived classically as well as quantum mechanically. For example, the quantum theoretical description of the single-particle states of the radiation field is identical with its classical description as far as the normalized eigenfunctions are concerned. The difference is not in the existence of normalized individual eigenfunctions but rather in the fact that classically a linear combination of eigenvalues would have been observable. Thus, the concept of quantization emerges from the 'normalization' (6.23) of the vector $\phi(x)$ and the interpretation of its coordinates, α_κ, as 'probability amplitudes'. It is this probabilistic interpretation of α_κ that is related to discrete numbers as the outcomes of individual physical observations. In other words the crucial step in the process of quantization seems already taken at the level of Equation (6.5) and its probabilistic interpretation. The process of field quantization via the action principle complicates the structure by bringing in multi-particle states but these are actually obtainable through an iterative process from the single-particle states. The field quantization may obscure the issue by the imposing superstructure of its multiparticles and their intricate statistics but these structures have to follow the interpretations previously given to their basic building blocks, the single-particle states. The conclusion seems to be that the process of quantization is essentially a single step which is contained in Equation (6.5); the so-called second quantization follows when the fields are considered as operators instead of ordinary numbers. Furthermore, when understood in this fashion the basic formula of quantization prevents the infinitely sharp Dirac functions and thereby avoids the problem of ultraviolet divergences. The price for this is the unlimited describability of nature. This almost perceptual statement is reflected in the entropy-information principle, and the fact that a single Lorentz frame is not sufficient to describe physical phenomena; in general we seem to need an infinite number of Lorentz frames to describe any process and therefore statistically diffuse the description. Despite the introduction of a new (possibly universal) ζ-parameter, and a fundamental change of attitude toward physical description, however, this revision is really a statistical generalization of special relativity and therefore does not violate any of the accepted principles of microscopic physics. Finally we may analyse the meaning of (6.10) and the consequent $\delta(x, x')$ which is not a sharp δ-function but a statistically diffused one. Evidently a further uncertainty (of a statistical kind) above and beyond the usual Heisenberg relations is here implied. For example according to the Heisenberg uncertainty, $\Delta p \Delta x \geqslant \frac{1}{2}$, if we are willing to let Δp to become infinitely large, then we could determine x with arbitrary accuracy. The present theory denies this possibility: "It is impossible, on principle, to design an apparatus which will yield arbitrarily detailed information about space and time." This in turn seems to imply that no elementary particle, useable as a probe can be heavier than some kinematical or statistical bound, $m \geqslant 1/|\zeta|$. In other words, a statistical uncertainty of the form $\Delta p \Delta \zeta \geqslant \kappa$ seems here to be involved. It is further to be noted that in the usual derivations of Pauli's master equation (that is the stochastic Chapman-Kolmogroff equation) from quantum mechanics one has to introduce the presence of a randomizing influence at each instant of time. This often takes the form of continuously imposed phase incoherence, prevention of exceptional states which are in their way of building up large statistical fluctuations, or attributing special randomizing properties to the interaction. Without such extraneous devices the irreversible macroscopic laws of physics cannot be derived from quantum mechanics.[10] The present theory seems to escape this difficulty by providing the ever present α_κ-coefficients which play the role of a randomizing influence. This, together with the presence of the unpredictable spontaneous emission inherent in the theory enables the statistical theory of fields to derive the master equation without further assumptions. The fact that this procedure arises from the basic condition of quantization, and is closely tied to the removal of field theory divergences would seem to make it less objectionable compared to the other methods utilized for the purpose.

The statistical theory of fields so formulated is essentially a theory of arbitrary objects in the sense that no further property of any object beyond the kinematical preliminaries is spelled out. On the other hand, if the theory is somehow valid to that extent it should then be possible to contemplate a physics in which (a) objects like elementary particles are to be described by restrictive impositions on the possible (statistical) representations of the theory. (b) The universe as a whole is to be understood as an assembly of representations (grand overall representation) restricted to exhibit certain self-consistency requirements and the experimentally observed regularities. In the next appendix we shall advance some arguments which, in our opinion, support a program so eloquently expressed by von Weizsäcker (vol. V, pp. 460–473 of the present series) on the possible overall unity of physics along these lines.

7. The Problem of the Unity of Physics

If one attempts to enlarge the scope of the conventional special relativity by considering reference frames more general than inertial, many confusing questions may arise. Consider, for instance, the case of a rotating disc. Let an observer at the rim of the disc send light rays forward and backward along the rim. Which one will come back (Figure 7.1) sooner to

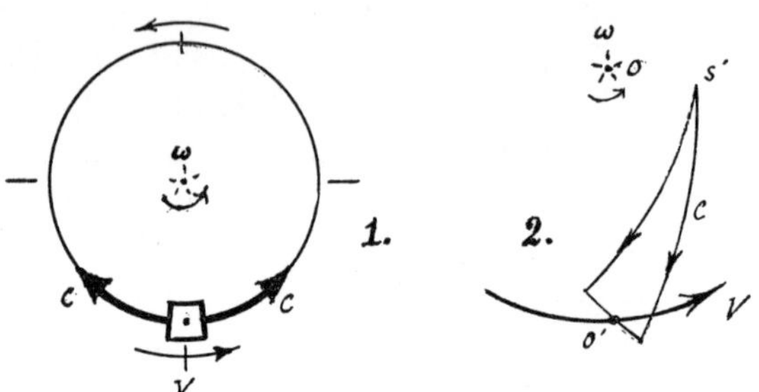

Figs. 7–1,2. Rotating disc and Sagnac experiment.

the observer? We can argue that the velocity of the observer at the rim relative to an observer on a non-rotating disc is immaterial and therefore forward and backward the velocity of light is c. Since this would be true for every point on the rim, the two rays should come back at the same time. This conclusion is, however, wrong because such an experiment was performed more than fifty years ago by Sagnac, who found that the waves sent backwards come back sooner by[1]

$$\Delta t = t_2 - t_1 \simeq \frac{4\omega\pi r^2}{c^2} \simeq \frac{2\pi r}{c - V} - \frac{2\pi r}{c + V}. \tag{7.1}$$

In other words, the velocity of light as measured by the observer at the rim seems not to be c but $c \pm V$. On the other hand this second conclusion must also be wrong because Michelson-Morley experiment, which originally suggested the invariance of the velocity of light was indeed performed at the rim of a rotating frame, the earth. The velocities due to the orbital motion (30 km s^{-1}) and rotatory motion (0.45 km s^{-1}) were both within the sensitivity of the experiments especially in their later versions and no influence from these motions on the velocity of light was ever detected[2]. So we have here a kind of paradox.

The generally accepted Langevin explanation of the Sagnac experiment seems to run as follows[3]: First transform the stationary line element into cylindrical coordinates and then transform it into a rotating frame by $\theta \rightarrow \theta - \omega t$

$$\text{(7.2)} \qquad \mathrm{d}s^2 = c^2\,\mathrm{d}t^2 - \mathrm{d}r^2 - r^2\,\mathrm{d}(\theta - \omega t)^2$$

where we have taken $z=0$. Since $\mathrm{d}s^2=0$ describes the motion of the light waves one writes, for the rays of light along the rim

$$\text{(7.3)} \qquad (c^2 - \omega^2 r^2)\,\mathrm{d}t^2 + 2r^2\omega\,\mathrm{d}\theta\,\mathrm{d}t - r^2\,\mathrm{d}\theta^2 = 0$$

which indeed reproduces the result of the Sagnac experiment. However this reproduction is not satisfactory, because if one solves Equation (7.3) for $\mathrm{d}t$ one obtains

$$\text{(7.4)} \qquad \mathrm{d}t = \frac{\mp r\,\mathrm{d}\theta\,(c \pm V)}{(c^2 - V^2)} = \frac{\pm r\,\mathrm{d}\theta}{c + V},$$

In other words, the line element (Equation (7.2)) is really a sophisticated way of saying that at the rim the velocity of light is $c \pm V$. It does not solve the problem.

In analyzing space-time relationships we find it much safer to argue directly on the basis of the wave-particle concept and not on the waves and particles separately. Thus if we remember the way we have derived the Lorentz transformations from the properties of rays and wave-fronts we can see how a pencil of light originating from the center of the disc will appear to the observer at the rim (Figure 7.2). The more reasonable assumption appears, then, to hold the velocity of light to be c as measured by an observer at the rim in agreement with Michelson-Morley experiment and require the Sagnac experiment to be explained differently than is done conventionally.

In order to attack the problem the wave-particle duality will now be generalized into non-inertial frames by the two following statements: (a) Physical observation is local. (b) The local value of the velocity of light is c. The latter statement is to be interpreted as the orthogonality of rays and wave-fronts. Then it becomes evident that the position of an observer will somehow be involved explicitly in the line element defining the space of that observer. In general relativity, which is the basis of the conventional explanation above (Equation (7.3)), this is apparently missing, and that is why (so it seems) we are having a problem. What we are trying to say is that Equation (7.3) is not the proper line-element for an observer situated at the rim of the disc. In order to see how the correct line element might be obtained, let us adjust the line element (7.2) so as to satisfy the local constancy of the velocity of light. We take[4]

$$\mathrm{d}s^2 = c^2\,\mathrm{d}t^2 - \mathrm{d}(\mathbf{r} - \mathbf{r}')^2 - (\mathbf{r} - \mathbf{r}')^2\,\mathrm{d}\,[(\phi/2) - \omega t]^2 \tag{7.5}$$

where $\mathbf{r}'$ is the position of the observer. Letting $\mathbf{r}=\mathbf{r}'$, $\mathrm{d}s^2=0$, we find the velocity of light locally to be c forward and backward along the rim for that observer. This result satisfies the second statement. Then, the first statement implies that for other points in "his space" the observer can only make certain inferences. Thus for a distant point of the rim the velocity of light may 'appear' different to him although if he were to go there and make a measurement he would again obtain c. These appearances, however, are not to be considered complete deceptions but the logical necessities of the description. Their consequences can in principle be tested by experiment. So to speak each observer has his 'own' geometry and apart from a requirement of mathematical consistency these geometries may be quite different[5]. Setting again $\mathrm{d}s^2=0$ and substituting $|\mathbf{r}-\mathbf{r}'|=2r\sin(\phi/2)$, and by solving for $\mathrm{d}t$ we now have

$$\mathrm{d}t = \frac{r\,\mathrm{d}\phi\,\{-\,2r\omega\sin^2(\phi/2) \pm \sqrt{c^2 - 4r^2\omega^2\sin^2(\phi/2)\cos^2(\phi/2)}\}}{c^2 - 4r^2\omega^2\sin^2(\phi/2)}. \tag{7.6}$$

Integrating this expression along the rim forward and backward and forming the time difference in question one obtains

$$\Delta t = t_2 - t_1 \simeq \frac{4r^2\omega}{c^2}\int_0^{2\pi} \sin^2(\phi/2)\,\mathrm{d}\phi = \frac{4\pi r^2\omega}{c^2} \tag{7.7}$$

which is equal to the result observed by Sagnac. Thus although the behaviors of (7.6) and (7.4) are very different locally, they lead to the same value for the Sagnac experiment. It goes without saying, of course, that the Langevin line element (7.2) may now be considered valid but only for an observer situated at the center of the rotating disc.

In general relativity, in dealing with the comparison of durations and lengths something similar to the above adjustment is made. For example from the sameness of ds^2 for two observers at different points one derives the red shift formula $\Delta\nu = GM/c^2\ (1/r - 1/r')\nu$ but apparently these are not sufficient to provide a full prescription in all cases. For example the bending of light rays is calculated on the basis that $c \rightarrow c(1 - GM/c^2 r)$ yet locally the velocity of light is supposed to be c everywhere. Such points are considered by many experts as implicit in the theory but we feel that they should be explicitly formulated in order for us to see the exact relationship between the 'observer' and the things he actually 'measures'. We suspect that the usual arguments based on the synchronization of clocks along the rim are not relevent here because in the Sagnac experiment we have only one clock, and this clock measures a time difference irrespective of whether or not there are other clocks at all. We shall therefore put the problem into a more general context, namely, that of the integration constants of the field equations of general relativity. These equations are second order partial differential equations and, in general, their solutions will involve a number of arbitrary functions and constants. Our two statements may then be said to require that, if the theory is consistent, it should always be possible to fix some of those arbitrary functions and constants so as to satisfy the local constancy of the velocity of light. Another and perhaps, somewhat better way of expressing the same idea is to carry Equations (5.8) and (5.11) over into general relativity and require the wave equation to be separable locally. It is known from Hamilton-Jacobi theory that such coordinates are singled out as our most natural frames of observations in that (for single particle situations, at least) they satisfy automatically the wave-particle duality. General relativity appears to have sufficient freedom to accommodate this requirement.

However, even if such a readjustment were to prove possible we shall probably run into other difficulties. This is because from a classical point of view a frame of reference can be a single solution of the gravitational field equations. With Equation (6.11) and its further elaboration in Appendix 6 one begins to realize, however, that in principle one needs an infinity of solutions to represent any frame of reference. Since the field equations of general relativity are non-linear we are here at a double disadvantage. In the first place we do not know infinite numbers of solutions. Secondly, the superposition principle no longer holds even in vacuum and we do not know how to combine solutions to make sense with the statistical interpretation. It seems that paradoxed which can easily arise, for example at $r \to \infty$ in a rotating frame of reference, problems arising in solutions intended for cosmology, problems associated with the Schwarzchild singularity, quantization, gravitational waves etc., seem to be indicative of the inadequacy of our conceptual framework to deal with them. In other words, the conventional general relativity, beautiful as it is, might not be the most convenient description of the phenomena concerning the non-inertial frames of reference although a workable reinterpretation of some kind is evidently a distinct possibility.

We shall now attempt a reformulation of general relativity which, in our opinion, establishes a logical contact with special relativistic quantum field theories and at least in a macroscopic, statistical sense preserves the unity of physics by allowing us to consider all of the basic principles of field theory (covariance, causality, positive energy, positive probability and equivalence) within the same, general, chain of reasoning. Although the resulting equations will be somewhat different we nevertheless hope that the effort will be worthwhile especially in view of von Weizsäcker's program on the possible unity and connectedness of physics. Our starting point is the above mentioned local sameness of the velocity of light and the consequent validity if a Lorentz invariant field theory locally in a gravitational field. We interpret this to mean that the gravitational field which we here denote with ϕ^{ν}_{μ} will be a locally special relativistic field and $g_{\mu\nu}$ will be a functional of this field as $g_{\mu\nu}(\phi^{\nu}_{\mu})$. It will then be possible to formulate the laws of gravitation as a direct relation between $g_{\mu\nu}$ and ϕ^{ν}_{μ}. We state this functional relation to be of the form

$$\text{(7.10)} \qquad \mathrm{d}g_{\mu\nu} = 2(g_{\mu\nu}\,\mathrm{d}\phi - g_{\mu\alpha}\,\mathrm{d}\phi^{\alpha}_{\nu} - \mathrm{d}\phi^{\alpha}_{\mu})\, g_{\alpha\nu}$$

where ϕ is the trace of ϕ^{ν}_{μ}. In accordance with our local requirements above, in a case where $\mathrm{d}\phi^{\alpha}_{\mu}=0$, namely, when the field differentials or their gradients vanish the $g_{\mu\nu}$ will reduce to the special relativistic form $\eta_{\mu\nu}=(1, -1, -1, -1)$. In other words in the absence of the fields or their variations the metric is assumed to go over into a Lorentz form which can be done without loss of generality. The fields ϕ^{ν}_{μ} themselves will then arise via the distributions of matter according to the locally special relativistic equations

$$\text{(7.11)} \qquad \phi^{\nu}_{\mu} = \int (\sigma u_{\mu} u^{\nu})' \,\frac{\mathrm{d}V'}{r'}$$

where prime (′) implies the usual retardation considerations on the integral. We shall therefore have the propagation equations

(7.12) $\partial^{\lambda}\partial_{\lambda}\phi^{\nu}_{\mu} = 4\pi\sigma u_{\mu}u^{\nu}$

where $\sigma u_{\mu}u^{\nu}$ is the usual 'matter' tensor and $\partial^{\lambda} = g^{\lambda\sigma}\partial_{\sigma}$. Note that (7.12) is already in the form (5.13) and hence satisfies the requirement of wave-particle duality in the sense of (5.8) and (5.11) for $m = 0$.

We may now point out that in a gravitational field the energy-momentum 'vector' of matter alone cannot be a conserved quantity and in fact it cannot be a true vector. We must add a 'field' stress-energy term t^{ν}_{μ} to (7.12) in order to obtain a conserved energy-momentum vector. We write

(7.13) $T^{\nu}_{\mu} = \sigma u_{\mu}u^{\nu} + \dfrac{1}{4\pi}\, t^{\nu}_{\mu}$

where t^{ν}_{μ} is not possible to determine uniquely from the usual special relativistic field theory arguments because the divergence of an arbitrary vector can be added to the Lagrangian and this will yield a different although equivalent expression. We shall, however, require that, as computed directly through (7.10) the geometric, divergence-free tensor $R^{\nu}_{\mu} - \frac{1}{2}\,\delta^{\nu}_{\mu}\, R$ will be equal to 8π times T^{ν}_{μ}. In other words we require the field equations to be of the form

(7.14) $R^{\nu}_{\mu} - \frac{1}{2}\delta^{\nu}_{\mu}R = 2(\partial^{\lambda}\partial_{\lambda}\phi^{\nu}_{\mu} + t^{\nu}_{\mu}) = 8\pi T^{\nu}_{\mu}\,.$

In this way we are able to guarantee the conservation of the energy momentum vector identically and at the same time determine the form of t^{ν}_{μ} essentially up to an additive constant. It turns out that in all cases so far examined via the exact solutions of (7.10) t^{ν}_{μ} is derivable from

(7.15) $L_f = -\,\partial^{\lambda}\phi^{\alpha}_{\beta}\,\partial_{\lambda}\phi^{\beta}_{\alpha} + \frac{1}{2}\,\partial^{\lambda}\phi\,\partial_{\lambda}\phi + \partial^{\alpha}J_{\alpha}$

which is indeed a familiar special relativistic stress-energy expression for the fields ϕ^{ν}_{μ} and ϕ.

It is to be noted in passing that in order for the second order derivatives of ϕ^{ν}_{μ} to combine properly into $\partial^{\lambda}\partial_{\lambda}\phi^{\nu}_{\mu}$ the field must satisfy a subsidiary condition

(7.16) $\partial^{\mu}\phi^{\nu}_{\mu} = 0$

which, contrary to a first impression, does not imply a separate conservation law for the matter part alone. It rather implies that locally the gravitational field will propagate with the velocity of light, c, in all directions.

Finally, we adopt, as usual, the geodesic equations of motion

$$(7.17)\qquad \frac{d^2x^\lambda}{ds^2} + \begin{Bmatrix} \lambda \\ \mu\nu \end{Bmatrix} u^\mu u^\nu = 0$$

as the equations governing the motions of test particles and light rays. We are able to do this because any line element that is a solution of (7.10) will also be a solution of (7.14). Since the geodesic equation above are a consequence of the field equations (7.14) under the subsidiary conditions $\partial^\nu(\sqrt{-g}\,\sigma\,u_\mu u^\nu) = 0$ (conservation of energy-momentum) we are justified in assuming that these are the correct equations of motion.

At this point it will be possible to start comparing the above theory with the standard Einstein theory to see that the two theories are very different conceptually and yet they give the same answers for the three classical tests of general relativity. For the time-delay of radar signals echoed from Venus and also for the experiment on the isotropy of inertia there seem to be certain differences in prediction or interpretation. The theoretical differences are mainly related to the fact that in Einstein's theory one assumes the right hand side of (7.14) to be the 'matter' tensor alone, hence t_μ^ν does not enter. Consequently in empty space where $\sigma = 0$ the matter tensor vanishes as well and one has the Einstenian law

$$(7.18)\qquad R_\mu^\nu = 0$$

By solving these equations for the static gravitational field of the Sun one obtains the usual Schwarzchild line elements. In the present theory this static case corresponds via (7.12) to $\phi_0^0 = \phi = M/r$, and putting this into (7.10) we immediately have the line element

$$(7.19)\qquad ds^2 = e^{-2\varphi}\,dt^2 - e^{2\varphi}(dx^2 + dy^2 + dz^2).$$

The reader could check easily that upon computing the tensor $R_\mu^\nu - \frac{1}{2}\delta_\mu^\nu R$ with this line element one obtains for t_μ^ν exactly the expression (7.15) with $\phi_\mu^\nu \to \phi_0^0 = \phi$. He could also repeat this process for other solutions given below, with the same end results. We can see that the two theories are infact very different. For example our solution is automatically isotropic and there are no non-isotropic solutions at all. At this point the main difference appears to be that in Einstein's theory the stress-energy components of the gravitational field itself do not contribute to $g_{\mu\nu}$ whereas in our theory they do because we now have, instead of $R^{\nu}_{\mu}=0$ the new empty-space law

$$(7.20)\quad R^{\nu}_{\mu} = 2(t^{\nu}_{\mu} - \tfrac{1}{2}\delta^{\nu}_{\mu}t).$$

We may point out immediately, however, that we do not have to solve these highly non-linear equations directly in order to obtain a line element. Instead we can simply deduce it from our functional Equations (7.10) as we have just done it to obtain the above line-element. As other examples of such exact solutions let all components of ϕ^{ν}_{μ} other than $\phi^{2}_{1}(t, z)$ vanish. Then the solution is $g_{00}=-g_{33}=1$, $g_{11}=g_{22}=-\cosh 4\,\phi^{2}_{1}$, $g_{12}=g_{21}==\sinh 4\phi^{2}_{1}$. If only $\phi^{3}_{0}(x, y)$ exists then $g_{11}=g_{22}=-1$, $g_{00}=-g_{33}==\cos 4\phi^{3}_{1}$, $g_{10}=g_{01}=\sin 4\phi^{3}_{1}$, etc. The latter solution with a slight extension is found to provide an explanation to the rotating disc and Sagnac type of experiments, and, former represents a plane gravitational wave which carries energy-momentum. There are other interesting exact solutions one of which seems to describe cylindrical gravitational waves. A general iteration solution of (7.10) with the boundary conditions that at the observation point $g_{\mu\nu}$ goes over into a Lorentz form $\eta_{\mu\nu}$ is given by

$$\begin{aligned}(7.21)\quad g_{\mu\nu} = \eta_{\mu\nu} &+ 2\,[\eta_{\mu\nu}\phi - \eta_{\mu\alpha}\phi^{\alpha}_{\nu} - \eta_{\nu\alpha}\phi^{\alpha}_{\mu}] \\ &+ 4\int [\eta_{\mu\nu}\phi - \eta_{\mu\alpha}\phi^{\alpha}_{\nu} - \eta_{\nu\alpha}\phi^{\alpha}_{\mu}]\,\mathrm{d}\phi \\ &+ 4\int [\eta_{\mu\lambda}\phi - \eta_{\mu\alpha}\phi^{\alpha}_{\lambda} - \eta_{\lambda\alpha}\phi^{\alpha}_{\mu}]\,\mathrm{d}\phi^{\lambda}_{\nu} \\ &+ 4\int [\eta_{\nu\lambda}\phi - \eta_{\nu\alpha}\phi^{\alpha}_{\lambda} - \eta_{\lambda\alpha}\phi^{\alpha}_{\nu}]\,\mathrm{d}\phi^{\lambda}_{\mu} + \cdots\end{aligned}$$

which, in a case where $\phi^{0}_{0}=\phi$ is the largest component, can be approximated by the simpler form

$$(7.22)\quad g_{\mu\nu} = \eta_{\mu\nu}(1 + 2\phi + 2\phi^{2}) - 4\,\eta_{\mu\lambda}\phi^{\lambda}_{\nu}.$$

Note that this approximation differs already by $(\tfrac{1}{2})\phi^{2}$ in g_{ii} from Einstein's theory.

Concerning the three classical tests of general relativity we may point out that the solution (7.20) agrees with the isotropic Schwarzchild line element in first and second order for g_{00} and in first order for g_{ii}. Since the three crucial tests are insensitive to higher terms in these components the present theory explains the three classical effects just as well as Einstein's theory.

With regard to the experiment on the isotropy of inertia the present theory seems to be in a better position than Einstein's theory because our line element is automatically isotropic. It may be argued that the principle of equivalence really implies isotropy of space because the gravitational energy is by definition only a function of position whereas the inertial energy could depend on direction if velocity of light were different in different directions. Thus the equality of inertial and gravitational energy (mass) seems to imply the isotropy of the velocity of light, hence the automatically isotropic functional solution (7.20) could be said to more appropriately represent the experiment on the isotropy of inertia.

For the time-delay of radar signals echoed from Venus there seems to be a curious difference in the prediction. This difference, however, does not arise from a difference in the terms or approximations of the line elements but rather from a difference in the interpretations of the two theories. In the usual theory one identifies the zeroth order special relativistic time-delay from the standard non-isotropic solution of Einstein's theory[4] whereas here there are only isotropic solutions and our correspondence procedure seems to imply that the zeroth order delay should be taken from the isotropic solution (3.2). It is then found that the time-delay seems to be $4M/c^3 \simeq 20\ \mu\text{s}$ larger than the value given by Ross and Schiff.[6] We believe the problem is worthy of further clarification especially since the correspondence procedure is a separate assumption.

It is interesting to note that Ross and Schiff have probably demonstrated clearly for the first time that Einstein's theory does not process a unique correspondence limit with regard to special relativity. For both the isotropic and non-isotropic solutions of Einstein's theory do reduce to special relativity at the limit of $M \to 0$ or $\phi \to 0$ but there exists no prescription as to which of them provides the correct zeroth order special relativistic time-delay. They found a 'plausible' way arguing, on the basis of a dynamical coincidence regarding the Keplerian laws of periods, that one should use the standard non-isotropic line elements for the purpose. But the argument is really rather weak because apart from the coincidental nature of the Keplerian analogy we would most likely take the zeroth order term from the isotropic line element if the non-isotropic solutions were never discovered hence end up predicting a delay 20 μs larger than they have stipulated.

It is important to emphasize that our procedure, based on the functional dependence $g_{\mu\nu}(\phi_{\mu}^{\nu})$ which might be constued as attaching importance to special coordinates (as in fact the procedure of Ross and Schiff also could) is a means of establishing correspondence with special relativity whereby the correct laws of nature are to be identified. Once this identification is made there seems to be no grave objection to transforming to more general coordinates where the interpretations of quantities are not referred to local field theory but are nevertheless possible in terms of appropriate reinterpretations.

It is further to be emphasized that this effect is also potentially capable of an independent check by a laser ranging of the Moon. Depending on whether the Moon is in the zenith or in the horizon the beam reflected from a metallic mirror on the Moon goes through the gravitational field of the Earth such that a difference $2M/c^3 \simeq 3 \times 10^{-5}$ μs is predicted by the above mentioned argument. This issue is in fact important from yet another point of view because as pointed out by Ross and Schiff the zeroth order time-delay is the only effect other than the perihelion motion of Mercury which is measurably sensitive to a second order term is g_{00}.

Concluding we may say that the present theory is a direct generalization of our locally Lorentz invariant field theory approach of 1958. It is essentially a tensor theory of spin 2 with vanishing rest mass. Locally it behaves similarly to a Pauli-Fierz type of tensor theory but the source term is here the stress-energy tensor of matter. It is expected that this new form, being much closer in nature to a standard Lorentz invariant field theory will be appealing to elementary particle physicists especially in its easier quantizability by the application of standard rules. Preliminary work has in fact already shown that there would be no self-energy divergence in the theory due to the exponential form of g_{00} in the limit of a static point mass. It can further be shown that in the present theory the vanishing covariant divergence of (7.14) and the geodesic equations of motion (7.17) are consistent only if $\partial_\nu(\sqrt{-g}\,\sigma\, u_\mu u^\nu) = 0$. This reduces to special relativistic conservation laws correctly.

The actual purpose of the last few paragraphs was to draw a logical connection between special relativistic quantum field theory and gravitation. Einstein's theory of gravitation, as we usually understand it, is quite isolated. It does not have anything common with other theories of physics such as electromagnetic or meson theories, has nothing to say on the problem of elementary particles, hyperons and quarks, not a single word on the non-conservation of parity or time reversal non-invariance. It is really a class by itself. The electromagnetic field, the electron field etc., have made great strides in the last forty years, whereas general relativity remained obscure, and proved almost sterile as far as the elementary constituents of matter are concerned. This situation is philosophically almost as unacceptable as it is truly deplorable. A common meeting ground, such as the requirement of a finite universe, as a means to remove the infrared divergences of field theory might stimulate a unified outlook for the physical theory as a whole, and help us in trying to find new ways of thinking about gravitation in relation to other branches of physics.

Following the same chain of reasoning we may further point out that a theory of elementary particles might be possible within the statistical theory of fields as a collection of special representations. Consider the scattering of two heavy particles. We know that $s=(p_1+p_2)^2$ and $t=(p_1-p_2)^2$ are the two relevant quantities which characterize the scattering. We may construct a distribution corresponding to such variables. Let K_0 be the operator associated with s and require a distribution which corresponds to a knowledge of $\langle K_0 \rangle$. As mentioned earlier we may write[5]

$$F_p^{(0)} = A^{(0)} e^{-\lambda^2 p^2/2} \tag{7.23}$$

where $p=p_1+p_2$ and λ is a statistical parameter which we shall treat as a constant. It is possible to regard this function as the zeroth member of a series of eigenfunctions constructed through a Gramm-Schmidt procedure. For $F_p^{(1)}, F_p^{(2)}, \ldots$ we shall have $A^{(1)}(p^2-\lambda^2)\, e^{-\lambda^2 p^2/2}$, $A^{(2)}(p^4 - -4p^{2\lambda 2}+2\lambda^4)\, e^{-x^2 p^2/2}$, so on. It can be verified that such functions satisfy the four dimensional harmonic oscillator equation[7]

$$K_0 F = sF, \qquad K_0 = \frac{1}{2}\left(-\frac{\partial^2}{\partial p^2} + \lambda^2 p^2\right). \tag{7.24}$$

It is then reasonable to assume that the eigenstates of this equation may act, in the s-channel, as statistical intermediate states of the scattering problem. These intermediate states may in turn be regarded as various unstable or metastable elementary particles. In fact the solution of Equation (7.24) by separation of variables reveals that the eigenvalue structure for the angular momentum is of the form $j = a + bs$, which is similar to linearly rising Regge trajectories. Furthermore, the three dimensional, polar coordinate solutions of Equation (7.24) possess SU_3 structure[8] which together with the above remark makes one wonder whether or not an SU_3 classification of elementary particles is conceivable along these lines.

Although what will be said below is considerably more speculative than the rest of our discussions in this assay, we nevertheless believe that it should be worthwhile to go through the arguments because with a proper extension, Equation (7.24) is found to lead to the recent, much discussed dual-resonance model of Veneziano and others. First we point out that the solutions of Equation (7.24) are stationary states, $F_n = |n\rangle$, $n = 0, 1, 2, \ldots$ corresponding to integer eigenvalues of K_0 and therefore the particles belonging to different states will be non-interacting. Since we know that particles do interact we augment K_0 by a perturbation term V and write

$$(7.25) \qquad KF = i\frac{\partial}{\partial\tau}F, \qquad K = K_0 + V.$$

Our intent will be to find, preferably deduce, a V which reproduces the dual resonance models. Below we indicate how this might be accomplished.

In order to solve Equation (7.25) we must have some information about the nature of K_0 and V. Two ingredients will be utilized: First, we remark that the interaction is very strong. We interpret this as the possibility of multiquanta states being created and annihilated. In other words V will contain not only the usual creation and annihilation operators a^+, a but also a_n^+, a_n which create and annihilate n-particle states at a given time. Such an assumption would require K_0 to be interpreted as

$$(7.26) \qquad K_0 = \sum_n n a_n^+ a_n + C.$$

We therefore generalize the oscillator commutation relations to read

$$(7.27) \qquad [a_n^+, a_{n'}] = i\delta_{nn'}, \qquad [a_n, a_{n'}] = [a_n^+, a_{n'}^+] = 0$$

where for simplicity the vector indices μ, μ' are omitted and a vacuum, $|0\rangle_n$, of all n is subsumed. Second, as part of the fundamental operator K, the term V itself must satisfy some invariant equation. We propose the simplest possible unitary invariant equation[9]

$$(7.28) \qquad a_n V = q_n' V = -i\frac{\delta}{\delta a_n^+}V$$

where q_n' is the eigenvalue of the annihilation operator a_n. We add immediately that although Equation (7.28) is highly plausible on account of its simplicity, its origin being the homogeneity of the displacements $a_n \to a_n + + \alpha_n$ where α_n is a c-number, its ultimate justification can only come from experiment. We shall therefore briefly outline a method of solution of these equations to show that this hope is justified to the extent that dual resonance models are supported by experiment.

Utilizing the commutations relations (7.27) we first solve Equation (7.28) for V

$$(7.29) \qquad V = \prod_{n=1}^{\infty} A_n e^{iq'_n (a^+{}_n + \alpha_n) + i\bar{q}'_n (a_n + \bar{\alpha}_n)}$$

where A_n are normalization factors and α_n, $\bar{\alpha}_n$ represent c-number displacements over the operators a_n^+ and respectively. We then set up, in analogy to the usual field theory, an interaction representation

$$(7.30) \qquad i\frac{\partial}{\partial\tau} F = V(\tau)\, F\,, \qquad V(\tau) = e^{iK_0\tau} V e^{-iK_0\tau}$$

and an (analogue) S-matrix

$$(7.31) \qquad F(\tau) = S(\tau, \tau_0)\, F(\tau_0)\,.$$

Through the usual iteration process we shall obtain

$$(7.32) \qquad S = \sum_{l=0}^{\infty} \frac{(-i)^l}{l!} \int_{\tau_1}^{\tau} \mathrm{d}\tau_1 \int_{\tau_0}^{\tau_1} \mathrm{d}\tau_2 \ldots \int_{\tau_0}^{\tau_{l-1}} \mathrm{d}\tau_l T\,\{V(\tau_1)\, V(\tau_2) \ldots V(\tau_l)\}$$

where T is the usual (but suitable extended) τ-ordering operator. We compute $V(\tau)$ from Equation (7.30) as

$$(7.33) \qquad V(\tau) = e^{iK_0\tau} V e^{-iK_0\tau} = \prod_{n=1}^{\infty} A_n e^{iq'_n a_n + e^{-in\tau} + i\bar{q}' a_n e^{in\tau}}$$

where, for the time being, a_n are absorbed into A_n (through the displacements α_n the solution in general allows fields of the form e^{ikx} as part of V. Note also that from the exponent of (7.33) one can drop $i\bar{q}_n' a_n e^{in\tau}$ if $V(\tau)$ is applied on $|0\rangle_n$). Equation (7.33) requires that the exponents, $\phi_n(\tau) = q_n' \, e^{-in\tau}$, are to be normalized like the Fourier components, ϕ_n, of a finite, elastic string problem

$$(7.34)\qquad -\frac{\partial^2}{\partial\tau^2}\phi_n = n^2\phi_n,\qquad q_n' \to q_n\sqrt{2/n}$$

where $\sqrt{2}$ is due to a choice of units. At this point all that remains to be pointed out is that our function, $V(\tau)$, as determined by the Equations (7.28, 33, 34) is the same functional conjectured by Fubini and Vneziano and also by Nambu[10]) in order to obtain the N-point Veneziano amplitude in a factorized form. In our present formalism these amplitudes arise from the lth term of the perturbation expansion (7.32). It may be noted that the Veneziano variable, z, corresponds to $e^{-i\tau}$. In order to fit the limits of integration of the Veneziano formulae one must have $\tau_0 \to -i\infty$, $\tau \to 0$ or else redefine $z = e^{-\tau}$ where τ is real. In this way Equation (7.32) may further be condensed into

$$(7.35)\qquad S = Te^{\int_0^\infty d\tau V(\tau)}\quad .$$

Through these relations a theory of elementary particles appears to be possible.

Of course we are well aware of the dangerous grounds we are treading in these considerations. We have no claim that resonance models shall succeed and yield a correct theory of elementary particles nor any reason to believe that physics is nearing its completion. It is nevertheless a matter of philosophic importance to behold as large a picture as we can at any given time, and we see that at present this may be possible along the lines advocated by Von Weizsäcker. Such a possibility in turn would strengthen our arguments in favor of the unity and relatedness of all science by inducing a unitary thinking in at least one of its disciplines, physics.[11]

8. Nature of a Physical Theory

Physical processes have been going on before man was evolved on Earth and raised the question of their description. Nature does not have to be described, nor can we be sure that it is inherently describable to unlimited detail. We have been able to devise a fairly efficient method of description we call physical theory, but how deeply this penetrates into the inner workings of nature and how far it can be carried is not known. That 'true' theoretical description of nature is possible must forever remain an assumption.

As far as we can see our theory is a kind of perception device (a subtle extension of elementary perception) and it has gradually evolved toward greater accuracy and scope. Evolution of such a system can proceed only by taking advantage of some regularities and correlations that exist in nature, and then only to the extent of those regularities and correlations. Therefore, it is essential that the basic statements of a theory be extracted from the realm of experience and establish correspondence with the ideas and theories which are known to be valid from previous experience. Moreover, the concepts and symbols used in the formulation of such statements must stand, as much as possible, in close correspondence with the experimental and observational operations. That is why we must try to formulate principles which are preferably macroscopic, directly testable and intuitively appealing. It is, however, not a necessity that every symbol used in a physical theory be a measurable or directly observable quantity. For science is not nature itself but only a perceptual construct. Some of the symbols which go into this construction may be there just to provide criteria (like guage functions and probability amplitudes) for the observability of other symbols – the former being essentially unobservable. Such concepts and symbols may refer to some inner, structural aspects of theory making, and need not have a meaning in reference to the external world. It would be conceivable that in an ultimate theory in which the observer's body and mind are explicitly represented in the equations such symbols may become representations of realities, but we are far from such a theory. Ultimately, of course, science is part of nature, and it can be conceived as nature perceiving itself in the form of man. Again we have as yet no way of contemplating a physics of such comprehensiveness except in poetical ecstasy.

Human logic (and therefore mathematics) as a formalized and highly refined outgrowth of ordinary language has the weakness that it is never completely decidable. Thus within any given set of symbols and postulates there will always be statements which are undecidable. Surprisingly the question of consistency to which we made much reference in our discussions turns out to be one of those undecidable concepts. It is possible to render some of these statements decidable by allowing ourselves to go beyond the original symbols and axioms but then there will be new statements which are undecidable.[1] Consequently, a given logico-mathematical system will cover a certain field of truth statements, leaving outside an infinite realm which is undecidable. This property of logic and mathematics is just what would be expected from an abstract perception device. It deals with an organized subset of a larger, conceivable set of possibilities and relationships, but within it there will be thought patterns, which are, so to speak, projections of the meaningful statement of the larger set. Relative to the smaller set these can be meaningless or undecidable. Now physical theory is a logico-mathematical construction intended to represent the relationships of the actual physical world. The question then arises as to what may be the criteria for the acceptability of a physical theory. The following heuristic position may be suggested: A theory should be considered adequate if, within its logic, its decidable numerical results are checked to available experimental accuracy and its undecidable numerical predictions are not known to occur in nature.[2] If we have two such theories the preference would then depend on simplicity, beauty, directness, overall goodness of fit, tradition, politics, personal and social taste, vested interest, emotional involvement, etc.

This digression into the nature of logic and mathematics is not intended to formulate an epistemological principle or to undermine the faith in the usefulness of mathematics in physical theory. But it is well to keep in mind that in constructing a theory we are not stating eternal truths, nor creating unambiguous worlds. Rather, we are evolving a little seeing eye on which we project a little image of the world. This image is not in one-to-one correspondence with nature, does not show her in exhaustive detail and has properties of its own, imparted to it by the very nature of the eye we have constructed.[3] The intuitive feelings of understanding, comprehension or grasp refer to our own impression of how direct, convenient consistent and non-redundant this image is rather than how true on real the relationships of our descriptions are. Likewise the equations we write down on paper do not describe nature; they describe our information about nature. This should not at all be surprising. In nature there does not exist such a thing as a ready made law or theory. Such phrases as true descriptions, absolute consistency, etc., are not of any real concern here. We never know whether we are right or wrong until the predictions of a theory are tested and then we will know more definitely if it is wrong. We can never find out if a theory is right although, of course, our confidence may increase every time we test a new prediction.

Scientific truth, in the form we know it, is a projection of ours. This projection evolves through long individual and social struggle of trial and error, prediction and testing, and through creative invention of concepts, toward greater objectivity, consistency and scope, and at the same time toward greater simplicity, economy and beauty – all these terms carrying only relative connotations and not absolute ones. In essence this is the same process through which our elementary perceptual powers came into being. In both spheres all we are trying to achieve is, in a very real sense, a representation as accurate as our observations require and a logical structure as simple as our minds can construct and comprehend. Beyond this scientific truth seems to have no operational meaning. We usually think that the laws are there for us to discover. This seems not quite so. Our active creation of a logico-mathematical framework in which facts acquire meaning seems to be an integral part of science and theory.

Science, as in perception, often possesses premature categorizations as a matter of expedient and our views of the world usually change in Gestaltic jumps from one categorization to another. If nature obeys a finite number of laws expressible in finite sentences, one of these categories could conceivably cover all that man will ever observe. On the other hand, it is entirely possible for nature to be such that no single theory constructable by the human mind is sufficient and that two or more independent or mutually exclusive theories are necessary to interpret it, in some logical, functional combination or on a purely statistical basis. Figuratively speaking, do we not need two separate eyes to really see, and do we not close them both to hear the music?

The mainspring of all cognitive adaptations seems to be the fact that in nature and in ourselves there exist certain regularities. In our struggle to survive we tend to adapt to some of these regularities and our adaptations manifest themselves as certain relationships between the sensed or measured quantities. Partly due to the existence of these external regularities independent of individual human beings, and partly due to the close similarity of all human beings, these relationships tend to have a subset which is common to a great many people so that it can be communicated. This subset is objective in the sense that up to a transformation depending on the frame of reference or frame of mind of the individual its meaning is the same or very nearly so among the communicants. There is great individual and social survival value in expressing, conveying, extending, categorizing and exploiting this subset because on the one hand it tends to lead to proper actions and responses in life situations, both individually and socially, and on the other hand, to a common ground of communications activities, in the form of languages, arts, sciences, games and sports and thereby makes life more pleasant and strengthens social bonds.[4] If not extended to all mankind this often results in special groups and their special follies and interests but, in the large, it eventually tends to become universal, at least in the case of man, due to the more efficient, communicative powers of man. Such a background is both the condition and the outcome of what is usually called psycho-social evolution of man. Physics as a science of external objects with measurements of ever greater precision and logical criteria of higher standards occupies a special position among all such activities. However, we must not forget that it is still only a member of man's larger domain of cognitive endeavour and its privileged position is due to the special nature of its problems rather than a basic distinction of its origins.

In conclusion the philosophy of science advocated here assumes science as part of the whole biological and psycho-social evolution subject to the same or analogous processes of selective retention. Such mechanisms range from genetic fixation to exasomatic cultural accumulation in more or less discernible forms of variation – struggle – selective retention themes of organisms, organs, organizations, powers, percepts, ideas and theories. The domain of cognitive theory is so defined that it does not become chemistry of molecules by lowering it to the level of viruses and does not lose its meaning by including all possible assumptions and hypotheses our minds cannot yet conceive or select. Somewhere between the two limits we chose perceptual organizations as the central theme from which we propose to descend to lower, sensory phenomena, to detection and differential reaction phenomena and in the other direction we ascend to gestalt, projection, and language processes until we reach the area of formalized physical theories. Our aim is, of course, to understand and, if possible, contribute to the healthy growth of knowledge. For this purpose we take the insights and statements concerning the theory of perception as a basis and use them to understand, possibly derive the properties of sensory, perceptual and conceptual organizations. Included in this program is a recasting of modern physical theories in the light of what we know about perception and its epistemology. This research is in a sense similar to Euclid's inferential efforts in geometry, as against the Egyptian accumulations of geometric rules. It is in itself a theory to be tested with further applications. It is recognized that such endeavor could lead, in view of decision pressures arising from the urgency of life situations and from the brevity of life itself, to some categorized alternatives or logical, statistical combinations thereof. The role of personal, social, aesthetic selection processes, economy, beauty, symmetry, generality, objectivity, analogy, fulfillment, vested interests etc., are not excluded. Man's niche, at least in the sense of indirect, observational and mental, inferential sense is defined as the whole observable universe and at any given stage of evolution and science his right to a unified world philosophy is recognized and in fact defended as a matter of need and challenge.

9. A THEORY OF PSYCHO-SOCIAL EVOLUTION

In this section a theory of the human phase of biological evolution – the so-called psycho-social evolution – is suggested. Our purpose is to formulate a small number of postulates, and possibly some subsidiary conditions, so as to lead logically to the features which typically belong to the human race and its society, as distinct from animals and animal societies.

According to Julian Huxley "there is as yet no single science of man in the sense of an organized branch of inquiry with a common body of postulates and ideas.[1] These sciences from psychology to history, from ethnology to economics are still groping for a central core of principles." He himself suggests the rather mystical concept of 'fulfillment' as a driving force for human progress. In this memorandum we suggest three simple principles which seem to lead in a straightforward manner to all human traits. In our proposed program we hope with suitable refinement and additions to formulate a fairly complete set of principles around which the main features of human sciences can be organized as a unified body of knowledge.

It seems to be generally agreed by many prominent biologists of our time that in man's mental and psycho-social behavior the biological evolution has entered a new and very significant phase.[2] This is man's method of evolution through cultural transmission. In this phase the main source of change and progress is no longer the Darwinian process of mutation and natural selection. Progress is achieved rather by the transmission of a common pool of human experience to subsequent generations through cultural beliefs, religions, arts, symbols, sciences, and transmissible skills. In this phase in addition to (or instead of) the genetic evolution we have to consider the self-reproduction and self-variation of mental activities operating via the medium of acquired cultural inheritance. The process here is typically Lamarckian instead of Darwinian. The possibility and explanation of such a new direction in evolution must of course be based on the nature of man and his brain organization; but the ultimate consequence of the process transcends the individual man, even shapes and determines his environment and the human destiny as a whole.

Thanks to the writings of Darwin and a century of subsequent workers (Neo-Darwinians, etc.) we have now a fairly complete understanding of biological evolution (differential reproduction through mutation and natural selection). But we do not as yet have a comparable theory of the second phase to explain the progress and the transformations of the human society. The purpose of this section is a search for a small number of fundamental principles through which the main features of human phase may be understood. In the next section we have postulated some preliminary principles which seem to lead logically to the over-all features of the human being and his social behavior as a distinct from animals and animal societies. The crucial postulates separating man from animal appear to be three: (a) the unbiased working brain, (b) the power of manipulation, and (c) the power of symbolic communication. Without these, all the remaining biological survival mechanisms would lead only to instincts and reflexes, devoid of the possibility of extensive learning, devoid of any sort of value judgment, and devoid of any real handing down of acquired experience to future generations.

Man, like other organisms on earth, has a long evolutionary genetic background based on survival of the race. In addition to this general postulate which is valid for all organisms, we propose the following three statements:

(1) The human brain tends to have no intrinsic bias.
(2) Man has the power of manipulation.
(3) Man has the ability to communicate through symbolic representation.

Before we indicate a possible avenue of postulational approach toward the understanding of the human phase, we may quickly summarize those features of man which are distinct from animals: In man's mental and social development two crucial features are speech and the creation of a common pool of organized experience. These are essentially social in nature but their realization must eventually rest on the brain organization of individual men. We therefore also enumerate characteristics which belong to the individual men.[1] Man is the only organism with a sense of right and wrong, good and evil. He is the only one with a power of abstraction and generalization. He alone has a notion of values and the idea of the ultimate end. Guilt and sin are also among the unique properties of man. Activities purposefully pursued and developed for their own sake exist only in man. Man alone devised sports and games with arbitrary rules. Religion and the feeling of awe and of sacredness are also among the unique features of man. Some animals seem to have a sense of beauty but man alone deliberately creates art and ardently pursues it. Man seems to be the only organism who can think of things in their absence. He alone possesses the power of imagination and grasp. Man is the only organism with a sense of past, present, and future. He alone has the idea of ideal and destiny. Man is constantly making conscious choices. He alone has what we call conscience, responsibility, and a feeling of rightness or wrongness. He is the only organism habitually subject to mental or emotional conflict. He alone has the power of symbolization, and he alone has conceived and pursued the idea of scientific truth. He has a greater degree of motivation than other organism.

The list may be extended considerably more but our purpose here is to indicate how a simple set of postulates can logically clarify and organize distinctly human traits into an understandable whole.

Next we present the attempted postulates and their possible origins and justifications. We do not pretend that these are complete or absolutely correct. Our aim here is to make a reasonable start. During the projected study we intend to examine, modify, and extend these postulates so as to lead more directly and more efficiently to an understanding of human phase.

These principles are not intended to be exact statements, like Newton's laws. They are rather Darwinian types of statements which enable us to follow the logical chain of developments. In fact, with proper constraints they may be looked upon as consequences of the evolutionary theory.

Man is a biological system evolved genetically through natural selection, and as such his over-all characteristics are directed toward the survival of the species. The evolution as a general principle needs hardly any explanation. In our context it enables us to classify all events and activities in man's life into three broad categories: (1) those which are survival-promoting, (2) those which are detrimental for survival, and (3) those which are neutral relative to immediate survival. We will use the evolutionary principle mostly in this sense of providing a reference frame for the classification of actual and potential events that can happen. We will see that without such a classification no system can have any value judgments.

The first postulate is a statement about the human brain. Of course we do not intend to say that man's brain is completely free of bias. We only intend to say that in man it is more pronounced when compared to animals and we need the property as part of our postulational structure to explain human traits. The need and the origin of this property of the brain can be understood as follows: When a certain stage of organization and complexity is reached in an organism's evolution, the number of situations to which the organism must respond greatly increases. Furthermore, each response becomes conditional to the presence or absence of various factors. This is especially true when the organism starts dealing with many objects simultaneously. The number of possible situations arising from only a mere permutation of ten objects is $10! \simeq 3 \times 10^6$. Since most possible situations and conditions would not occur sufficiently often in the lifetime of an organism, it is evident that evolution of a reflex for each possible situation would be out of the question. Therefore an organ which has no intrinsic bias but enables an organism to respond as the situations arise becomes a necessity. Such an organ resembles a blackboard as against a carved stone. Things to which we respond enter consciousness as needed. They are effaced when we turn our attention elsewhere. We have, so to speak, a continuum, like a blank paper, and a means to elucidate an image of some part of the external world when it is necessary. When things are not equally probable, a learning giving greater weight to the more probable situations must be possible. This brings us to learning through experience. In this context learning is related to the ability of organizing the unbiased brain so as to assign proper weights of various situations. Without this implementation of learning the unbiased brain would be powerless as an instrument of action. On the other hand inborn bias would be closer to reflex or instinct – an animal trait.

The second principle, manipulation, is the expression of the pronounced active nature of man with regard to his environment. Man has the capacity to change, rearrange, and transform objects. In this way he can modify his environment to a much greater extent than any other organism and does this consciously. Manipulation leads also to creation of situations rather than waiting for them to occur. In this way a larger number of possibilities can be realized. This gives man the opportunity to learn more, to check and to create situations which normally do not occur in our particular physical environment. The eventual effect of such a process is the extension of the habitat, creation of new environments. Creation of new environments and experimentation with them leads to extension of knowledge.

The last principle, the symbolic communication between individuals, enables man to learn through the experience of other individuals. Communication between individuals exist in other organisms to some extent but theirs is not symbolic. A cat shows the sign of anger, a dog displays its intentions, but our words 'anger' and 'intention' are not signs but symbols. As sounds they have essentially nothing to do with situations they represent and they are different in every language. Through the power of manipulation man can represent the objects and situations in permanent material symbols. Thus we can carve the words 'anger' and 'intention' into stone or write them on sheepskin. In this way communication of experience with individuals of other places and times becomes possible. The result of this process is the spread and the accumulation of a common pool of human experience.

The need for symbolic representation lies mainly in economy. It would be impractical to bring a lion in when we want to communicate the fact that lions can be dangerous. It is a social need. But in man the realization of it is not instinctual as it is in the case of some interesting communication in bees. It is learned consciously and is related to the extensive manipulative power of the human brain.

Now we would like to show that all the human characteristics enumerated in Section 5 can essentially be understood in terms of these principles. Since evolutionary background is common to all animals, the difference between animal and man rests on the missing of one or more of the properties expressed in the other three principles. For example, monkeys seem to lack only the symbolic means of communication. Bees are able to symbolize and communicate to a certain extent with their dances, but they lack the manipulation of objects and the unbiased brain. Their actions are controlled by reflexes and instincts. Since bees do not manipulate objects, their communication is only with other bees, near in space and time. They cannot accumulate culture.

An immediate consequence of shifting from reflex to an unbiased brain is decision. As the situations arise, an unbiased brain considers various possibilities before making a choice for action. But the outcome of actions are new situations for the brain to consider. In this way a sequence of situations will arise. Now some actions are survival-promoting for the individual or the species as a whole, some are detrimental. Since in the domain of unbiased brain activity there is no reflexive or instinctive mechanism to control useful and harmful actions, the idea of right or wrong as conscious judgments will arise. These words are used here with the implication of good or evil. Actions which are on the whole neutral relative to survival are neither good nor bad. This, coupled with the power of manipulation, leads to responsibility for conscious actions and thus to the idea of guilt and sin. The concept of correct and incorrect refers more to the external world and is related essentially to the manipulation postulate. When we manipulate things, certain arrangements work, certain others do not work. If we superimpose two triangles, they either fit or do not fit. The fitting may not have anything to do directly with survival although the knowledge of correct and incorrect combinations can eventually be turned to evolutionary significance. Here we see that without the power of manipulation the idea of truth or development of scientific attitude would be practically impossible. Likewise some categorization processes and also aspects of human motivational characteristics are consequences of manipulative power.[3]

The ability of thinking of things in their absence follows from the ability of symbolic representation of things and situations. This simply operates in terms of manipulating the symbols in place of things that they represent. Imagination is also related to the manipulation of symbols – it is the construction mentally of whole processes and systems in terms of mental images. This property, coupled with the successive outcomes of actions as new situations, leads to the concepts of sequences, series and the idea of the future and past. The idea of ultimate end and destiny follows.

The arts, sports, and hobbies seem to be among the most intricate to understand. They fall under the classification of things which are neither immediately beneficial nor detrimental to survival. But an active unbiased brain tends to scan various possibilities. It wanders around among possibilities. Through the power of manipulation some of these possibilities can also be actually realized. Since through the biological background man has a certain built-in structure, some of these constructions will be more interesting to his nature. Through the power of communication such constructions can be communicated and eventually become a medium of expression. They can also become a medium of relaxation or a means of social bond.

All adaptation can only take place by conforming with the laws and regularities of the external world. In this, man either adapts to the external world or manipulates the external world to a smaller or greater extent to adapt it to himself. Animals do not in general manipulate nature and therefore their adaptation is one-sided. Animal adaptation can take place through natural selection and the laws of nature do not have to come into consciousness directly. In a large scale adaptation of nature to man's need,s a conscious feeling of the regularities of nature is necessary. Such a feeling may originally develop by trial and error, especially through the idea of correct and incorrect. Then manipulation of things with a feeling of knowledge of what works and what does not work can quickly become a powerful means of survival. From this the origin of pure and applied science may be understood.

Through the symbolic means of communication, the sum total of all human experience can be transmitted to future generations, thus leading, as a grand over-all effect, to the mechanism of human progress and evolution via cultural transmission.

The process of evolution through cultural transmission, that is, the human phase of progress, is discussed by various authors. Pierre Auger draws analogies between 'ideas' and 'organisms'. According to him, ideas breed and multiply in the brains of individuals when the idea is right or useful. When wrong or harmful they die. The sum total of good and useful ideas that survive through a process of selection becomes the repertoire of culture for the human race. Julian Huxley suggests that the driving force behind the psycho-social evolution may be a sense of 'fulfillment'. According to him, in the human phase, struggle for existence has been largely superseded by the struggle for fulfillment. The satisfaction of needs, spiritual as well as material, the emergence of new qualities of experience to be enjoyed are achieved through struggle. But when achieved, one experiences a feeling of fulfillment. Thus he too seems to draw an analogy to Darwinian mechanism of evolution through the idea of struggle for fulfillment.[4]

It is interesting that our principles too seem to lead to a kind of analogy between the psycho-social and Darwinian evolution. The unbiased brain is continually considering all kinds of new possibilities and through the power of manipulation some of these are actually realizable. Not all of these constructions are good or useful, however. The ones which are found to be obviously harmful for the survival of the race as a whole are inhibited through law, tradition, or the concept of sin. The ones which are useful are selected and transmitted to future generations through the symbolic means of communications, tales, rituals, written messages, and books. The following crude analogy to Darwinian mechanism can therefore be drawn: mutation-imagination, selection-validation, transmission by genes-transmission through cultural inheritance. Note, however, that this analogy is not to be taken too literally for, as has been mentioned above, the psycho-social evolution is a Lamarckian and not a Darwinian process. Nonetheless the process has further analogues in the scientific method and intelligent behavior in general.[5]

Perception Technology Corporation,
Winchester, Mass.

NOTES

Appendix 1. Nature of a Perceptual Theory

[1] H. Yilmaz, 'Color Vision and New Approach to General Perception', in E. E. Bernard and M. R. Kare (eds), *Biological Prototypes and Synthetic Systems* **1**, Plenum Press, New York, 1962, pp. 126–41.

[2] Formally Equation (1.1) is similar to the optimum principles of mechanics such as the principles of least action and the principles of optimum control. The common ingredient of all such principles seems to be that anything permanent or at least quasi-permanent in time must satisfy a stability condition which in turn implies certain relationships or equations among the constituent variables. A variational principle is a way of deriving all such equations from the optimality of a single suitably defined cost. Note, however, that there may be equivalent but nevertheless different costs leading to equivalent theories with different interpretations.

[3] An example of such exploration via the statistical fluctuations is the process by which a physical system reaches thermal equilibrium, H. B. Callen, *Thermodynamics*, John Wiley & Sons, Inc., New York, 1963, pp. 267–92.

[4] The ancient 'principle of optimal design' seems to have been formulated clearly for the first time by N. Rashevski (*Mathematical Biophysics*, Vol. II, Dover, New York, 1960, p. 292) and applied by him and his students to many biological problems; R. Rosen, '*Optimality Principles in Biology*', Plenum Press, New York and London, 1967. Note that Rashevski implies not only the optimization of the physical configuration of design but also functional behavior of the organ or organism with respect to the environment and other organisms in the environment.

[5] We may argue that the Earth, in the presence of the energy flow from the Sun, is in a steady-state condition so that its negentropy content and the rate of production of entropy, on the average, both remain constant. We write, $\eta = \mathrm{d}S/\mathrm{d}t = -\mathrm{d}N/\mathrm{d}t$, $N = -\eta t + K$, where K is the finite entropy content available for order and organization and η is the flow rate of entropy available for the maintenance of various physical and chemical processes. The amount ηt is, of course, removed continually by the cooling effect of heat radiation by the Earth into space. Thus, in an oversimplified picture where the effect of radioactive decay, heat flow from the interior of the Earth, etc., are omitted, the total negentropy, K, and the total flow, η, remain fairly constant and finite. It is clear that an organized system which reproduces itself in large numbers eventually has to come into conflict with the limited supply (actually only a fraction of K and η) leading to competition. Since through mutations, fluctuations, etc., there are considerable variations in the reproductions it is also clear that only those systems which are able to sustain stability with the least amount of accessible K and η can exist in largest numbers. With minor reservations on transients, fluctuations and occasional storage of K in the form of fossil fuels, etc., we may then identify the overall cost, A, per individual, say, within the community of a species, as

$$A = \eta' + \lambda K'$$

where η' and K' are the flow and the negentropy content represented by the organism and λ is a Lagrangian parameter. In essence this expression seems to cover the barest essentials of design at a most idealized functional level and must be supplemented with subsidiary conditions involving energy, material properties, etc. Then the statement of optimality on A which applies over the whole lifetimes and over the whole body of an organism may often be broken up into parts which apply over shorter periods and over the component organs including the faculties of perception and conceptualization. On the other hand when applied over the whole biological span it would seem to imply that adaptive radiation into new niches, social structuralization, interspecies adjustments and ultimately the selection, design and evolution of genetic material itself ought to come under its jurisdiction. At the present stage of biophysical research the latter question is hardly on anybody's mind but eventually is bound to become an issue of major importance.

In this connection a point of fundamental importance is the nature of the method to be applied in the course of optimization. It is clear from the above that ultimately the cost function has to be expressed in terms of probability and information and this makes the whole process statistical. The method is therefore a statistical one although one usually is able to find parameters which can be treated as though they were continuous classical variables. The analogy here is something like the usual thermo-dynamic variables as against the statistical mechanics which provides a firm foundation to the former and, in addition, is able to treat also phenomena of variance and fluctuation. Recent studies of sensory processes from the point of view of detection and statistical decision theory should be considered as positive steps in this direction.

Appendix 2. The Psychophysical Law

[1] The principle that percepts tend to be invariant against the variations of the environmental situations appears to have been conceived by many people. For example, as early as Seventeenth Century Descartes introduced a similar argument in the perception of shapes. However, a clearer exposition and the idea of combining it with the concept of homological modeling and thereby deriving the psychological power law from them seems to be a recent event; H. Yilmaz, 'Perceptual Invariance and the Psychophysical Law', *Perception and Psychophysics* **2** II (1967), 533–38.

[2] S. S. Stevens, 'On the Psychophysical Law', *Psychol. Rev.* **64** (1957), 153–81. This paper, for the first time, established the power law with solid experimental evidence and replaced the century-old logarithmic law of Fechner. Weber's law, on the other hand, remains empirically justifiable.

[3] E. H. Land, 'Color Vision and the Natural Image', *Proc. Nat. Acad. Sci.* **45** (1959), 115–29. This and subsequent papers of Land made popular some old but little appreciated experiments in color vision which show that all is not well with classical color theory.

[4] R. Plutchik, '*The Emotions*', Random House, New York, 1962. This work, apart from being a highly imaginative approach to the problem of emotions, is interesting also from the point of view of methodology. The author does not make use of the analogues of the transformation formulae in color vision. With their inclusion some of the problems in emotional adjustment would seem to be possible to treat in the framework of the theory. In very general terms the perception theory seems to imply that there exists a state of equilibrium of perception which serves as a reference and deviations from this equilibrium are perceived as certain sensations feelings and perceptions. If the space of parameters have isotropy with respect to the equilibrium, as would be expected in most instances, then there are complementarities in perceptions and sensations. If the state of equilibrium is a result of competing processes then there are trans-

formations in the perception etc. This perceptual equilibrium, however, should not be thought of as chemical or physical kind of condition which takes place on its own accord but rather as an active organizational property of the sensory devices which is evolved and structured by the information content of the environment to be perceived.

Appendix 3. Perception of Light and Color

1 With minor modifications in the way of simplification this exposition is essentially that of Note 1, Appendix 1. For simplicity we have, however, left out some important sections such as the simultaneous contrast and experimental predictions of the theory which were later substatiated by experiments.

2 For example, the fact that the human eye is also a pattern recognizer implies that a large portion of the channel capacity of the eye has to be devoted to pattern resolutions. Since pattern acuity is very important for the organism, only very few of the channels can be devoted to color. As a matter of fact in the fovea, where pattern resolution is greatest, the color vision happens to be quite deficient; E. N. Wilmer, in Julian Huxley, A. C. Hardy, and E. B. Ford (eds.), *Color Vision and its Evolution in the Vertebrates*, pp. 306–25, Collier Books, New York, 1963. The necessity of invariance under a given set of illuminants also plays a restricting role upon the number of Fourier terms. For example if the set of natural alluminants are so smooth it can be represented by the first three of these components, then there would be no way of implementing transformations for the higher eigenfunctions. This means they cannot be made inveariant hence not useful.

3 Many of the consequences of the theory can be easily tested by an inexpensive cardboard box and a set of, preferably, wide-band plastic color filters. For this purpose perform the two room experiment of Figure 3.3 by building it in the form of a cardboard box in such a manner that one eye looks into one room and the other eye into the other room. Through the hole in the separating wall the left room is visible by the right eye and the right room by the left eye. In the beginning of the experiment the hole is closed. Each of the rooms receive colored lights through windows on which one of the color filters is placed. To give an example, let one of the rooms be lighted by a red-orange filter and the other by a yellow filter. Each room will nevertheless appear achromatic to its respective observers eye. When the hole is opened one finds that the room which receives red-orange light appears violet, and the room which receives yellow light appears green. The violet and the green observed by the two eyes are such that they are complementary to each other. Latter statement is true for any choice of two filters (representing literally hundreds of different choices). Furthermore, when the two rooms are arranged to be equal in brightness, each eye perceives the other room brighter than its own room. This effect is quite striking in the case of two quite saturated and roughly complementary filters, because then the hole appears so bright as to take on a luminous quality. These experiments show that the human color space tends to operate quite similarly to Special theory of relativity, hence the geometry involved is Lobachevskian and not Euclidean in nature. Recent studies seem to show in fact that the classical von Kries Law of coefficients has to be revised in favor of the present theory.

4 Thus despite the traditional rivalry between the two theories the Hering and Young-Helmholtz constructions represent one and the same structure. As a transformation from cartesian to polar coordinates does not carry us into a new theory in physics so in color science a mere substitution of variables cannot lead to a new theory. In fact it

becomes now quite clear that these so-called older theories of color are really not theories in the modern sense of the word, but only certain crude models which need further explanations of their own existence.

Appendix 4. Perception of Voice and Music

[1] The residue pitch was first discovered by J. F. Schouten in 1937 and fascinated scores of researchers and interested laymen under the name of the 'perception of missing fundamental'. Three of the clearest articles on the subject are: J. F. Schouten, R. J. Ritsma, and B. L. Cardozo, *J.A.S.A.* **34** (1962), 1418. R. J. Ritsma and B. L. Cardozo, *Philips Technical Rev.* **25** (1964), 37, and R. J. Ritsma, *J.A.S.A.* **34** (1962), 1224.
[2] Originally a non-linear mechanism in the ear was thought to explain the residue phenomenon. However, this ran into difficulties for reasons given in R. J. Ritsma and B. L. Cardozo, *ibid.* A time domain mechanism proposed by E. de Boer 'Residue in Hearing', thesis, Amsterdam, 1956, explains some of the gross properties of the phenomenon as indicated in the above references but this too runs into difficulties because of the existence of other phenomena known under the name of the 'second effect of pitch'.
[3] H. Yilmaz, 'On the Pitch of the Residue', Technical Report No. 41, I. P. O., Eindhoven, 1964.
[4] It might occur to the reader that similarity of color theory to relativity and the analogy of residue theory to quantum mechanics seem quite unexpected. Evidently perception theories cannot always be analogous to microscopic physical theories. The reason for the two coincidences is that we have chosen two of the sensory theories, which happen to be rather analogous. For example theory of psychophysical law has little analogue in physics and the theory of kinethetics has no similarity to quantum jumps although the theory of the phenomene has some crude resemblence to the theory of elementary particles.
[5] To obtain the existence region in terms of f we plot $\bar{n}$ against f as follows: If for a given ν, the number of harmonics implemented is $\bar{n}$, then for this ν, up to the frequency $f=\bar{n}\nu$ the residue is perceptible. Therefore in (4.10) substitute $\nu=f/\bar{n}$ and plot the implicit function $\bar{n}$ against f. This plot will give the existence region for a large modulation depth. At lower modulation depths smaller numbers of harmonics are excited and the existence region gets correspondingly narrower. R. J. Ritsma, 'Existence Region of the Tonal Residue I', *J.A.S.A.* **34** (1962), 1224. A striking confirmation of the residue effect is recently found in frequency-inverted speech. If frequency is inverted at, say, 1200 Hz, speech sounds will be inverted but not the pitch. Similarly a whistled melody will be unrecognizably inverted whereas the same tune sung by human voice will be heard as the same melody.

Appendix 5. Theory of Space and Time

[1] Galilean transformations as originally derived from the classical relativistic behavior of mechanical objects were admirably suited to the study of point particles and their motions. After the discovery that light was a wave propogation the unity of kinematical description was lost because wave motion was not consistent with mechanical relativity. Einstein in 1905 revised Galilean transformations so as to make them applicable to wave and particle equations simultaneously. So in a sense Lorentz transformations imply a

wave-particle duality at least for light, which was further exploited by Einstein in his theory of photoelectric effect. Our discussions in this appendix will show, among other things, that conceivably, the discovery of Lorentz transformations could have been made as early as 1728 if Huygens' wave theory of light were taken seriously by some imaginative theorist.

[2] The wave-particle duality implicit in the photoelectric (photon) theory of Einstein was extended by Louis de Broglie in 1924 to material particles, especially to electron, and thereby modern quantum theory was initiated. The rigorous relation between the orthogonality of rays and wave-fronts and the relativity of simultaneity also shows that Potier's 1874 hypothesis, despite the denial of first order effects of ether drag could hardly be considered as a forerunner of special relativity since it does not take into account the full reciprocity of reference frames. M. A. Potier, *J. Phys.* (Paris), **3** (1874), 201. See also R. G. Newburgh and T. E. Phipps, Jr. (submitted to *A.J.P.*). It could, however, be considered as a forerunner of the Michelson-Morley experiment, in first order. More important, Potier implicitly regarded the Fresnel drag as a time transformation instead of a velocity effect in the space part. This was real advance in the direction of modern theory of relativity.

[3] It seems that a reasonable understanding of wave and particle duality is lacking, among the majority of physicists, even to this day. This is evidenced by the confusion which reigns in many textbooks and research papers on the subject of quantum mechanics and its interpretations.

[4] The idea that space and time concepts should arise from the existence of material elements goes back to Ernst Mach. His ideas, however, were along the line of global requirements which eventually gave rise to the so-called 'Mach's principle' much discussed in general relativity and cosmology. Our intention is strictly on the local side as will be evident a little further. On a level more elementary and geometrical the perceptual character of special relativity was argued by D. Bohm, *Special Theory of Relativity*, Benjamin, New York–Amsterdam, 1962 (see Appendix, p. 185) and the present writer, *Introduction to the Theory of Relativity*, Blaisdell, New York, 1965 (see Introduction, p. xiii).

[5] Although only trivially different in appearance this new way of writing the content of special relativity changes profoundly the meaning of (5.7). This is because (5.8) is an eigenvalue equation which has radically different properties than (5.7). In fact it is precisely this difference that allows wave and particle concepts into a logically coherent entity.

[6] Both (5.10) and (5.12) may be said to be consequences of the fact that Equation (5.8) is implied by (5.9) only up to an additive constant K in x, namely, the resulting space-time concept possesses homogeneity.

[7] The real ingredient of this unity (and also the essential novelty over the classical physics) is the fact that Equation (5.9) is an eigenvalue equation. This forces us to keep track of two things simultaneously. The quantities such as $p_\mu = i\partial_\mu$ which presumably have direct physical relevance and the operands P, ϕ which condition the spectrum of eigenvalues actually available for observation. A priori these two kinds of objects do not even have to obey the same kind of logic. It becomes necessary to refer to experiment in order to see what the conceptual status is. Classically the coefficients of the eigenstates in a superposition $\phi(x) = \alpha_1\phi_1 + \alpha_2\phi_2$ are ordinary numbers so that a linear combination $a_1k_1 + a_2k_2$ of the eigenvalues k_1 and k_2 would be a result of an observation. Experiment denies this contention. It sides with quantum mechanics which states that only one or the other of the two eigenvalues k_1 and k_2 can be the result of an obser-

vation. The coefficients α_1 and α_2 are therefore not classical vector components; they are, instead, probability amplitudes. It is precisely here that lies the fundamental novelty one now has to engineer into the framework of physics. In Appendix 6 an attempt in this direction will be presented.

[8] M. Born, *Atomic Physics*, Hefner Publishing Co., 1936, p. 94.

[9] The more intuitive form of the principle of causality considers time as positive definite (orthochronous Lorentz frame). Then Equation (5.17) can be written $\Delta(x) = 0$ if $t < |x|/c$ where x is the spatial position vector. This simply means that no physical influence reaches from origin to a point x faster than the signal velocity c and before the arrival of the signal there. This way of formulating the principle of causality, however, has a minor efficiency because the usual Feynman inter-pretation of antiparticles as the original particles running backward in time cannot be retained. The alternative is to replace $p_0 > 0$ with $p_0^2 > 0$ and allow negative energy particles. This can be formulated consistently as in Dirac's original positron theory but the formalism becomes somewhat complicated.

Appendix 6. Statistical Theory of Fields

[1] Since space-time is homogeneous and the origin of coordinates are undefined the α-coefficients may in general have phase factors.

[2] This conclusion about the kinematical background of fields and particles including the reinterpretation of special relativity implicit in Equations (6.3) and (6.5) was reached as early as 1963 in a report 'On the Theory of S-Matrix', on the basis of the Equation (5.8). The publication of the consequent Statistical Theory of Fields, however, was delayed several years, mainly because of the unconventional nature of the presentation containing perceptual ingredients. Eventually it was possible to find a presentation of the theory along the lines of familiar formulations of quantum field theory; H. Yilmaz, 'Statistical Theory of Fields' in H. Feschbach and U. Ingard (eds.), *In Honor of Philip M. Morse*, MIT Press, Cambridge, 1969, pp. 170–80.

[3] Usually this point is not clearly discussed in the exposition of variational techniques. See, however, H. Yilmaz, *Introduction to the Theory of Relativity*, Blaisdell Publishing Co., New York, 1965, p. 48.

[4] L. Pauling and E. B. Wilson, *Introduction to Quantum Mechanics*, McGraw-Hill Publishing Co., New York, 1935, p. 3996. We apply the principle here in the context of the quantum theory of fields.

[5] J. M. Jauch, and R. Rohrlich, *Theory of Photons and Electrons*, Addison-Weslley Publ. Co., Inc., Reading, Mass. 1955, pp. 22–23.

[6] The action principle here uses the stationarity of the action alone and the minimality is not required. In quantum theory the principle of least action is valid only in the sense of expectation values and, precisely speaking, only in the classical limit.

[7] R. Rohrlich, *Classical Charged Particles*, Addison-Weslley Publ. Co., Inc., Reading, Mass. 1965,p. 238. Professor Rohrlich seems to have recognized clearly the deficiency of the conventional action principles but he attributes the difficultiy to the operator character of the fields solely. In our view, this is in fact part of the problem but it can be overcome if the kinematical preliminaries are understood and separated properly.

[8] For a neutral scalar field interacting with fixed c-number sources, for example we have $L = L_0 + L_I$, $L_I = g\,\Sigma i\delta(x - x_i)\phi(x_i)$, $\phi(x) = \Sigma k\alpha_k\phi_k(x)$, $\phi_k(x) = (2\omega_k V)^{-1/2} (a_k e^{-ikx} + a_k^* e^{ikx})$, $\alpha_k = N^{1/2} e^{-\zeta \cdot k}$. Then one finds the propagator

$$\Delta(x) = \frac{N}{(2\pi)^4} \int_c d^4k\, e^{-2\zeta\cdot k} \frac{e^{ik(x-x')}}{k^2 + m^2}.$$

A straightforward perturbation theory calculation then shows that the interaction energy between two sources separated by a distance r is

$$E_{\text{int}} = -\frac{\pi}{2}\frac{Ng^2\, e^{-m_a}}{a} - \frac{g^2 N}{4\pi}\frac{\pi}{2}\frac{e^{-m|r+ia|}}{|r+ia|}$$

which is finite. For a single source the self-energy, $E_s \simeq -(\pi/2)(Ng^2/2a)\, e^{-ma} = \delta M$ and the expressions

$$g_{\text{exp}} = N^{1/2} g, \qquad M_{\text{exp}} = M - \frac{\pi}{2}\frac{g^2_{\text{exp}}}{2a}\, e^{-ma}, \quad ma \ll 1$$

where $M_{\text{exp}} = M + \delta M$ corresponds to a renormalization. For the present theory the renormalization is not mandatory, however; this is because we may just as well assume $a \simeq \pi|\zeta|$ to be determined from some high energy experiments. Since g_{exp}, and M_{exp} are known from low energy limits, g and M which are the parameters that go into the unrenormalized Lagrangian are then calculable. The function $Y(r) = e^{-m|r+ia|}/|r+ia|$ and the electromagnetic analogue $V(r) = 1/|r+ia|$ derivable by a similar method to that of Note 4, p. 84 replace the Yukawa and Coulomb potentials respectively. These formulae would really apply even if a source theory point of view like Schwinger's were adopted as long as the statistical assumption and the interpretation of Lorentz frames are retained.

[9] The reason simply is that $\zeta = 0$ leads to equal probabilities for all $\phi_k(x)$ which then violates the requirement (6.7). Without such a requirement, on the other hand, integrals of field theory are not well defined (L. Streit, in Paul Urban (ed), *Quamtum Electrodynamics*, Springer Verlag, Inc., New York, 1965, p. 14.) For similar reasons theories with indefinite metric are not satisfactory although they may ameliorate some of the divergences. It seems that a theory with indefinite metric would lead at most to an asymptotic behavior $\sim \Delta(k)/k^2$ hence logarithmically divergent charge renormalization although the mass and the vertex renormalizations are finite. In such a theory there are further problems associated with causality and zero-point energy which is also divergent.

[10] N. G. van Kampen, in E. G. D. Cohen (ed.), *Fundamental Problems in Statistical Mechanics,* North-Holland Publ. Co., Amsterdam, 1962, pp. 173–202.

Appendix 7. The Problem of the Unity of Physics

[1]Nowadays the Sagnac effect is used as a laser gyro to detect inertial motion W. M. Macek and T. M. Davis, Jr., *Appl. Phys. Letters* **2** (1963), 67.

[2] Michelson-Morley experiment was recently repeated with great accuracy with laser techniques by A. Arna, T. S. Jasea and C. H. Townes. See the summary article by A. L. Schawlow, *Scientific American* **209** (1963), 2.

[3] P. Langevin, *Comp. Rend.* **173** (1921) 831.

[4] Note that the redefinition we adopted $t - |\mathbf{r}|/c \rightarrow t$ and from the law of sines $|\mathbf{r}|\sin\theta = |\mathbf{r}-\mathbf{r}'|\sin\theta'$, we have, foı $|\mathbf{r}| = |\mathbf{r}'|$, $\sin\theta' = \cos\theta/2$, hence $\theta' = (\theta+\pi)/2 \rightarrow \phi/2$.

[4] An attempt in formulating the general theory of relativity within such a context was earlier made by the author; H. Yilmaz, *Phys. Rev.* **111,** (1958) 1417–26. Since

in the normal journal publications so much of the philosophical background which led to certain physical theories is inhibited we shall here take the liberty of elucidating some of the early thinking which were essentially perceptual in nature and depended heavily on wave-particle duality.

Let a kinematically accelerated frame of reference $\Gamma(\alpha)$ be put into a one-to-one correspondence with an infinite set of Lorentz frames $L(\alpha)$ such that the fixed observers at various points of Γ are at momentary rest relative to the observers at the origins of the Lorentz observers of L. The arrangement can alternatively be thought of Γ being stationary but having a gravitational field in it. Then the Lorentz frames in question are the freely falling material frames whose origins happen to be everywhere at rest relative to the points of Γ. The existence of this one-to-one correspondence will here be considered as an expression of the principle of equivalence. This correspondence parametrizes the observers in terms of the group parameter α of L. We know that various Lorentz observers are connected by a relative velocity $v = c \tan k\alpha$. For example the Doppler shift between two Lorentz observers is $\nu = \nu_0(1 - v/c)/\sqrt{1 - v^2/c^2} = \nu_0 e^{-\alpha}$. On the other hand due to the local validity of special relativity the same parameter α can be computed in $\Gamma(\alpha)$ in terms of the proper-time, τ, and the locally measurable acceleration, γ, as

$$\alpha = \int (\gamma/c)\, \mathrm{d}\tau = \int (1/c)\, \gamma_i\, \mathrm{d}x^i = \phi/c^2$$

where we used the fact that distances are measured locally as they would be measured in special relativity, namely, $|\mathrm{d}r| = c\mathrm{d}\tau$ via the local sameness of the velocity of light. In view of the above Doppler formula this means that in $\Gamma(\alpha)$ we shall have (in units $c = 1$)

$$\mathrm{d}t = \mathrm{d}t_0 e^{-\phi}.$$

In this way we know how the time differential transforms in $\Gamma(\alpha)$. Concerning the space differentials we require that light rays will appear to fall in the space of $\Gamma(\alpha)$ according to the principle of equivalence. This means that the wave will appear as though it is transformed by a Lorentz transformation (namely, it will appear falling with the Lorentz frame)

$$\mathrm{d}r = \mathrm{d}r_0(1 + v/c)/\sqrt{1 - v^2/c^2} = \mathrm{d}r_0 e^{\phi}.$$

Contrary to a first impression the factor e^{ϕ} is to be imposed on $\mathrm{d}r = (\mathrm{d}x^2 + \mathrm{d}y^2 + \mathrm{d}z^2)^{1/2}$ isotropically and not for the direction of $\mathrm{d}r$ alone. This is because locally the velocity of light is isotropic and this implies the isotropy of inertia. If we do not require this isotropy of inertia we would (a) destroy the local validity of special relativity and (b) make inertia non-isotropic and destroy the equivalence of inertial and gravitational mass, since the gravitational energy is by definition only position dependent and therefore isotropic. Accordingly the line element corresponding the geometry of $\Gamma(\alpha)$ is of the form

$$\mathrm{d}s^2 = e^{-2\phi}\, \mathrm{d}t^2 - e^{2\phi}(\mathrm{d}x^2 + \mathrm{d}y^2 + \mathrm{d}z^2).$$

This simple line-element has the amazing property (which is actually to be expected) that if one computes the Einstein tensor $R_\mu{}^\nu - \frac{1}{2}\delta_\mu{}^\nu R$ with it one finds

$$R_\mu{}^\nu - \tfrac{1}{2}\delta_\mu{}^\nu R = (\sqrt{-g})^{-1}(-\nabla^2\phi\delta_0{}^0 + t_\mu{}^\nu)$$
$$t_\mu{}^\nu = -\,\partial_\mu\phi\partial^\nu\phi + \tfrac{1}{2}\delta_\mu{}^\nu\partial_\lambda\phi\partial^\lambda\phi.$$

In other words the reference frame $\Gamma(\alpha)$ represents a curved space in which there exists a locally special relativistic field with energy-momentum tensor $t_\mu{}^\nu$ and a source part which can be taken as

$$\nabla^2\phi = - 4\pi \sum_j M_j \delta(x - x^j).$$

The local acceleration γ_i is then nothing other than the Newtonian field of acceleration $\gamma_i = \partial\phi/\partial x^i$. Through the property of equivalence we can thus establish the connection between the local special relativistic field theory and the Newtonian theory of gravitation. It simply means that the local field theory considered is correctly transformed into a curved-space theory of gravitation. Note also that on account of our $t^\mu{}_\nu$ the present theory reduces at the correspondence limit $v/c \to 0$ to a special relativistic field theory whereas Einstein's theory does not.

Next we inquire the form of the equations of motion of test particles in the theory. In $\Gamma(\alpha)$ we have for test particles at rest the Newtonian acceleration $\gamma_i = \partial\phi/\partial x^i$. However, for light rays we must have, as a generalization of the special relativistic equation, $\mathrm{d}s^2 = 0$. This leads in $\Gamma(\alpha)$ to $\mathrm{d}r/\mathrm{d}t = ce^{-2\phi}$. It implies that light waves behave as though they are propogating in a medium with refractive index $e^{2\phi}$. But this means that if the light wave is considered as a test particle (wave-particle duality) a doubling of the local acceleration occurs. This doubling is the result of the two formulae concerning the time and space transformations above and results in the doubling of the bending of light rays compared to its Newtonian value based on a particle concept alone. Clearly the local field theory is not strictly the usual Newtonian theory. We must consider locally the motion to take place with the equations

$$-\frac{\mathrm{d}^2 x_\nu}{\mathrm{d}t^2} = \frac{\partial\phi}{\partial x_\nu}\left(1 + \frac{v^2}{c^2}\right), \qquad \partial_0\phi = 0$$

which for small velocities $v^2/c^2 \ll 1$ reduces to the Newtonian equations but for $v^2/c^2 = 1$ the acceleration doubles. Powers other than $(v/c)^2$ are not acceptable because either the principle of equivalence or the second order character of $\mathrm{d}t^2$ will be violated. Forms like $1 + Av_x{}^2 + Bv_y{}^2 + Cv_z{}^2$ are not acceptable because the isotropy requirement, hence the principle of equivalence, will be violated. That this deceptively simple equation represents locally the correct law of motion can be demonstrated by substituting $\mathrm{d}t \to \mathrm{d}te^{-\phi}$, $\mathrm{d}x^i \to \mathrm{d}x^i e^{\phi}$ and carrying out the differentiation. This process gives

$$-\frac{\mathrm{d}^2 x^\lambda}{\mathrm{d}t^2} = \frac{\partial\phi}{\partial x^\lambda}\left(e^{-4\phi} + \frac{v^2}{c^2}\right) - 2\frac{\partial\phi}{\partial x^\sigma}\frac{\mathrm{d}x^\sigma}{\mathrm{d}t}\frac{\mathrm{d}x^\lambda}{\mathrm{d}t} = +\left\{{\lambda \atop \mu\nu}\right\}\frac{\mathrm{d}x^\mu}{\mathrm{d}t}\frac{\mathrm{d}x^\nu}{\mathrm{d}t}.$$

Rewriting this expression in terms of the proper time, $\mathrm{d}s = e^{-2\phi}\,\mathrm{d}t$, we have exactly the geodesic equations of motion in $\Gamma(\alpha)$, namely,

$$\frac{\mathrm{d}^2 x^\lambda}{\mathrm{d}t^2} + \left\{{\lambda \atop \mu\nu}\right\}\frac{\mathrm{d}x^\mu}{\mathrm{d}s}\frac{\mathrm{d}x^\nu}{\mathrm{d}s} = 0.$$

Realizing that we have also the correct laws of motion, locally, corresponding to our line element we write the results of our special relativistic analysis in the differential form $\mathrm{d}g_{00} = -2g_{00}\,\mathrm{d}\phi$, and $\mathrm{d}g_{ii} = 2g_{ii}\,\mathrm{d}\phi$. Together with considerations concerning the (a) rotating frames of reference (coriolis and centrifugal forces) and (b) gravitational radiation and the local sameness of the velocity, c, of the massless fields in general, (c) the fact that the ten field components $\phi_\mu{}^\nu$ are to be a representation of the six group parameters of the Lorentz transformations, we write,

$\tilde{g} = \tilde{\eta} \cdot \exp[2(\phi - 2\tilde{\phi})]$, where $\tilde{\phi}$ is the field matrix, hence find the general differential relation in the form

$$dg_{\mu\nu} = 2(g_{\mu\nu}\, d\phi - g_{\mu\alpha}\, d\phi_\nu{}^\alpha - d\phi_\mu{}^\alpha g_{\alpha\nu}),\ \partial^\mu \phi_\mu{}^\nu = 0$$

which carries us to the opening statement (7.10) and the subsequent developments presented in outline.

The theory so constructed can be quantized locally and the results can be translated to the curved space where the factor $e^{-\phi}$ acts as a convergence factor. This factor is in fact found to suppress not only the gravitational divergences but also the quantum electrodynamic divergences without violating the gauge invariance and locality. The same factor seems, however, introduceable in the local quantized theory as a statistical factor $e^{-2\zeta k}$, $k \leftrightarrow 1/r$. The meaning of the statistical factor may then be made in relation to curved-space concept as being the condition of normalizability of the field as in (6.5), hence the scarcity of virtual gravitons at small distances near the point particles. For large scale matter distributions $e^{-\phi}$ gets small for another reason, hence both for $r \to 0$ and $r \to \infty$ the divergences occurring in quantum field theory (ultraviolet and infrared divergences) are removed by the universal presence of gravition.

From an observational point of view the above theory seems to have some definite advantages over the conventional theory. This is because (a) the solution here is automatically isotropic and there are no non-isotropic solutions. This seems to do justice to the experiments of Hughes and Drever which show that inertia is isotropic to the incredible accuracy of 10^{-23}. (b) The isotropic line-element above accounts correctly the outcomes of the three classical tests and gives a unique answer to time-delay due to our unique correspondence procedure. In the conventional general relativity one usually calculates the time-delay in the non-isotropic Schwarzchild line element and believes that it will provide the correct correspondence limit. This, however, is not clear. There are in fact people who believe that (Fock and his school, this author, Brillouin and others) the two line elements are not equivalent. It is therefore extremely important to see what the time-delay experiment eventually will say when it reaches an accuracy high enough to make a definite distinction. In view of our arguments above and in view of the isotropy experiments of Hughes and Drever we tend to believe that the result should confirm the isotropic result for δt. (D. K. Ross and L. I. Schiff, *Phys. Rev.* **141** (1965) 1215.) (c) However, according to the present theory there is an objection to the isotropic Schwarzchild line element also. This is because the second order terms in the expansions

$$e^{-2\phi} = 1 - 2\phi + 2\phi^2 + \cdots, \qquad e^{2\phi} = 1 + 2\phi + 2\phi^2 + \cdots$$

arise from the transverse Doppler and the standard Lorentz-Fitzgerald contraction of the usual special relativity and these terms must enter here symmetrically conforming to special relativity. In the isotropic Schwarzchild line-element the corresponding expansions are

$$g_{00} = 1 - 2\phi + 2\phi^2 + \cdots, \qquad g_{\alpha\alpha} = 1 + 2\phi + \tfrac{3}{2}\phi^2 + \cdots$$

and the necessary symmetry is violated in second order. Since the symmetry in question is a solidly established part in the rest of physics this seems to imply that Einstein's original field equations

$$R_\mu{}^\nu - \tfrac{1}{2}\delta_\mu{}^\nu R = T_\mu{}^\nu(M)$$

where $T_\mu{}^\nu(M)$ is the 'matter tensor' alone (without the containing the stress-energy tensor of the field) a slight omission is made leading to a violation in second order of the principle of equivalence as formulated above. That formulation, however, is a rigorous statement of the usual principle of equivalence implemented in orders beyond first, whereas, the original Einstenian implementations was only in first order. The question at issue is very crucial because aside from its extreme importance from an experimental point of view (transverse Doppler effects are observed everyday in high energy accelerators and the principle of equivalence is supposed to translate them into gravitational field correctly) the theoretically important result that, locally, the theory is a Lorentz invariant tensor theory of the Pauli-Fierz type depends entirely on this rigorous implementation of the principle of equivalence. (d) In fact this problem with Doppler symmetry in Einstein's theory is even worse than it is described above. For example for a uniformly accelerated frame of reference the Einstein's equations $R_{\mu\nu}=0$ yield $g_{00}=e^{-2\phi}$, $g_{22}=g_{33}=-e^{4\phi}$, $g_{11}=-e^{6\phi}$ with $\phi=\gamma x$. In other words Doppler symmetry is here violated in first order as well as in second. Consequently, light rays fall 4 times the Newtonian fall instead of the expected 2 times that value! (e) In the usual theory the definition of energy is quite ambiguous so that one does not know whether the gravitational waves carry energy. In fact due to the generality of the coordinate problem one does not know whether the gravitational waves propagate with the same velocity as the velocity of light. In the present theory, where there is the locally valid special relativistic field theory, the local sameness of the velocity of light is guaranteed for all fields. The ambiguity of the definition of energy does not arise because the energy is locally the same as it would be in special relativity and at infinity where $\phi\to 0$ the space reverts to the Lorentzian form. In fact in the present theory the field equations and the geodesic equations of motion are consistent under $\partial_\nu(\sqrt{-g}\sigma u_\mu u^\nu) = 0$ whereas in Einstein's theory they are consistent under $\partial_\nu(\sqrt{-g}\sigma u^\nu) = 0$. The latter implies a conservation of rest-mass which violates special relativity. Thus many of the problems of the usual general relativity would be overcome if the present approach is considered as a proper formulation of the theory of gravitation. It is furthermore justifiable to say that the present formulation of the theory of relativity both in Appendix 5 and 7 is along the line of the philosophy of science of Ernst Mach. As is well known the original Einstein theory was heavily based on Mach's ideas in mechanics but did not pay much attention to his ideas on the theory of knowledge. The present approach is, as we have already acknowledged in the main text of the lecture, to be considered a serious effort towards completing Mach's overall program both in physics and in biopsychology.

5 See for example W. H. Louisell, *Radiation and Noise in Quantum Electronics* for similar arguments on distributions not belonging to conserved quantities, McGraw-Hill Inc., 1964, pp. 221–252.

6 In elementary particle physics a linear harmonic oscillator equation was first introduced by Born in relation to his reciprocity principle, M. Born, *Rev. Mod. Phys.* **21** (1949), 463. Its extension by Equations (7.15–18) and their subsequent solution leading to the factorised Veneziano amplitudes was presented by the author. Since the validity of reciprocity as a general law of nature is quite doubtful it is probably more desirable to base it on other grounds such as the one we have indicated here. The possible connection between a linear harmonic oscillator and the Veneziano amplitude was also suggested by Suskind who conceived it along the more conventional lines of quark-antiquark interactions (L. Suskind, *Phys. Rev. Letters.* **23** (1969), 545). Venziano, Fubini and Nambu (see note 9, Appendix 7) arrived at the form of V from a study of the problem of factorization of the Veneziano amplitudes, whereas

our intent has been to determine K_0 and V from a reasonable theoretical basis and then to look for solutions that reproduce the amplitudes in a rigorous or in a sufficiently accurate fashion.

[7] H. Lipkin, *Groups for Pedestrians*, Wiley, 1965, pp. 57–68.

[8] Note for example that a_n^+ does not satisfy any eigenvalue equation. As we shall see the solutions of Equation (7.28) correspond to coherent, minimum-uncertainty wave packetts and represent mutually displaced ground-state solutions of undisturbed oscillators given by K_0. In this sense K_0 and V are intimately related to each other.

[9] S. Fubini and G. Veneziano, *Duality in Operator Formalism*, MIT preprint, CTP #114, 1969; Y. Nambu, University of Chicago, preprint, 1969.

[10] With regard to the unification process we may (a) Proceed in the sense of Weizsäcker and conceive a unified physics of quantum mechanics, elementary particles and cosmology. In the process revise general relativity. (b) Conceive theories as sets of categorized conceptualizations similar to a perceptual process. In the process revise logic, possibly as proposed by Finkelstein and others, (c) Unify all science in the sense of Jakobson by considering science as a language-based communication process where linguistics forms a bridge between positive sciences and the humanities, (d) Unify all cognitive processes based on the biological evolution and optimization in the sense of Mach, Rashevski, Popper and others.

Appendix 8. Nature of a Physical Theory

[1] R. Blanché, *Axiomatics*, The Free Press, New York, 1962, pp. 56–64.

[2] Thus if a theory is regarded as valid, what is undecidable becomes forbidden and what is decidable becomes compulsory. Physicists' dictum: 'in nature' what is not forbidden is compulsory reflects a faith in the essential correctness of his contemporary approach to physical theory.

[3] We would like to comment on an important paper by D. Finkelstein (Vol. V, pp. 199–215, in this series. See also the succeeding paper by H. Putnam.) where he argues that (a) logic is empirical, (b) quantum mechanics calls for a revision in logic, (c) the fracture is in the distributive law. While we fully agree on the first two of these statements we may not necessarily adopt the third. This is because in his demonstration of the breakdown of the distributive law, $E \cdot (L \vee R) = (E \cdot L) \vee (E \cdot R)$, he uses the incompatibility of even (E) and left (L), but the way he uses it the argument would seem to apply not only to wave functions but also to statistical distributions of a classical kind. We prefer the interpretation that a wave function or a statistical distribution represents our information about the system so that an even function is not necessarily contradictory to the leftness or rightness but merely indicates that the particle is expected at left and at right with equal probability. In this case the distributive law could be preserved but the fracture is shifted to the meaning of logical symbols which now signify eventualities rather than circumstances. The next problem – in our opinion the really crucial one – would be how to distinguish between classical and quantum mechanical information. This is analogous to Finkelstein's second step in reconstructing quantum mechanics from its logic. He assumes, besides the above mentioned fracture in the distributive law, that the sets associated with quantum mechanics have the property of coherence. He does not provide a prescription for the partially coherent case. We believe this can be done in our interpretation through a suitable definition of information. We replace the classical definition of information

(essentially the negentropy), $I = -\Sigma_k w_k \ln w_k$ where $\Sigma_k w_k = 1$, by the expression $I = -\text{trace}\,(\rho \ln \rho)$ where ρ is the density operator, $\text{trace}\,(\rho) = 1$. Implicit in this step is the assumption that the microscopic observables are operators (the c-numbers we encounter in our microscopic observations are eigenvalues of these operators) and that the objects carrying the information are themselves subject to the same laws. In other words I contains all possible information (the carrier being included as part of the total, completed system). Essentially that is why it is not proper to ask which hole the electron went through unless we had an object (say, a photon) that was actually emitted and carried some information. Even then the answer depends on whether or not the information was of a nature which warrants the conclusion. Now notice that $\Lambda = \text{trace}\,(\rho^2)$ lies between 1 and 0, such that $\Lambda = 1$ corresponds to fully coherent and $\Lambda = 0$ to fully incoherent distributions. The intermediate situations, $1 > \Lambda > 0$, which are included in our definition, describe partially coherent distributions. In other words, Finkelstein's fourth premise, (d) 'After fracture comes flow' might be easier to visualize in this form than in his original direction, although the two approaches may in the end, be equivalent. In any case the more important conclusion here seems to be that logic, as intended for the real world of physics, is empirical, and evolves in time like any other cognitive structure which claims at least a partial contact with reality.

[4] At this point comparing again the faculties of perception and theories of physics we find no essential difference between them, in principle. Both can be described as a modeling activity and both are conditioned with requirements of ever greater efficiency, economy, significance and scope. They can both be understood as part of the overall adaptation that living beings try to make to the realities of the world. In this conception, the distinctions made on the basis of what is learned what is inherited should not matter. What is inherited should be considered as something learned in the past by the species. As far as we can see there is no real obstacle in conceiving man's cognitive development as a single process of adaptations. However, we must emphasize that there are areas where further efforts are needed for clarification. For example it seems to us that language behavior and linguistics as a transitional area between perception and science, and between language and logic need careful consideration. Linguistics seems to occupy a central position between natural sciences and the sciences of man (humanities). It seems clear that a detailed study of linguistics, both as a perception and as a science, would contribute decisively toward a unitary conception of all science: See the penetrating analysis by R. Jakobson, 'Linguistics in Its Relation to Other Sciences', *Actes du Xe Congres International des Linguistes*, Editions Bucarest, 1969, pp. 75–122. In this connection we would like to mention two articles by the present author on the subject of speech perception which, in our opinion, treats some of the more elementary aspects of human speech in a form closely consistent with these goals₃ H. Yilmaz, 'A Theory of Speech Perception, I–II', *Bulletin of Mathematical Biophysics* **29** (1967), 793–825, and **30** (1968), 455–79. Further work in the direction of a statistical theory of speech perception is now in progress and will be reported separately at a future occasion. This theory explains on a purely deductive manner the results obtained by Li House and Hughes in *JASA* **46** (1969), 1019 and has similarities to the statistical theory of fields presented in Appendix 6. A theory of Psycho-social evolution of the author (unpublished, 1963) analyses the essentially Lamarckian-Spencerian process of cultural transmission in the light of language processes and identifies the essentially selection retention nature of this evolution in view of the unbiased, active mental powers of man and his manipulative,

tool-making characteristics. A science of man, leading to essentially all the distinctive human characteristic are thereby inferable at least in an outline form. In all these activities we are in close resonance with Michael Polanyi (to appear in this series of essays) and K. R. Popper (*Logic of Scientific Discovery*, Basic Books, N.Y., 1963; *Conjectures and Refutations*, Basic Books, N.Y., 1963; *Of Clouds and Clocks*, Arthur Holly Compton Memorial Lecture, Washington University, 1965) although there are also noticeable differences, especially in our greater willingness to found physical theory firmly on cognitive and perceptual propositions of a tractable nature. With regard to social status of science we find ourselves in essential agreement with the writing of D. T. Campbell (*Objectivity and Social Locus of Scientific Knowledge*, to appear in this series of essays) and others with some variations and extensions.

Appendix 9. A Theory of Psychosocial Evolution

1 J. S. Huxley, *Evolution in Action*, Harper, N.Y., 1953.
2 T. Dobzhansky, *Mankind Evolving*, Bantam Books, 1962.
3 H. Yilmaz, *Categorization and Multistability*, Unpublished, 1964.
4 J. S. Huxley, *ibid.*
5 We believe this kind of parallel was epitomized most eloquently by Karl R. Popper in his *Conjectures and Refutations*, Routledge and Kegan Paul, London, 1963, as the most and the only intelligent behavior in the face of the unknown. It is interesting to note that it is essentially the process of acquiring information by trial and error, and continual revisions in view of new information similarly gathered.

JEFFREY BUB AND WILLIAM DEMOPOULOS

THE INTERPRETATION OF QUANTUM MECHANICS

> ... no solution of the problem is possible as long as in adherence to the tendencies of Huyghens and Mach one disregards the structure of the world.
>
> Hermann Weyl ([21] p. 105)

INTRODUCTION

The quantum theory is interpreted in the technical (semantical) sense. By an interpretation of quantum mechanics we mean something much less precise. As a rough approximation, an interpretation of a theory should show in what fundamental respects the theory is related to preceding theories. In the case of the quantum theory this means understanding the transition from classical mechanics to elementary (i.e. non-relativistic) quantum mechanics.

To begin with, we distinguish between two types of physical theory: 'principle' theories and 'constructive' theories.[1] The difference is basically this: In the case of constructive theories the idea is to reduce a wide class of diverse systems to component systems of a particular kind. The existence claims to which theories of this type have led are well-known – especially in the case of the molecular hypothesis of the kinetic theory. The classical discussions of the reality of theoretical concepts have focussed on theories of this type. Principle theories have a different aim. These theories introduce abstract, structural constraints which events are held to satisfy.[2]

In this paper, classical and quantum mechanics are represented as a particular type of principle theory. We call theories of this type 'phase space theories' or 'theories of logical structure' since the type of structural constraint they introduce concerns the logical structure of events, and this is given by the phase space of the theory.[3] The logical structure of a physical system is understood as imposing the most general constraint on the occurrence and non-occurrence of events.

In Section 1 we consider the concept of an abstract structural constraint in the more familiar context of space-time theories. This motivates the preliminary discussion, in Section 2, of our concept of logical structure.

Section 3 presents an elementary characterization of the imbeddability relations between the phase space structures of classical and quantum mechanics. We mean this in the technical sense, i.e., in terms of the validity of elementary or first-order propositional formulae. The characterization depends on Kochen and Specker's crucial Theorem 4.[4]

As is well-known, the set S of statistical states of the quantum theory does not contain any states that are dispersion free. In Section 4 we discuss this in the light of Kochen and Specker's Theorem 1. In this section, we also compare the quantum theory with classical statistical mechanics. We conclude this section by relating our interpretation of the quantum theory in terms of logical structure to the original proposal of Birkhoff and von Neumann [2].

The mathematical discussion of this paper is based on Kochen and Specker's concept of a partial Boolean algebra. In Section 5, we consider two alternative representations of Hilbert space: orthomodular partially ordered sets (posets) and orthomodular lattices.

1. Space-time theories

The following discussion is based on the formulations of Anderson [1] and Trautman [18, 19]. However the central idea of this section, viz. the distinction between coordinate transformations and symmetries, originates with Weyl. (See, e.g. [21], Section 13.)

Denote by E the set of all possible histories of physical systems, and by F the subset of physical histories that may actually occur; i.e., the histories in F are dynamically possible or allowed by the physical laws. The structural constraints of a space-time theory may be understood in terms of the concept of a symmetry. By this we mean an automorphism of E on E which carries F into F. Symmetries thus preserve the laws of motion, which determine the subset F of E. The space-time structure of a physical theory is given by the invariants of its symmetry group.

It is clearly possible to restrict F on the assumption that a particular symmetry group obtains; we are then restricting the character of physical laws on the basis of space-time structure. For example, the transition from Newtonian mechanics to special relativity consists in a modification of the symmetry group from the inhomogeneous Galilean group to the Poincaré group (inhomogeneous Lorentz group). The two fundamental invariants of the classical symmetry group, the Euclidian metric, and absolute time, are dropped altogether. Space-time structure is determined by the Maxwell-Lorentz theory, and the classical laws are modified: new law of motion for rapidly moving mass points.

A theory of space-time structure, if correct, tells us something perfectly objective about the world. There is an important misconception of the relation of relativity principles to the role of coordinates in physics which suggests that at least the choice of space-time *metric* is largely, if not wholly, a matter of convention or descriptive simplicity. We discuss this matter here since it anticipates issues which will arise later in connection with the quantum theory.

To see what is involved, it suffices to consider the principle of general relativity. To begin with, we distinguish this principle from the principle of general covariance. As we use the term (cf. Anderson [1], Section 4.2), generally covariant theories have the property that the transform of a solution of an equation is the solution of the transformed equation for 'arbitrary' coordinate transformations. That is to say, general covariance requires that the coordinates should not occur essentially in the formulation of physical laws. Roughly speaking, the possibility of a generally covariant formulation of physical laws is mainly a mathematical development, which was initiated by Minkowski.

In going from special to general relativity, the symmetry group is enlarged to include all diffeomorphisms, i.e. all maps which preserve the topological and differential structure of space-time. In the general theory metrical structures do not occur among the invariants of the space-time symmetry group. This is expressed by saying that the metric is not an absolute element of the general theory, but a dynamical variable which appears as a component in the histories in F. The local validity of the special theory requires that for infinitesimal regions of space-time the metric must assume (flat) Minkowskian values.

The important point to notice in the transition from special to general relativity theory is that it concerns the symmetry group, not merely the covariance group of the theory. In contrast to general relativity, the principle of general covariance is compatible with the existence of a symmetry group which is properly included in the set of all diffeomorphisms. For example, in a generally covariant formulation of special relativity, we may replace the coordinates by their curvilinear transforms. But this leaves invariant the Minkowski tensor which represents an absolute element of the special theory. In general relativity, there are no absolute metrical elements.

So long as these two principles are not kept clearly distinct, the generalization of a relativity principle, together with the corresponding change in space-time structure which this induces, will appear to be a purely formal development. This is because it seems plausible to view a change in the covariance group of a theory as largely a matter of mathematical convenience. But even if this were true, it would be irrelevant to the interpretation of relativity principles; since they concern the *symmetry group*, not the covariance group of the theory. Similarly, hypotheses concerning the metric depend on relativity principles, and thus on the symmetry group, not the covariance group of the theory. But the character of the symmetry group, and therefore the metrical structure of space-time, is independent of how we describe the dynamically possible histories in E. So even if it were possible to show that the choice of covariance group is conventional, nothing would follow concerning the choice of symmetry group.

(Notice, even the claim that the covariance group is conventional cannot be completely correct, since the requirement of general covariance restricts the class of mathematical objects which may represent physical magnitudes, and, to this extent, restricts the actual content of the theory. On this see Trautman's discussion.)

2. Phase Space Theories[5]

The fundamental problem for a phase space theory is the *representation problem*. It is required to find a phase space structure and a probability algorithm which correctly represents the totality of all possible events associated with a certain class of physical systems. In classical particle mechanics an event is represented by a point in a subset Ω of $6N$ dimensional Euclidian space, where the $6N$-tuple $(q_1, \ldots, q_{3N}, p_1, \ldots, p_{3N})$ of real numbers denotes the coordinates of position and momentum of the N components. In quantum mechanics an event is represented by a ray in a separable Hilbert space $\mathscr{H}$. In this section, we confine our attention to phase space structures.

Consider the set of all intervals of the real line, R, half-open (on the right). The Borel subsets of R are the sets contained in the σ-ring generated by this set. A *theoretical proposition* about the system asserts that the value of a physical magnitude lies in one of these intervals. In the case of *basic propositions* (i.e., basic theoretical propositions) the intervals are atoms in the field of Borel subsets of R.

We may imagine that the propositions of a phase space theory express the result of ideal – i.e., non-interfering – measurements. In the case of basic propositions, the measurements are also infinitely precise.

Now consider the system at a particular instant. The greatest lower bound $\wedge_i\{a_i\}$, $i \in I$ (I is just some index set) of the set of all the basic propositions true of the system at that instant is called an *atom* or *atomic proposition*. Each atomic proposition determines an ultrafilter in the algebra of theoretical propositions.

Notice, theoretical propositions are algebraic objects and the structure of theoretical propositions is an algebraic structure of a certain kind. For example, in the commutative algebra R^{Ω}, theoretical propositions are associated with the characteristic functions of the Borel subsets of Ω; in the set of self-adjoint operators on a separable Hilbert space, theoretical propositions correspond to the projection operators. We adopt the following notational convention: T denotes the algebra of theoretical propositions of an arbitrary phase space theory; C, the algebra of propositions of classical mechanics, and Q, of quantum mechanics.

The phase space of the theory provides an alternative way of viewing this structure in terms of the topology of the space. For example, in the case of classical mechanics, the points in Ω correspond one-to-one with maximally consistent sets of theoretical propositions, i.e., with ultrafilters in the Boolean algebra of theoretical propositions. Now let $S(C)$ denote the Stone space of C (the set of all ultrafilters in C). The Stone isomorphism $h: C \to S(C)$ which maps a theoretical proposition onto the set of ultrafilters which contain it preserves the structure of C. Because of the correspondence between the points of Ω and the ultrafilters of $S(C)$, we may replace the Stone space of C by Ω. Then Ω is the Boolean space of C; and h is an isomorphism of C onto the perfect and reduced field $\mathscr{F}(\Omega)$ of simultaneously open and closed subsets of Ω. Under this mapping the image of a consistent set of propositions (i.e., a proper filter in C) is a non-empty closed subset of Ω. An ultrafilter in C corresponds to a singleton subset $\{\omega\}$ of Ω. The unit filter in C is associated with the whole space, and the dual of the unit filter, the zero ideal, with the empty set.

In classical mechanics, atomic events in the history of a system are represented by the points in Ω, or the ultrafilters in C, so that the algebra $\mathscr{B}$ of all possible events associated with a physical system is a Boolean algebra.

The theoretical propositions of quantum mechanics form a partial Boolean algebra. This structure may be viewed as a collection $Q = \{Q_i\}_{i\in I}$ of Boolean algebras such that for every $i, j \in I$ there is a $k \in I$ such that $Q_i \cap Q_j = Q_k$; and if $a_1, \ldots, a_n$ are elements of $Q = \bigcup_{i\in I}\{Q_i\}$ such that any two of them lie in a common Q_i, then there is a $k \in I$ such that $a_1, \ldots, a_n \in Q_k$. Q is a partial Boolean algebra if we restrict the algebraic operations to elements in Q which lie in a common Boolean algebra Q_i. For the quantum theory, Q is taken to be isomorphic to the partial Boolean algebra of linear subspaces of a suitable Hilbert space. A partial Boolean algebra may be pictured as 'built up' from its maximal Boolean subalgebras. In terms of this representation, the phase space of the quantum theory is just the isomorphic collection of Boolean spaces corresponding to the Boolean algebras Q_i. Just as in classical mechanics, an atomic event is represented by an ultrafilter in Q or the image of this filter in the collection of Boolean spaces corresponding to Q, so that the algebra $\mathscr{A}$ of all possible events is a particular type of partial Boolean algebra.

For our purposes, it is sufficient to distinguish between Boolean and non-Boolean systems of events. The logical structure of an individual event is Boolean or non-Boolean according to whether the physical system to which it belongs is or is not Boolean. The distinction depends on a reflexive and symmetric binary relation of compatibility. Let $\hat{A}$ denote the set of all possible *atomic* events which a phase space theory associates with a physical system. If the relation of compatibility is transitive in $\hat{A}$, the system is Boolean. This is the case in classical mechanics. The quantum theory generalizes the logical structures of classical physics by introducing a relation of compatibility which is not transitive in $\hat{A}$. This leads to a class of event structures which differ strongly from classical logical structures in the sense that they are not even imbeddable into a Boolean algebra.

This distinction between classical and non-classical logical structures does not coincide with the distinction between classical and non-classical formal logics. That is, non-imbeddability into a Boolean algebra is a necessary condition for the logical structure of a system of events to be considered non-classical. But the non-classical logics usually considered in the literature determine classical logical structures. Each formal logic is associated with a characteristic algebra: the Lindenbaum-Tarski algebra of the logic. This coincides with the logical structure determined by the formal logic. The Pseudo-Boolean algebras [6] associated with Intuitionist logic and the modal systems of Lewis coincide with distributive lattices. So the theorem of MacNiele [14] applies: for each such algebra, there is a Boolean imbedding. And by a result of Peremans [15], the imbedding is constructive.

The fact that there exist strongly different theoretical conceptions of the logical structure of a physical system indicates that this is as objective a component of the world as the event themselves. At least this is the major consideration in favor of realism elsewhere in science. If it is maintained that logical structure is conventional it must be possible to show that there is something which the question of logical structure does not share with other theoretical issues which would justify such an interpretation. For example, it is generally required that conventions be dispensible. Hence, if the choice of non-Boolean logical structure were conventional, it should be possible to reformulate the theory without this choice. But the logical structure of the quantum theory does not have this character.

Notice, in this connection, that we maintain a sharp distinction between logical structure in the sense of the phase space structure and the syntax and semantics of the formal language $\mathscr{L}$ in which the propositions of T are reconstructed. The choice of phase space is directly related to the representation problem, and therefore, to the quantum theory. The syntax and semantics of $\mathscr{L}$ raises a completely different set of problems. The events which the propositions of T describe are represented in a certain algebraic structure. This is given by the phase space which the theory associates with the physical system considered. Now on any reconstruction of the theory, this structure is retained, whatever the choice of $\mathscr{L}$. Thus the syntax and semantics of $\mathscr{L}$ – i.e., logical structure in the conventional sense – is not theoretically important.

There is another respect in which our approach differs from accounts which use the concept of a theoretical proposition: Usually the concept of a theoretical proposition is introduced in order to identify a phase space theory with the pair (T, S) consisting of the system T of its theoretical propositions, and set S of its statistical states. Now the properties asserted by theoretical propositions are elementary or first order properties, so they are expressible in an elementary language. But most theoretically interesting properties are not even *general first-order properties*; i.e., properties P such that a structure has P if and only if it is a model of some (possibly infinite) set Σ of first order sentences. For example, the property of being a Euclidean space is not a general first order property. This is true of other space-time properties as well. Since, in our interpretation, such properties are an essential component of principle theories, the identification of a theory with a first order reconstruction of the system of its theoretical propositions is not justified.

Finally, it is necessary to consider the objection that the concept of logical structure introduced here involves an unjustifiable extension of 'logic'. Insofar as this is not a completely verbal issue, it overlooks several important considerations.

(A) It is possible to characterize the difference between the classical and quantum mechanical phase spaces in terms of the validity and refutability of classical tautologies, the so-called 'propositions of logic'. The concept of validity employed in this characterization is a generalization of the classical concept of validity in a straight-forward sense. Both of these points will be explained in detail in the discussion of Kochen and Specker's work.

(B) The phase space structures with which we are concerned are Boolean algebras or generalizations of Boolean algebras. From a mathematical point of view, classical propositional logic is essentially a Boolean algebra when equivalent sentences are 'identified'. There is also the well-known equivalence of the representation theory of Boolean algebras with the metatheory of classical logic.

(C) The identification of logic, or logical structure with the syntax and semantics of formal languages is by no means a necessary delimitation of the subject but is due to a particular point of view: viz., Formalism. Thus, in part, this paper may be viewed as a rejection of Formalism as an adequate theory of the application of logic in physics.

3. Validity and Imbeddability

This section is essentially an exposition and clarification of the work of Kochen and Specker. For the most part, we adopt their notation and terminology.

A *partial algebra* over a field K is a set A with a reflexive and symmetric binary relation $\leftrightarrow$ (termed 'compatibility'), closed under the operations of addition and multiplication, which are defined only from $\leftrightarrow$ to A, and the operation of scalar multiplication from $K \times A$ to A. That is:

(i) $\leftrightarrow \subseteq A \times A$

(ii) every element of A is compatible with itself

(iii) if a is compatible with b, then b is compatible with a, for all $a, b \in A$

(iv) if any $a, b, c \in A$ are mutually compatible, then $(a+b) \leftrightarrow c$, $ab \leftrightarrow c$, and $\lambda a \leftrightarrow b$ for all $\lambda \in K$.

In addition, there is a unit element 1 which is compatible with every element of A, and if a, b, c are mutually compatible, then the values of the polynomials in a, b, c form a commutative algebra over the field K.

A partial algebra over the field Z_2 of two elements is termed a *partial Boolean algebra*. The Boolean operations $\wedge$, $\vee$, and $'$ may be defined in terms of the ring operations in the usual way:

$$a \wedge b = ab$$
$$a \vee b = a + b - ab$$
$$a' = 1 - a.$$

If a, b, c are mutually compatible, then the values of the polynomials in a, b, c form a Boolean algebra.

Clearly, if B is a set of mutually compatible elements in a partial algebra $\mathscr{A}$, then B generates a commutative sub-algebra in $\mathscr{A}$; and in the case of a partial Boolean algebra $\mathscr{A}$, B generates a Boolean sub-algebra in $\mathscr{A}$. Just as the set of idempotent elements of a commutative algebra forms a Boolean algebra, so the set of idempotents of a partial algebra forms a partial Boolean algebra. A partial Boolean algebra may also be defined directly in terms of the Boolean operations $\wedge$, $\vee$, and $'$. A partial Boolean algebra associated with a Hibbert space may be regarded as a partially ordered set with a reflexive and symmetric relation of compatibility, such that each maximal compatible subset is a Boolean algebra.

We restrict the discussion now to partial *Boolean* algebras. A *homomorphism*, *h*, between two partial Boolean algebras, $\mathscr{A}$ and $\mathscr{A}'$, is a map $h:\mathscr{A}\to\mathscr{A}'$ which preserves the algebraic operations, i.e. for all compatible $a, b\in\mathscr{A}$:

$$h(a)\leftrightarrow h(b)$$
$$h(a+b)=h(a)+h(b)$$
$$h(ab)=h(a)\,h(b)$$
$$h(1)=1.$$

A homomorphism is an *imbedding* if it is one-to-one, and into.

A *weak imbedding* is a homomorphism which is an imbedding on Boolean sub-algebras of $\mathscr{A}$. More precisely, a homomorphism, *h*, of $\mathscr{A}$ into $\mathscr{A}'$ is a weak imbedding if $h(a)\neq h(b)$ whenever $a\leftrightarrow b$ and $a\neq b$ in $\mathscr{A}$.

A necessary and sufficient condition for the imbeddability of a partial Boolean algebra $\mathscr{A}$ into a *Boolean algebra* $\mathscr{B}$, is that for every pair of distinct elements $a, b\in\mathscr{A}$ there exists a homomorphism $h:\mathscr{A}\to Z_2$ which separates them in Z_2, i.e. such that $h(a)\neq h(b)$ in Z_2. This is Kochen and Specker's Theorem O. The result depends on the semi-simplicity property of Boolean algebras, i.e. essentially, the homomorphism or ultrafilter theorem.

The counterpart of Theorem O for weak imbeddability is the following: A necessary and sufficient condition for the weak imbeddability of a partial Boolean algebra $\mathscr{A}$ into a Boolean algebra $\mathscr{B}$ is that for every non-zero element $a\in\mathscr{A}$ there exists a homomorphism $h:\mathscr{A}\to Z_2$ such that $h(a)\neq 0$.

A propositional or Boolean function $\Phi(x_1,\ldots,x_n)$ may be regarded as a polynomial over Z_2. To say that a particular propositional function, e.g. the function

$$x_1\wedge(x_2\wedge x_3)\equiv(x_1\wedge x_2)\wedge x_3$$

is a *classical tautology*, is to say that every substitution of elements from a *Boolean algebra* $\mathscr{B}$ for the variables x_1, x_2, x_3 yields the unit element in $\mathscr{B}$. We seek a generalization of this classical notion of validity to include substitutions from *partial Boolean algebras*. Kochen and Specker propose that a propositional function such as the above is valid in a partial

Boolean algebra $\mathscr{A}$ if every 'meaningful' substitution of elements from $\mathscr{A}$ yields the unit element in $\mathscr{A}$. A 'meaningful' substitution is one which satisfies the compatibility relations; otherwise the partial operations are undefined in $\mathscr{A}$. In this particular case, we require that the elements a_1, a_2, a_3 of $\mathscr{A}$ substituted for x_1, x_2, x_3 satisfy the conditions:

$$a_2 \leftrightarrow a_3$$
$$a_1 \leftrightarrow a_2$$
$$a_1 \leftrightarrow a_2 \wedge a_3$$
$$a_1 \wedge a_2 \leftrightarrow a_3$$
$$a_1 \wedge (a_2 \wedge a_3) \leftrightarrow (a_1 \wedge a_2) \wedge a_3.$$

This notion is formalized in the following definition:

Let $a = \langle a_1, \ldots, a_n \rangle$ be an element in $\mathscr{A}^n$, the n-fold Cartesian product $\mathscr{A} \times \cdots \times \mathscr{A}$ of the partial Boolean algebra $\mathscr{A}$. The *domain*, D_Φ, in $\mathscr{A}$ of a propositional function $\Phi(x_1, \ldots, x_n)$ is defined recursively, together with a recursive definition of a map Φ^* (corresponding to Φ) from D_Φ into $\mathscr{A}$, as follows:

(1) If Φ is the polynomial 1, then $D_\Phi = \mathscr{A}^n$ and $\Phi^*(a) = 1$.

(2) If Φ is the polynomial $x_i (i = 1, \ldots, n)$, then $D_\Phi = \mathscr{A}^n$, and $\Phi^*(a) = a_i$.

If $\Phi = \Psi \otimes \chi$ (where $\otimes$ is either + or .), then D_Φ consists of those sequences a which belong to the intersection of the domains of Ψ and χ (i.e. $a \in D_\Psi \cap D_\chi$), and also satisfy the compatibility condition $\Psi^*(a) \leftrightarrow \chi^*(a)$. The map $\Phi^*(a)$ is defined by $\Phi^*(a) = \Psi^*(a) \otimes \chi^*(a)$.

The definition of the domain of a propositional function in a given partial Boolean algebra $\mathscr{A}$ serves to make precise the notion of a 'meaningful' substitution, while the map Φ^* defines the value of the polynomial in $\mathscr{A}$ for each such substitution.

The statement that the identity

$$\Phi(x_1, \ldots, x_n) = 1$$

holds in $\mathscr{A}$ is to be understood in the sense that

$$\Phi^*(a) = 1$$

for all $a \in D_\Phi$.

The statement that the identity

$$\Phi(x_1,\ldots,x_n) = \Psi(x_1,\ldots,x_n)$$

holds in $\mathscr{A}$ is to be understood in the sense that

$$\Phi^*(a) = \Psi^*(a)$$

for all $a \in D_\Phi \cap D_\Psi$.

Now, the generalized definition of validity is this:

A propositional function $\Phi(x_1,\ldots,x_n)$ (i.e. a Boolean function – a polynomial over Z_2) is valid in the *partial Boolean algebra* $\mathscr{A}$ if the identity $\Phi = 1$ holds in $\mathscr{A}$.

Φ is refutable in $\mathscr{A}$ if for some $a \in D_\Phi$, $\Phi^*(a) = 0$ in $\mathscr{A}$.

Φ is logically valid in the generalized sense, i.e. *Q-valid*, if Φ is valid in every partial Boolean algebra $\mathscr{A}$. If the choice of $\mathscr{A}$ is restricted to Boolean algebras, this definition of validity coincides with the usual definition: the set of valid propositional formulae is just the set of classical tautologies. Thus the recursive definition of the domain of a propositional function coupled with the recursive definition of the map Φ^* just generalizes the classical, Boolean interpretation of Φ.

It is important to appreciate the distinction between the validity of a Boolean function

$$\Psi \equiv \chi$$

in a partial Boolean algebra $\mathscr{A}$, and the holding of the identity

$$\Psi = \chi$$

in $\mathscr{A}$. To say that $\Psi \equiv \chi$ is valid in $\mathscr{A}$ is to say that

$$(\Psi \equiv \chi) = 1$$

in $\mathscr{A}$, i.e. writing $\Phi = (\Psi \equiv \chi)$, we require that

$$\Phi^*(a) = 1$$

for every sequence $a \in D_\Psi \cap D_\chi$ satisfying the additional compatibility condition

$$\Psi^*(a) \leftrightarrow \chi^*(a).$$

But, for the identity $\Psi = \chi$ to hold in $\mathscr{A}$, we require that $\Psi^*(a) = \chi^*(a)$

for *every* sequence $a \in D_\Psi \cap D_\chi$, not only those sequences satisfying the additional compatibility condition $\Psi^*(a) \leftrightarrow \chi^*(a)$. Thus, the set of admissable sequences $a \in \mathscr{A}^n$ is *smaller* in the case of the validity of the biconditional than in the case of the identity. If the identity holds in $\mathscr{A}$, then certainly the biconditional is valid in $\mathscr{A}$, but the converse is not in general true. The validity of the biconditional amounts to the holding of the identity for the restricted set of sequences which satisfy the compatibility condition $\Psi^*(a) \leftrightarrow \chi^*(a)$.

For example, let $\Phi = (\Psi \equiv \chi)$ be the classical tautology:

$$x_1 \wedge (x_2 \vee x_3) \equiv (x_1 \wedge x_2) \vee (x_1 \wedge x_3).$$

Φ is not only valid in every partial Boolean algebra, it is also the case that the identity

$$x_1 \wedge (x_2 \vee x_3) = (x_1 \wedge x_2) \vee (x_1 \wedge x_3)$$

holds in every $\mathscr{A}$. For if $\mathbf{a} = \langle a_1, a_2, a_3 \rangle \in \mathrm{D}_\Phi$

$$\begin{aligned} &a_2 \leftrightarrow a_3 \\ &a_1 \leftrightarrow a_2 \\ &a_1 \leftrightarrow a_3 . \end{aligned}$$

But then a_1, a_2, a_3, generate a Boolean algebra. It follows that

$$a_1 \wedge (a_2 \vee a_3) = (a_1 \wedge a_2) \vee (a_1 \wedge a_3),$$

and hence

$$a_1 \wedge (a_2 \vee a_3) \leftrightarrow (a_1 \wedge a_2) \vee (a_1 \wedge a_3).$$

Thus, every sequence $a \in D_\Psi \cap D_\chi$ automatically satisfies the compatibility condition $\Psi^*(a) \leftrightarrow \chi^*(a)$.

In the case of a partial Boolean algebra $\mathscr{A}$ *imbeddable* into a Boolean algebra, the validity of the biconditional $\Psi \equiv \chi$ in Z_2 (i.e. the classical tautologousness of the biconditional) entails the holding of the identity $\Psi = \chi$ in $\mathscr{A}$. Thus, in the case of imbeddability (and only in this case):

$$\Psi \equiv \chi \text{ is valid in } Z_2$$

is equivalent to

$$\Psi = \chi \text{ holds in } \mathscr{A}.$$

This is a consequence of Kochen and Specker's Theorem 4, to which we now turn.

Kochen and Specker's Theorem 4 establishes an elementary condition for the imbeddability of a partial Boolean algebra into a Boolean algebra. This clarifies the relationship between the validity of classical tautologies in a partial Boolean algebra $\mathscr{A}$ and the imbeddability of $\mathscr{A}$ into a Boolean algebra. The statement of the theorem is as follows:

(1) A necessary and sufficient condition for the imbeddability of a partial Boolean algebra $\mathscr{A}$ into a Boolean algebra is the holding of the corresponding identity $\Psi = \chi$ in $\mathscr{A}$ for every classical tautology of the form $\Psi \equiv \chi$.

(2) A necessary and sufficient condition for the *weak imbeddability* of a partial Boolean algebra $\mathscr{A}$ into a Boolean algebra is the validity in $\mathscr{A}$ of every classical tautology.

(3) A necessary and sufficient condition for the existence of a *homomorphism* from a partial Boolean algebra $\mathscr{A}$ into a Boolean algebra is the irrefutability in $\mathscr{A}$ of every classical tautology.

The first part of the theorem states that $\mathscr{A}$ is imbeddable into a Boolean algebra if and only if, for every Boolean function of the form $\Psi \equiv \chi$ which is *valid in* Z_2 (i.e. for which the identity $(\Psi \equiv \chi) = 1$ holds in Z_2), $\Psi = \chi$ is valid in $\mathscr{A}$.

If $\mathscr{A}$ is imbeddable into a Boolean algebra and Φ is a propositional formula not valid in $\mathscr{A}$, i.e.

$$\Phi^*(a) \neq 1$$

for some $a \in D_\Phi$ in $\mathscr{A}$, then by Theorem O there is homomorphism onto Z_2 such that

$$h(\Phi^*(a)) \neq h(1)$$

i.e.

$$\Phi^*(h(a)) \neq 1 \text{ in } Z_2,$$

hence Φ is not valid in Z_2. Thus if $\mathscr{A}$ is imbeddable into a Boolean algebra, *all* classical tautologies are valid in $\mathscr{A}$.

(If we assume that the theorem holds, this may be proved directly, as follows: Suppose that for every tautology of the form $\Phi \equiv \Psi$, $\Phi = \Psi$ is valid in $\mathscr{A}$. Now let χ be a classical tautology. Then

$$\chi \equiv 1$$

is a classical tautology, where 1 is the constant Boolean function. It follows that

$$\chi = 1$$

holds in $\mathscr{A}$, i.e. that χ is valid in $\mathscr{A}$.)

The difference between weak imbeddability and (strong) imbeddability for the set of functions valid in $\mathscr{A}$ is just this: In the case of weak imbeddability, all the classical tautologies are valid in $\mathscr{A}$ (and in general there are also functions valid in $\mathscr{A}$ which are not classical tautologies). In the case of (strong) imbeddability, all the classical tautologies are valid in $\mathscr{A}$. There may also be functions valid in $\mathscr{A}$ which are not classical tautologies. But here we know in addition that if $\Psi \equiv \chi$ is a classical tautology, then $\Psi = \chi$ holds in $\mathscr{A}$.

Thus, for weak imbeddability, if $\Psi \equiv \chi$ is a classical tautology (i.e. if $\Psi \equiv \chi$ is valid in Z_2), we know that $\Psi \equiv \chi$ is valid in $\mathscr{A}$ (by the second part of the theorem), but we cannot conclude that $\Psi = \chi$ holds in $\mathscr{A}$. In the case of (strong) imbeddability, this inference is legitimate, i.e. from the validity of a biconditional in Z_2, we may infer that the corresponding *identity* holds in $\mathscr{A}$. This means that in the case of imbeddability we may infer the holding of the identity

$$\Psi = \chi \text{ in } \mathscr{A}$$

from the validity of the biconditional; i.e. from

$$(\Psi \equiv \chi) = 1 \text{ in } \mathscr{A},$$

whenever $\Psi \equiv \chi$ is a classical tautology, as well as the converse (which follows immediately from the definition of validity for an identity).

Notice that we cannot conclude that *only* the classical tautologies are valid in $\mathscr{A}$ if $\mathscr{A}$ is imbeddable into a Boolean algebra; since it does not follow that if $\mathscr{A}$ is imbeddable, and Φ is valid in $\mathscr{A}$, then Φ is valid in Z_2. For, to say that Φ is valid in $\mathscr{A}$ is to say that

$$\Phi^*(a) = 1 \text{ in } \mathscr{A}$$

for every $a \in D_\Phi$ in $\mathscr{A}$, and to say that Φ is valid in Z_2 is to say that

$$\Phi^*(a) = 1 \text{ in } Z_2$$

for every $a \in D_\Phi$ in Z_2. But the imbedding into $\mathscr{B}$ may only use a proper subset of the sequences in Z_2^ω associated with the elements of $\mathscr{B}$. We can conclude that all and *only* the classical tautologies are valid in $\mathscr{A}$ only if the imbedding is an isomorphism.

We give an exposition here only of the proof of the first part of the theorem.

The necessity of the condition is relatively easy to prove. We are required to show that the holding of the corresponding identity $\Psi = \chi$ in $\mathscr{A}$ for every classical tautology of the form $\Psi \equiv \chi$ is a necessary condition for the imbeddability of $\mathscr{A}$ into a Boolean algebra. In other words, we are required to show that if $\mathscr{A}$ is imbeddable, then for every biconditional $\Psi \equiv \chi$ which is a classical tautology (i.e. which is valid in Z_2), the corresponding identity $\Psi = \chi$ holds in $\mathscr{A}$.

Suppose $\mathscr{A}$ is imbeddable into a Boolean algebra, and that $\Psi \equiv \chi$ is a classical tautology. We must show that this entails that the identity $\Psi = \chi$ holds in $\mathscr{A}$. We show this by proving that

$$\Psi \neq \chi \text{ in } \mathscr{A}$$

leads to a contradiction.

If $\Psi \neq \chi$ in A, then for some $a \in D_\Psi \cap D_\chi$:

$$\Psi^*(a) \neq \chi^*(a).$$

Now, by Theorem O, since $\mathscr{A}$ is imbeddable into a Boolean algebra, for each $b, c \in \mathscr{A}\,(b \neq c)$ there exists a homomorphism $h: \mathscr{A} \to Z_2$ such that

$$h(b) \neq h(c)$$

and so there exists a homomorphism $h: \mathscr{A} \to Z_2$ such that:

$$h(\Psi^*(a)) \neq h(\chi^*(a))$$

or

$$\Psi^*(h(a_1), \ldots, h(a_n)) \neq \chi^*(h(a_1), \ldots, h(a_n)).$$

In other words, $\langle h(a_1), \ldots, h(a_n)\rangle$ is an admissable sequence in Z_2^n such that

$$\Psi^*(h(a_1), \ldots, h(a_n)) \neq \chi^*(h(a_1), \ldots, h(a_n)).$$

This means that $\Psi \neq \chi$ in Z_2, and so the biconditional $\Psi \equiv \chi$ is not valid in Z_2, i.e. $\Psi \equiv \chi$ is not a classical tautology, contrary to our original assumption.

(Notice, it would not in general be permissable to infer the non-validity of the biconditional $\Psi \equiv \chi$ from the fact that the identity $\Psi = \chi$ failed to hold in a partial Boolean algebra. This inference is, however, obviously legitimate in Z_2.)

To prove the sufficiency of the condition, we must show that the holding of the corresponding identity $\Psi = \chi$ in $\mathscr{A}$ for every classical tautology entails the imbeddability of $\mathscr{A}$ into a Boolean Algebra. Kochen and Specker prove the contrapositive; if $\mathscr{A}$ is not imbeddable into a Boolean algebra, then there exists a classical tautology $\Psi \equiv \chi$ such that for some $a \in D_\Phi \cap D_\Psi$, $\Psi = \chi$ is not valid in $\mathscr{A}$.

Let K_1 denote the set of positive statements from the diagram of $\mathscr{A}$, i.e. the sentences formulated in some first-order language $\mathscr{L}$ which describe all equations of the form $\alpha + \beta = \gamma$ or $\xi\eta = \zeta$ which subsist among elements of $\mathscr{A}$.

Let K_2 be the set of sentences formulated in $\mathscr{L}$ describing the class of Boolean algebras.

Write $K = K_1 \cup K_2$.

It is very important to bear in mind throughout the following that K is a subset of the sentences in $\mathscr{L}$, the first-order language in which the Boolean axioms are formulated, and in which relations of the form: $\alpha + \beta = \gamma$ and $\xi\eta = \zeta$ which subsist among elements of $\mathscr{A}$ are formulated.

Now, the models of the set of sentences K_1 are all homomorphic images of $\mathscr{A}$. Hence, the class of all models of K comprises all homomorphic images of $\mathscr{A}$ which are Boolean algebras.

If $\mathscr{A}$ is *not* imbeddable into a Boolean algebra, then, by Theorem O, there exists a pair of elements $a, b \in \mathscr{A}$ such that *no homomorphism onto* Z_2 *will separate them*. That is, a and b are two distinct elements in $\mathscr{A}$ which are identified by every homomorphism onto Z_2.

If a and b are not separated by any homomorphism onto Z_2, then they cannot be separated by a homomorphism into any Boolean algebra (by the homomorphism theorem, or the semi-simplicity property of Boolean algebras). That is to say, a and b are identified in every model of K (since the models of K arc just a class of Boolean algebras, viz. those which are homomorphic images of $\mathscr{A}$).

Thus:

$$K \models_{h(\mathscr{A})} a = b$$

(where $h(\mathscr{A})$ is the class of Boolean algebras which are homomorphic images of $\mathscr{A}$). Note that $a=b$ is to be understood here as a sentence in the first order language $\mathscr{L}$.

By the completeness of $\mathscr{L}$, we have:

$$K \vdash a = b.$$

By the (syntactic) compactness of $\mathscr{L}$, $a=b$ follows from a finite subset L of K_1, where

$$L = \{\alpha_j + \beta_j = \gamma_j,\ \xi_k\eta_k = \zeta_k \mid 1 \leqslant j \leqslant n,\ 1 \leqslant k \leqslant m\}.$$

Hence:

$$K_2 \cup L \vdash a = b$$

or

$$K_2, \{\textstyle\bigwedge_{j,k} L\} \vdash a = b$$

where

$$\textstyle\bigwedge_{j,k} L$$

is a finite conjunction of sentences of $\mathscr{L}$. Hence by the deduction theorem for $\mathscr{L}$:

$$K_2 \vdash \textstyle\bigwedge_{j,k} L \to a = b.$$

Clearly, $\bigwedge_{j,k} L$ is logically equivalent to the conjunction

$$\textstyle\bigwedge_{j,k} \{\alpha_j + \beta_j + \gamma_j = 0,\ \xi_k\eta_k + \zeta_k = 0 \mid 1 \leqslant j \leqslant n, 1 \leqslant k \leqslant m\}$$

of sentences of $\mathscr{L}$, which is logically equivalent to the sentence:

$$\textstyle\bigvee_{j,k} \{\alpha_j + \beta_j + \gamma_j,\ \xi_k\eta_k + \zeta_k \mid 1 \leqslant j \leqslant n, 1 \leqslant k \leqslant m\} = 0$$

where the sign $\bigvee$ is to be understood as the supremum or least upper bound *in the partial Boolean algebra* $\mathscr{A}$. That is,

$$\textstyle\bigvee_{j,k} \{\alpha_j + \beta_j + \gamma_j,\ \xi_k\eta_k + \zeta_k \mid 1 \leqslant j \leqslant n, 1 \leqslant k \leqslant m\}$$

denotes an *element in* $\mathscr{A}$: the least upper bound of all the elements of the form $\alpha_j+\beta_j+\gamma_j$, and $\xi_k\eta_k+\zeta_k$. The sentence asserting that the least upper bound of all these elements is zero is equivalent to the conjunction of all the sentences asserting separately that $\alpha_j+\beta_j+\gamma_j=0$ $(1\leqslant j\leqslant n)$ and $\xi_k\eta_k+\zeta_k=0$ $(1\leqslant k\leqslant m)$.

It follows immediately that:

$$K_2 \vdash (\textstyle\bigvee_{j,k} \{\alpha_j + \beta_j + \gamma_j,\ \xi_k\eta_k + \zeta_k \mid 1 \leqslant j \leqslant n, 1 \leqslant k \leqslant m\} = \\ = 0) \to a = b.$$

Write:

$$\rho(\alpha_1, \ldots, \zeta_m) =$$
$$= \bigvee_{j,k} \{\alpha_j + \beta_j + \gamma_j, \xi_k \eta_k + \zeta_k \mid 1 \leqslant j \leqslant n, 1 \leqslant k \leqslant m\}.$$

Then:

$$K_2 \vdash \rho(\alpha_1, \ldots, \zeta_m) = 0 \rightarrow a = b.$$

Since the constants $\alpha_1, \ldots, \zeta_m, a, b$ do not occur in K_2, they may be replaced by variables $x_1, \ldots, x_m, x, y$ to obtain:

$$K_2 \vdash \rho(x_1, \ldots, x_m) = 0 \rightarrow x = y.$$

We have now shown that the conditional:

$$\rho(x_1, \ldots, x_m) = 0 \rightarrow x = y,$$

which is to be understood as a formula in $\mathscr{L}$, is *valid in all Boolean algebras.*

Let

$$\Psi \text{ denote } x \rightarrow \rho$$
$$\chi \text{ denote } y \rightarrow \rho$$

i.e. Ψ and χ are Boolean functions, explicitly:

$$\Psi \text{ is the function } 1 - \chi + \rho - (1 - x)\rho$$
$$\chi \text{ is the function } 1 - y + \rho - (1 - y)\rho.$$

Since:

$$\rho = 0 \rightarrow x = y$$

is valid in all Boolean algebras, it follows that the identity

$$\Psi = \chi$$

holds in Z_2, i.e.

$$\Psi^*(a) = \chi^*(a)$$

for every sequence $a \in D_\Psi \cap D_\chi$. For, suppose under some substitution for the variables $x_1, \ldots, x_m$, that $\rho^*(a) = 0$. Then because $\rho = 0 \rightarrow x = y$ is valid in Z_2, we have:

$$\Psi^*(a) = \chi^*(a).$$

If, under some substitution $\rho^*(a) = 1$, we have:

$$\Psi^*(a) = 1 = \chi^*(a).$$

Since the identity

$$\Psi = \chi$$

holds in Z_2, it follows that

$$\Psi \equiv \chi$$

is valid in Z_2, i.e. that $\Psi \equiv \chi$ is a *classical tautology*. But $\Psi = \chi$ *does not hold in* $\mathscr{A}$. For, substituting the sequence $\langle \alpha_1, \ldots, \zeta_m, a, b \rangle$ of elements from $\mathscr{A}$ for the variables $(x_1, \ldots, x_m, x, y)$ yields the value $1-a$ for Ψ and $1-b$ for χ. That is, under this valuation for Ψ and χ in $\mathscr{A}$, we have

$$\Psi^* = 1 - a$$

and

$$\chi^* = 1 - b$$

with $a \neq b$.

Thus, in the case of a partial Boolean algebra $\mathscr{A}$ for which there is no Boolean imbedding, there is a biconditional, $\Psi \equiv \chi$, which is a classical tautology, and a sequence $\langle \alpha_1, \ldots, \zeta_m, a, b \rangle \in D_\Psi \cap D_\chi$ under which the corresponding identity, $\Psi = \chi$, does not hold in $\mathscr{A}$. This proves the theorem.

4. The Basic Problem

The set S of statistical states of quantum mechanics does not contain states which are dispersion free. This is the property of the quantum theory which generates the problem of 'interpretation': i.e., the problem is to understand the absence of dispersion free states.

A statistical state $\psi \in S$ is a map $\psi : T \to [0, 1]$ such that $\psi(1) = 1$ and $\psi(\vee_i \{a_i\}) = \sum_i \psi(a_i)$ if $\{a_i\}$ is a disjoint sequence. That is, a statistical state is a generalized probability assignment to the theoretical propositions of the theory which satisfies the usual conditions for a probability measure on each maximal compatible subset of the partial Boolean algebra T. The probability algorithm of a phase space theory is a function P which assigns to each magnitude A and each state ψ, a probability measure on R. $P_{A\psi}(U)$ denotes the probability that in the state ψ the value of the magnitude A lies in U. For dispersion-free states, the probability assigned to each magnitude reduces to an atomic measure concentrated on the value of A. It is not difficult to show that ψ is dispersion free if and only if

ψ is a homomorphism of T onto Z_2. (See e.g., Gudder [8] for a proof.)

In classical mechanics the algebra $\mathscr{B}$ of events is a Boolean algebra, so there is a one-to-one correspondence between atomic events $a \in \hat{B}$ and two-valued homomorphisms on $\mathscr{B}$. This property is preserved in the algebra of theoretical propositions. Thus when $\mathscr{B}$ is a Boolean algebra, each atomic event determines a two-valued homomorphism, and hence, a dispersion free state, on T.

Each event in $\hat{B}$ is represented bi-uniquely by a point $\omega \in \Omega$, since the field of subsets of Ω is perfect and reduced. Each point $\omega \in \Omega$ determines a two-valued homomorphism on $\mathscr{F}(\Omega)$, and therefore, a dispersion free state on C. Conversely, since C is isomorphic to $\mathscr{F}(\Omega)$, $\psi \circ h^{-1}$, where h is the Stone isomorphism, is a statistical state on Ω.

A dispersion free state on Ω may be replaced by the point $\omega \in \Omega$ which determines it. A *classical mechanical state* is just the phase point which determines ψ when ψ is dispersion free. Thus a classical mechanical state corresponds to an atomic proposition in C. Because C is a Boolean algebra, each atom in C determines a homomorphism onto Z_2, and hence a dispersion free statistical state.

In a Boolean algebra, each $a \in \hat{B}$ determines a maximal proper filter F in $\mathscr{B}$; similarly, in a partial Boolean algebra each $a \in \hat{A}$, determines a maximal proper filter F in $\mathscr{A}$. Notice, in case $\mathscr{A}$ is a Boolean algebra, each maximal proper filter in $\mathscr{A}$ may be used to define a homomorphism onto Z_2 by the condition, $h(a)=1$ if $a \in F$ and $h(a)=0$ if $a \notin F$. This possibility depends on the distributivity of Boolean algebras, for this implies that maximal proper filters in $\mathscr{A}$ are prime filters.[8] If the ultrafilters in $\mathscr{A}$ are not prime, we may have $a \notin F$ and $a' \notin F$, but $1 \in F$, and therefore, $a \vee a' \in F$. Hence, the correspondence between ultrafilters in $\mathscr{A}$ and two valued homomorphisms on $\mathscr{A}$ breaks down. This means that the correspondence between atomic propositions (or events) and two-valued homomorphisms breaks down.

Notice, even if the ultrafilters in $\mathscr{A}$ are not prime, $\mathscr{A}$ may be the homomorphic image of a Boolean algebra. If this is the case, there exist homomorphisms $h: \mathscr{A} \to Z_2$, however, these are not in general determined by maximal proper filters in $\mathscr{A}$.

In general, event structures which determine dispersion free states on the associated algebra of theoretical propositions include all those that may be mapped homomorphically into a Boolean algebra, since, by the homomorphism theorem, every Boolean algebra admits homomorphisms onto Z_2. All the event structures isomorphic to or containing the partial Boolean algebra $B(\mathscr{H}_3)$ of linear subspaces of a three dimensional Hilbert space[9] fall outside of this class. By Kochen and Specker's Theorem 1, for each such $\mathscr{A}$ there are no homomorphisms onto Z_2, hence, no two-valued homomorphisms on the partial Boolean algebra Q of theoretical propositions associated with $\mathscr{A}$. Because of the equivalence between two-valued homomorphisms and dispersion free states, there are no dispersion free states on Q.

Thus the absence of dispersion free states on Q is a direct consequence of the fact that $\mathscr{A}$ is a particular type of partial Boolean algebra, just as the existence of dispersion free states on C is a consequence of fact that $\mathscr{B}$ is a Boolean algebra.[10]

The fact that there are no homomorphisms $h: Q \to Z_2$ must be sharply distinguished from the question of the bi-valence of the language $\mathscr{L}$ in which the theoretical propositions are formulated. It is trivially possible to make bi-valent assignments of truth values to the propositions of Q, and therefore, to the corresponding sentences of $\mathscr{L}$: Let every proposition associated with an event in an ultrafilter in $\mathscr{A}$ be true, and every proposition associated with an event outside the filter, false. Since the ultrafilters in $\mathscr{A}$ are not prime filters, the bivalent assignment of truth values to $\mathscr{L}$ is not induced by a homomorphism of $\mathscr{A}$ onto Z_2. But since the homomorphism theorem is equivalent to Stone's Representation Theorem, this is to be expected, if $\mathscr{A}$ is strongly non-Boolean.

In the view advanced here, events $b_i \in \mathscr{A}$ incompatible with an event $a \in \mathscr{A}$ are excluded by the logical structure of the system; the chief advantage of the bi-valent truth value assignment defined above is that it makes this fact explicit. This definition has the consequence that it is a sufficient but not a necessary condition for the truth of a disjunction that one of the disjuncts be true. For example, the propositional formula

$$x_i + x_j + x_k - x_i x_j x_k$$

in $\mathscr{L}$ is true whenever it is interpreted over three mutually compatible events a_i, a_j, $a_k \in \hat{A}$ since

$$a_i \vee a_j \vee a_k = 1.$$

Now the trivial event (i.e., 1) is compatible with every event and is a member of every filter. But it does not follow that exactly one of every triple of mutually compatible atomic propositions is true.

It may be objected that such a truth value assignment is 'unintuitive'. But this is surely a pseudo-problem. For if the quantum theory is assumed, the models M of $\mathscr{L}$ are isomorphic to $B(\mathscr{H}_3)$. Therefore the ultrafilters in M are not prime filters. Thus, so far as the event structures are concerned, this property is presewed in any reconstruction of the theory. The objection is interesting only if it is coupled with a classical solution to the representation problem. But this problem is left untouched by the choice of $\mathscr{L}$.

In the remainder of this section we compare the quantum theory with classical statistical mechanics, since for this theory it is also true that dispersion free states are not theoretically fundamental. This is done in two stages. We begin by considering why dispersion free states are not theoretically important in statistical mechanics. Next, we examine the sense in which the description of statistical mechanics is incomplete relative to the Newtonian description. We conclude this section with some remarks on the interpretation of Birkhoff and von Neumann.

Classical mechanics and classical statistical mechanics share the same phase space as well as the same dynamics. Thus for statistical mechanics a dispersion free state is determined by the classical mechanical state of the system. Just as in classical mechanics, physical magnitudes $A \in \mathcal{O}$ are associated with functions in $R^{\Omega} = \{f_A : \Omega \to R\}$ from Ω into the Borel subsets of R. Each A is associated with a family of subsets of Ω by the inverse image

$$f_A^{-1}(U) = \{\omega \mid f_A(\omega) \in U\}$$

of the Borel subsets of R under the map corresponding to A. In statistical mechanics Ω is also a sample space, so that a point $\omega \in \Omega$ is also interpreted as the phase point of a sample system in an ensemble of similar systems.

Now for a certain class $\mathscr{K}$ of regions of the sample space, there are *macroscopic magnitudes*, i.e. properties of the ensemble, with values concentrated on a small subset of R.[11] These magnitudes obey the phenomenological equations of classical thermodynamics. For some $\Lambda \in \mathscr{K}$, and any distribution of phase points in Λ, the values of each macroscopic magnitude remain concentrated on a small subset of R. This is to be expected; for the basic Newtonian laws are symmetric with respect to time, hence the law of motion of an individual system is time-symmetric. But the phenomenological laws are irreversible; if the macroscopic magnitudes were not independent of the precise location of the phase point, the macroscopic laws would be reversible, not irreversible. Therefore the theoretical unimportance of the classical mechanical state is a necessary condition for the successful application of statistical mechanics to thermodynamic systems.[12]

The application of the laws of Newtonian physics to a thermodynamic system requires too fine a specification of the classical mechanical state. The slightest discrepancy amplifies very rapidly and renders the initial specification theoretically useless. Because of this difficulty, we forgo a complete description in terms of the classical mechanical state in favor of an incomplete description involving a proper subset S' of the statistical states $\psi : \mathscr{F}(\Omega) \rightarrow [0, 1]$. Relative to this description, there exist magnitudes $A, B \in \mathcal{O}$ such that

$$P_{A\Psi}(U) = P_{B\Psi}(U)$$

for every Borel set $U \subset R$ and every $\psi \in S'$. A and B are equivalent with respect to the set S' of statistical states. Therefore, C contains theoretical propositions of the form

$$f_A(\omega) \in U$$

and

$$f_B(\omega) \in U$$

which are equivalent with respect to S'. By extending S' to S it is possible to distinguish these magnitudes together with the theoretical propositions corresponding to them. Thus, the statistical description in terms of S' is incomplete relative to the classical description in the sense that the extension to S leads to an imbedding of this description into the classical description. For this reason, the absence of dispersion free states may be taken to mean that, relative to classical physics, the description of statistical mechanics is based on incomplete knowledge of the exact classical state of the system.

Now in the case of quantum mechanics, the absence of dispersion free states cannot be understood in this way. For if $\mathscr{A}$ is a partial Boolean algebra of events corresponding to a Hilbert space of at least three dimensions, there is no imbedding of $\mathscr{A}$ into a Boolean algebra; hence, there is no imbedding of the partial Boolean algebra Q of theoretical propositions associated with $\mathscr{A}$ into a classical (i.e., Boolean) description. This follows from Kochen and Specker's Theorem 1, because the existence of two-valued homomorphisms is a necessary condition for the imbeddability of Q into a Boolean algebra. There even exist finite subalgebras D of Q which are not imbeddable into a Boolean algebra. D contains pairs of propositions such that $a \neq b$ but $h(a) = h(b)$ for every homomorphism $h: D \rightarrow Z_2$. If D is weakly imbeddable, a and b will have to be incompatible. This follows immediately from the definitions of strong and weak imbeddability, or, more precisely, from Kochen and Specker's Theorem O and its counterpart for weak imbeddings.

It is often suggested that the quantum theory is more vague than classical physics in the sense that there are distinctions which can be made in classical physics which are ill-defined in the quantum theory. The opposite is the case – there are finer distinctions possible in the quantum mechanical case than in the classical case. The a and b above are in a sense distinguishable quantum mechanically, but not classically, i.e.[13] not in terms of homomorphisms onto Z_2. Intuitively, there exist completely symmetric but distinct elements in a non-imbeddable $\mathscr{A}$.

Though Birkhoff and von Neumann recognize that the algebra of theoretical propositions of quantum mechanics is not a Boolean algebra, they do not consider the possibility of imbedding Q into a Boolean algebra, and thus, into a classical description. This is due to their conception of the role of the classical mechanical state in statistical mechanics.

In their view there are basically two reasons for ignoring the exact classical description in statistical mechanics: first, it is 'convenient' to do so, and secondly, knowledge of the phase point requires a degree of precision which it is impossible to obtain experimentally. While this is certainly true, it is not an analysis of the relationship between the two theories. Statistical mechanics can ignore the classical state because it deals with irreversible processes, and these must be independent of the exact phase point of the system. This account is combined with a particularly naive confusion of reference with evidence. Birkhoff and von Neumann are thus led to the view that it is meaningless to suppose that the system is always in a state corresponding to a point in Ω; i.e., the exact classical description is not relevant to statistical mechanics on largely independent, epistemological grounds. But blurring the distinction between the world, and our knowledge of the world, makes it impossible to distinguish a Boolean description based on incomplete knowledge from a complete non-Boolean description. Since the concept of the exact state of a system is considered fundamentally incoherent, the non-Boolean character of Q may merely express the impossibility of knowing the complete classical state. Because of this unclarity, Birkhoff and von Neumann have only succeeded in reformulating the orthodox interpretation.[14] Though their discussion is couched in terms of the logical structure of quantum propositions it suffers from all the ambiguities of the conventional view.

5. Alternative Representations

In this section we consider alternative representations of the Hilbert space structure of quantum mechanics, viz.: orthomodular posets, and orthomodular lattices.[15] It is clear that the partial Boolean algebra of subspaces of a Hilbert space may be extended to an orthomodular poset by simply defining the order relation in each maximal Boolean sub-algebra in the usual way; and even to an orthomodular lattice by defining g.l.b. and l.u.b. for incompatible elements. Thus the mathematical differences are not essential.

In the lattice and poset approaches there are basically two properties that are held to distinguish Q from C: non-distributivity, and the existence of incompatible pairs. Neither corresponds to non-imbeddability. Because of this, interpretations based on these representations suffer from the same ambiguity as the view of Birkhoff and von Neumann.

The orthomodular lattice[16] $\mathscr{H}_2$ of linear subspaces of a two dimensional Hilbert space is isomorphic to the lattice of subspaces through a point in ordinary two-dimensional Euclidian space. In this representation, compatibility corresponds to orthogonality, i.e. two linear subspaces are compatible if and only if they are orthogonal in the sense of elementary geometry. (Thus $a \leftrightarrow b$, if a is a subspace of b.) Joins, meets, and complements correspond to spans, intersections, and orthocomplements. The unit of the lattice is the whole space, and the zero is associated with the zero-dimensional subspace or origin. It is obvious from Kochen and Specker's Theorem O that the partial Boolean algebra $B(\mathscr{H}_2)$ associated with $\mathscr{H}_2$ is *imbeddable* into a Boolean algebra.

Since $B(\mathscr{H}_2)$ is imbeddable into a Boolean algebra, the distributive law

$$a \wedge (b \vee c) \equiv (a \wedge b) \vee (a \wedge c)$$

of classical logic holds. In fact the identity holds in $B(\mathscr{H}_2)$. This is not peculiar to $B(\mathscr{H}_2)$, for the distributive identity is Q-valid, i.e. valid in all partial Boolean algebras.[17] Clearly, non-distributivity, in the sense of failure of the distributive law of classical logic, depends on the definition of validity. More interestingly: even when the propositional (i.e. Boolean) functions in $\mathscr{L}$ are interpreted as lattice polynomials, the failure of the distributive law does not correspond to non-imbeddability.

Similarly, there are clearly incompatible elements in $\mathscr{H}_2$; yet $\mathscr{H}_2$ may be imbedded into a Boolean algebra. So the existence of incompatible pairs must be distinguished from non-imbeddability.

The work of Zierler and Schlesinger [22] shows that there always exists a map $h:\mathscr{A} \to \mathscr{B}$ from an orthomodular poset $\mathscr{A}$ into a Boolean algebra which preserves the ordering and orthocomplementation. That is,

(i) if $a \leqslant b$, then $h(a) \leqslant h(b)$
(ii) $h(a') = h(a)'$.

The map is also monomorphic so that

(iii) if $h(a) \leqslant h(b)$ then $a \leqslant b$.

Notice it does not follow that such a map preserves lattice meets and joins, for although the ordering of elements above and below a and b is preserved in the image $h(\mathscr{A})$ of $\mathscr{A}$ in $\mathscr{B}$, there may be an element of $\mathscr{B}$ smaller (in $\mathscr{B}$) than the image of any element above a and b in $\mathscr{A}$, so that *this* element would qualify as $h(a) \vee h(b)$ and not $h(a \vee b)$. That is, for any $x \in \mathscr{A}$, such that $a \vee b \leqslant x$ in $\mathscr{A}$,

$$h(a \vee b) \leqslant h(x),$$

but the smallest element above $h(a)$ and $h(b)$ might be an element in $\mathscr{B}$ which is not the image of any element in $\mathscr{A}$.

Zierler and Schlesinger show further that there does not in general exist a map satisfying conditions (i)–(iii) which also preserves the lattice operations for compatible elements. Now this is already clear from the work of Kochen and Specker, since independently of the question of the preservation of order, meets and joins cannot be preserved in the case of partial Boolean algebras associated with Hilbert spaces of three or more dimensions. On any representation, what is fundamental about the non-Boolean structures of quantum mechanics is that they are not imbeddable into a Boolean algebra, and this depends on the fact that $\leftrightarrow$ is not transitive in $\hat{A}$. In this sense, the order structure is redundant.

(Notice non-transitivity of $\leftrightarrow$ in $\hat{A}$ is not the same as non-transitivity of $\leftrightarrow$ in $\mathscr{A}$. For every element in $\mathscr{A}$ is compatible with the unit. Thus if compatibility is transitive in $\mathscr{A}$, then every element is compatible with every other: $a \leftrightarrow 1$ and $1 \leftrightarrow b$ implies $a \leftrightarrow b$. That is, there are no incompatible pairs. Conversely if $\leftrightarrow = A \times A$, $\leftrightarrow$ is obviously transitive. Thus transitivity of $\leftrightarrow$ in $\mathscr{A}$ is equivalent to $\leftrightarrow = A \times A$.)

One final point: It seems natural to understand the generalization of the order relation in a Boolean algebra as a generalized implication. However this leads to difficulties. By a result of Fây [6], implication cannot be defined as in classical logic by

$$a \Rightarrow b \text{ if and only if } a' \vee b = 1.$$

For in an orthomodular poset or orthomodular lattice, if the relation

$$a' \vee b = 1$$

is transitive, the poset or lattice is a Boolean algebra. For this reason it has been argued (e.g. by Gudder and Greechie [9]) that the transition from classical to quantum mechanics is not properly concerned with logic. But in view of Kochen and Specker's Theorem 4, this is obviously a purely verbal issue.

6. Conclusion

It remains to be shown how the discussion of this paper leads to a solution of the 'measurement problem'. This subject, together with a complete discussion of the role of probabilities on non-Boolean event structures, will be dealt with in a separate paper. The present paper clarifies what is required of an interpretation of the quantum theory: The problem is to explain the transition from classical mechanics to quantum mechanics, given that the set S of statistical states of the quantum theory does not contain dispersion free states. It also explains the sense in which quantum mechanics and classical mechanics are theories of the world's logical structure. This, in conjunction with Theorems 1 and 4 of Kochen and Specker, completely solves the problem of interpretation. Clarification of the problem of hidden variables, in the sense of the importance of Gleason's Theorem and its corollaries, is immediate: such results have the character of completeness theorems for the logical structures of quantum mechanics. The whole discussion rests on the distinction between logical structure in the sense of the syntax and semantics of a formal language, and the logical structure of events. This distinction is completely analogous to the one drawn in Section 1 between coordinate transformations and symmetries. Only the first component of each pair involves conventional elements. Logical structure and space-time symmetries are objective structural properties of the world.

University of Tel Aviv and University of Western Ontario (J. B.)
and
University of Western Ontario and University of New Brunswick. (W. D.)

NOTES

[1] This distinction is suggested by Einstein [4]. We have retained his terminology. Cf. also [5], pp. 53ff.

[2] There are theories which are clearly both constructive and principle theories, e.g. classical statistical mechanics.

[3] Notice that in Bub [3], 'phase space theory' denotes a *classical* phase space theory. We extend the use of the term here to include any theory in which the concept of logical structure occurs explicitly.

[4] This theorem is proved in [13]. Unless otherwise indicated, all references to Kochen and Specker, are to this paper.

[5] This paper assumes some acquaintance with the representation theory of Boolean algebras. See, e.g. Sikorski [17], Chapter I.

[6] For a characterization of these algebras, see Rasiowa and Sikorski [16].

[7] Kochen and Specker use the term 'commeasurability' to refer to this relation, clearly suggesting that the relation should be understood in terms of simultaneous measurability. This is at least misleading, since the simultaneous measurability of two magnitudes is a *consequence* of the fact that they are compatible; but compatibility is not operationally definable in terms of simultaneous measurability. [Cf. the discussion below (i.e. Section 4) of Birkhoff and von Neumann.]

[8] See Rasiowa and Sikorski [16], Chapter I, Section 9, for a discussion of this point.

[9] For definiteness, we restrict the discussion of this section to partial Boolean algebras of this class.

[10] In general, for a Hilbert space of three or more dimensions, all possible statistical states on the partial Boolean algebra of linear subspaces are generated by the statistical operators according to the algorithm of the quantum theory. That is to say, the probability algorithm of the theory generates *all* possible statistical states on Q. (This is essentially the content of Gleason's Theorem [7].) Yet the set S of statistical states does not contain states which are dispersion free. So, by the equivalence between

two-valued homomorphisms and dispersion free states, an extension of the theory which recovers the correspondence between events and two-valued homomorphisms does not exist.

11 $\mathscr{K}$ is the class of Borel subsets of Ω modulo Borel sets of Lebesgue measure zero. This class is identical with the class of Lebesgue measurable subsets of Ω modulo sets of Lebesgue measure zero (see e.g. Halmos [10], Section 15).

12 This has the character of a randomness assumption. It is also a sufficient condition for applying statistical mechanics to thermodynamic systems. For a thorough discussion see van Kampen [20], Chapter 1.

13 By Stone's Representation Theorem, every pair of distinct elements in a Boolean algebra must be distinguishable by a pair of homomorphisms onto Z_2.

14 In its original form, Heisenberg's interpretation was compatible with the existence of a classical mechanical state. This assumption was later rejected by Heisenberg and Bohr and replaced by the thesis that an atomic system cannot be significantly described independently of a measurement process.

15 Birkhoff and von Neumann assume an orthocomplemented, modular lattice. This assumes more structure than an orthomodular lattice. (See Jauch [11], Chapter 5, Section 6, for a more detailed discussion of this point.) For our purposes, the difference is not important, and everything said concerning orthomodular lattices may be extended to the lattice of Birkhoff and von Neumann.

16 For simplicity of exposition we restrict the discussion to lattices.

17 Cf. Section 3, above. For a generalization, see Kochen and Specker [12], Section 6.

BIBLIOGRAPHY

[1] J. L. Anderson, *Principles of Relativity Physics*, Academic Press, 1967.

[2] G. Birkhoff and J. Von Neumann, 'The Logic of Quantum Mechanics', *Ann. Math.* **37** (1936) 823–843.

[3] J. Bub, 'On the Possibility of a Phase Space Reconstruction of the Quantum Statistics: a Refutation of the Bell-Wigner Locality Argument', to appear in *Found. of Phys.* **3** (1973) 29–44.

[4] A. Einstein, 'What is the Theory of Relativity?', (1919) *Essays in Science*, Phil. Lib., 1934.

[5] A. Einstein, *Autobiographical Notes* (1949), *Albert Einstein: Philosopher-Scientist* (ed. by P. A. Schilpp), Harper, 1959.

[6] G. Fây, 'Transitivity of Implication in Orthomodular Lattices', *Acta Sci. Math.* (Szeged) **28** (1967) 267–270.

[7] A. Gleason, 'Measures on Closed Subspaces of Hilbert Space', *J. of Math. and Mech.* **6** (1957) 885–893.

[8] S. Gudder, 'On Hidden-Variable Theories', *J. of Math. Phys.* **11** (1970) 431–436.

[9] S. Gudder and R. Greechie, 'Is Quantum Logic a Logic?', *Helv. Phys. Acta.* **44** (1971) 238–240.

[10] P. Halmos, *Lectures on Boolean Algebras*, Van Nostrand, 1963.

[11] J. Jauch, *Foundations of Quantum Mechanics*, Addison-Wesley, 1968.

[12] S. Kochen and E. P. Specker, 'Logical Structures Arising in Quantum Theory', *The Theory of Models, 1963 Symposium at Berkeley*, compiled by J. Addison *et al.* North-Holland, 1965.

[13] S. Kochen and E. P. Specker, 'The Problem of Hidden Variables in Quantum Mechanics', *J. of Math. and Mech.* **17** (1967) 59–87.

[14] H. M. MacNeille, 'Partially Ordered Sets', *Trans. Am. Math. Soc.* **42** (1937) 416–460.
[15] W. Peremans, 'Embedding of a Distributive Lattice into a Boolean Algebra', *Nederl. Akad. Wetensch. Indag. Math.* **19** (1957) 73–81.
[16] H. Rasiowa and R. Sikorski, *The Mathematics of Metamathematics*, P.W.N., 1963.
[17] R. Sikorski, *Boolean Algebras* (2nd ed.), Springer-Verlag, 1964.
[18] A. Trautman, *Lectures on General Relativity*, Brandeis Summer Institute, 1964, Prentice Hall, 1965.
[19] A. Trautman, *The General Theory of Relativity*, Nuclear Energy Information Center of the Polish Government, 1968.
[20] N. G. van Kampen, 'Fundamental Problems in the Statistical Mechanics of Irreversible Processes', *Fundamental Problems in Statistical Mechanics, Proceedings of the NUFFIC International Summer Course in Science at Nijenrode Castle, The Netherlands*, 1961, compiled by E. G. D. Cohen, North-Holland, 1962.
[21] H. Weyl, *Philosophy of Mathematics and Natural Science* (Rev. ed.), Princeton, 1949.
[22] N. Zierler and M. Schlessinger, 'Boolean Embeddings of Ortho-Modular Sets and Quantum Logic', *Duke J.* **32** (1965) 251–262.

C.A. HOOKER

DEFENSE OF A NON-CONVENTIONALIST INTERPRETATION OF CLASSICAL MECHANICS*

I

1. *Introduction*

In an earlier paper [15] I have outlined and defended a version of Scientific Realism. According to that doctrine, theories are imaginative attempts to grasp the true nature of a universe apprehended very imperfectly via the senses; they are *intended* literal truths (though very likely all false in fact); the correlative epistemology is naturalistic, being informed by, and in turn informing, the science of human beings. In that paper I described an alternative theory of science, which I called (after the historical tradition) Conventionalism. This theory holds that (i) there is a sharp, objective observational/theoretical dichotomy within the terms of science and (ii) all sentences in science with certain theoretical terms are to be regarded as expressing conventions (or partial conventions) for the linking together of observational terms. Epistemologically, the purely observational level offers the only empirical content of science. These two positions are in utter opposition ontologically, epistemologically and semantically.

It is to be expected that Scientific Realism and Conventionalism will take different approaches to scientific theories. The Conventionalist will be out to show that a given theory may *fruitfully* be viewed as a system of conventions for manipulating observables. The Realist is not committed, of course, to denying the occurrence of conventions in science (that would be absurd) but he is committed to providing a coherent ontological account of the world and in so doing to take very seriously what theories say there is. In general the two will conflict for the one will be wanting to write off as conventional sentences which the other will want to understand as describing a realm of unobservable entities which are the real constituents of things. (Of no real entity or property can it be said that the situations in which it does or does not occur are conventionally assignable; such entities must be fictional, they cannot belong to the ontology of the theory in question.[1])

The motives for adopting Conventionalism seem to be very similar to those for adopting Empiricism (cf. [15]), roughly they centre around the desire to achieve epistemological security by restricting the content of science to the observable. Often it is hard to distinguish the two. Ernst Mach is a good case in point; what we find here is an attempt to show that unobservable entities may be eliminated from the corpus of science without impairing it.[2] One of the ways in which this goal can be achieved is to argue that the offending theoretical term in question has no troublesome ontological import either because it is definable in observable terms or because it represents only a conventional shorthand description of a class of purely observable circumstances. Such, eg. was Mach's approach to *mass* and *force* respectively in Classical Mechanics.

From the time of Mach forward, it has become increasingly fashionable, and often mandatory, to analyze the foundations of Classical Mechancis with an eye to the large degree of conventionality which is alleged to be contained within it. One of the invariable consequences of such a critical rephrasing of Classical Mechanics has been the sacrificing of a purely empirical status for one or more of the laws of motion and of an ontologically serious status for forces. This approach to Classical Mechanics has such universal currency today that he who would defend the non-conventionality of the fundamental laws of that system and would take at their ontological face value the terms contained therein finds himself quite definitely on the defensive (cf., for example, [22], [23], [25]).

There are of course other ways to arrive at these same conclusions for Classical Mechanics besides the approach taken by Mach. Adolf Grünbaum arrives there from an analysis of the consequences of the conventionality of the assignment of a metric to space-time (assumed a continuum) and Brian Ellis from a consideration of the relation of forces to effects of forces (cf. [5], [7] and below). What all these approaches share in common is the conviction that forcefree circumstances are conventionally assignable in a relevant sense and hence either that forces are not part of the fundamental ontology or that one or more of the usual laws of mechanics is without empirical content but rather conventionally or definitionally true.

I happen to hold the view that a more coherent (and more interesting) ontology is obtained if forces (and masses) are conceded real features of the world and I am going to attempt a defense of that view here. Along the way a number of other philosophically interesting issues will arise and this will be an added bonus of the discussion, as will the defense of the historical understanding of Classical Mechanics with its 'full' empirical content.

The reader is warned in advance that in the earlier essay [15] I stressed the systematic differences between philosophies of science, these differences will appear frequently in the form of argument which I choose and the rebuttals I will accept, none of which may be acceptable to a philosophical opponent.

I might also point out that I have concentrated here on a discursive argument concerning the role of conventionality in Classical Mechanics rather than on the presentation of a formal treatment of the theory which would 'demonstrate' my case. There are two, closely related, reasons for this: (i) many of the arguments revolve around general philosophical doctrines and I want to exhibit the pattern of this dependency and (ii) formal treatments tend to build in from the start the crucial assumptions rather than self-consciously discussing them and I believe that with sufficient ingenuity one could construct a system to support any of the philosophical positions. This is not to say that a more formal investigation is not worthwhile, to the contrary, a careful, vigorous examination of what is formally possible by way of reconstructions of Classical Mechanics would be very important. This task will prove to be an exceedingly complex one, as two of the best recent attempts of this kind show (see [1], [30], cf. [21]).

2. *Three Kinds of Conventionality*

Before proceeding, a few comments on what is intended by conventionality in this essay are in place. First of all, what is *not* intended by conventionality is reference to those decisions which link a particular semantic content to a particular sound or typographical mark.[3] In contra-distinction to this kind of conventionality, the conventionality with which this essay is concerned arises because of the *factual absence of conditions in the world, or features of the world, sufficient to uniquely determine the extension of one or more scientific terms.* It is not a question at all of assigning meanings to sounds or other symbols and it is not even primarily concerned with the meanings of terms. It is that kind of conventionality which arises when a decision must be made between several alternatives and *there is no factual (objective, non-pragmatic) basis on which to make that decision.* Under these circumstances we are forced to make a decision on other than empirical grounds and, since no decision can be distinguished empirically from any other possible decision, the decision is conventional in the sense of the term here intended. This is not to say, of course, that there may not be pragmatic, aesthetic, emotional or other such factors which serve to determine us on one particular alternative rather than another. But such factors have no *empirical* significance for the particular choice in question.

It is important to distinguish here between an epistemic and an ontological version of this type of conventionality. The world may be such that, whilst there are empirical features of it which would determine various alternative decisions in a given case, or the truth values of the alternative answers to a given question, we either do not know, or cannot know, of those features. Then the decision would be epistemically conventional *for us*, but not conventional in Grünbaum's ontological sense. On the other hand, whilst the factual absence of such features certainly entails that we have no corresponding knowledge of them, in this case the decision would be conventional in Grünbaum's ontological sense. Hereafter, whenever the term 'conventional' is used I shall be using it to refer to Grünbaum's stronger, ontological sense of conventionality. I shall mark appearances of the other sort of conventionality by the phrase 'epistemic conventionality'.

There is, to repeat, an important difference between epistemic and ontological conventionality. For from the epistemological conventionality of a given term it does *not* follow that the term cannot be interpreted realistically; all that follows is that we do not, or cannot, *know* that it is to be so understood.

There is another issue which ought to be raised here and which is hard to disentangle from the foregoing. In a certain sense, we can imagine there being alternative conceptual resources to our own, so that reality is dissected in different ways. (Derivatively this is linked to the first, trivial, conventionality through the fact that we may, in many cases, be able to choose differing primitive and defined terms for a theory and hence develop it using different conceptual patterns.) In an obvious way, it is often felt that this introduces a pervasive element of conventionality into any position, Empiricist or Realist. (Often, however, the Empiricist will feel comfortable with this argument because he takes it, not to apply to observational concepts but only to theoretical concepts and this probably because of the cases where theories may be developed in alternative ways, though neither the argument nor any other considerations concerning perception suggest that this might be a valid assumption.) I do not propose to discuss here the difficult notion of alternative conceptual repetoires in general; I shall simply assert that, according to Realism, any conceptual scheme would have to reproduce the ontological commitments of a true theory and also its specifications of the behaviour of that ontology, so that from this point of view there will be a mapping between the two. Since I am largely concerned with the ontology of the world according to Classical Mechanics, this reply suffices. If there is surplus conceptual richness beyond these theoretical agreements it may perhaps be conventionally chosen, one would have to judge each case on its merit and any additional criteria proposed (Ockham's razor etc.). Equally, the choice among such theoretically equivalent descriptions may be described as conventional (though again I would want to examine each case in the light of such criteria as ontological coherence before committing myself), but that is of no moment here, it is rather akin to the trivial semantic conventionality first mentioned. In any event the Realist position is quite consistent with allowing that there may be differing descriptions of the one reality, so long as everything of importance is preserved, and no relevant element of conventionality is involved. Issues of epistemic and ontological conventionality concern the status of distinctions that can be made *within* a given conceptual scheme.

It is my intention in this paper to weigh some of the major contemporary arguments for and against the maximally empirical interpretation of Classical Mechanics. In the first place (Sections II-VI), I shall aim to show that, several new and powerful arguments in favour of a conventionalist position notwithstanding, a non-conventionalist approach to the foundations of mechanics may still be *consistently* adhered to and that *there are arguments in favour of doing so.* Secondly (Sections VII-VIII) I shall argue that there are some distinctive advantages, which are not balanced by corresponding disadvantages, in the adherance to the non conventionalist option.

II

3. *Prima Facie Considerations in Favour of an Ontically Serious Status for Forces: The Argument from Sensory Experience*

If we can convince ourselves that forces are a real part of the 'furniture of the earth' we shall be a long way along the road to eliminating any serious element of conventionality from Classical Mechanics. Under normal conditions we ordinarily grant that, whatever be the relations between sensory experiences and the world, there are genuine objects of perception. We perceive, by touching, that a rock is round, that it is hard and so forth. We also have, *prima facie* such sensory experience of forces. We perceive pressures and strains on the limbs of the body, or over entire areas of the body. We perceive sharp blows at impact and so forth.

Prima facie, therefore, we may be said to perceive the actions of forces in these situations. Insofar as untutored common experience informs us, forces are a real and ubiquitous part of our everyday world. There is as little reason to disown them as there is to disown the objects which are their sources. Indeed, if they are to be ontically disowned then some alternative account of our 'force perceptions' must be offered.

Of course, our force perceptions are crude indeed. But this feature does not of itself make them either illusory or misinformed. There is, in particular, one objection to regarding forces seriously which whould be dealt with immediately. In the ordinary run of life we encounter forces in many situations, in some where accelerations are produced, in some where constant velocities are produced and in some where no movement whatsoever is produced. *All* of these situations present us with forces and force perceptions. It might then be demanded to know why, if mechanics is to find experience at all relevant to its founding, we cannot perceptually distinguish amongst these three situations. Alternatively, the objection may be put in this manner: if mechanics is based on experience at all, should it not have *three types of force*, one corresponding to each of the types of situations mentioned above, and not just a single variety?

In reply to this objection we simple affirm that we *can* distinguish between these three situations on the basis of experience. We experience balanced forces in the pressure of hand upon hand, and know that *this same pressure* will produce an acceleration in other circumstances *if not otherwise prevented* (by us). We experience the gravitational force and we can conduct Galileo's famous inclined plane experiment and so forth. If we set for ourselves the criterion of obtaining a consistent and unified explanation of all our experiences, then we will eventually be led to adopt the Newtonian-style account of forces. (Though as the drawn-out fate of the Aristotelian alternative so vividly shows, such a process of decision may be a long and complex one. Moreover the possibility of the General Relativistic account of forces, especially the gravitational force, must also be squarely faced.) And all of this without any denial that our force perceptions are reliable.

Nor does the employment of experience necessarily introduce any undue anthropomorphic element into the notion of forces. We may perceive correctly and objectively on the basis, if you like, of feelings. If all such perceptions were suspect we should have no cause to trust our senses at all. And yet it is in the fidelity of the tactile sense, which figures so largely in our perceptions of force, that we place the most abiding trust.

Unfortunately these considerations do not decide the issue, for there is the possibility of an alternative theoretical account of our sensory experience in these respects from that suggested by untutored commonsense (cf. Section IV); all that we have done is laid out some evidence most simply understood as commonsense suggests. The question for Classical Mechanics at this point is how we are really to interpret our experiences here. Are we on to some genuine feature of the real physical world, or can these experiences be explained without the appeal to any such entities as forces? We shall only have an adequate view of the issue when we reach the last stage of the arguments now to be embarked upon.

III. THE ARGUMENT FROM THE NON-CONVENTIONAL COMPLETEABILITY OF CLASSICAL MECHANICS

4. *Introduction*

My aim is to argue that one can defensibly adopt a non-conventionalist account of Classical Mechanics. One of the necessary conditions for establishing this is to show that the form of the laws of that theory at least *permit* such an interpretation – after that its *defense* can be undertaken. This first task is doubly worthwhile because it has been often regarded as impossible and because the argument reveals some philosophical subtleties often (or always) overlooked.

I might remark, incidentally, that the account of Classical Mechanics that I will offer would seem to do most justice to the sense and order of the original development found within Newton's *Principia* [24]. I am not interested here, however, in usurping the roles of either historian or Latin scholar. Rather, I shall be content to arrive at a fair outline of the general order and development of the fundamental laws of mechanics as found in Newton's writings, expressed in a more or less contemporary garb.[4]

The central role of forces in common perceptual experience certainly permeated Newton's construction of his mechanics. There seems little doubt that we should assume that Newton would subscribe to the following statement concerning forces:

(F) Forces are assumed to be real entities that are identifiable independently of their effects.

The approach to forces characteristic of the statement F seems to underly what Newton has to say of them. The relevant definitions and laws I shall now discuss in order and comment upon them in the process of building up a non-Conventionalist account of them.

5. *Definition 4*

Newton's remarks, pertinent to the nature of forces, begin with Definition 4. In Definition 4 forces are defined, in the case of *external* forces, to be sufficient for producing accelerations.

(Definition 4: An impressed force is an action exerted upon a body, in order to change its state, either of rest, or of uniform motion in a right line, Newton [26], p. 2.)

Comment: The statement F and Definition 4 may seem to be in conflict unless carefully understood. For example, if the concept of a force is taken to be *merely* the concept of something sufficient for its effects, then surely forces cannot be identified independently of their effects as F would allege.[5] Now the question of whether this is the correct concept of a force is crucial to the ultimate evaluation of the status of forces. For to characterize forces as not identifiable independently of their effects is to open the way to a Conventionalist evaluation of their status; (cf., for example, Section IV below). Certainly one can understand how definition 4 has suggested *that* particular characterization of the concept of a force.

But we must ask whether this is the only – or even the correct – way to view the matter.

We shall come to see, I think, that it is too glib a rendering of Definition 4, if we consider the question of *balanced forces.* If two forces offset one another, then the effects of *neither* are present, where 'effect' is here understood to be the behaviour arising from each force acting singly. In this case and in this sense forces are obviously *not* 'sufficient for their effects'. For the effect which the force would have had acting singly does not always occur whenever the force is present. Though this is not the only way to view the matter, the view which I will defend holds that these 'null effects' or 'joint effects' mean that we cannot, and do not, identify forces purely on the basis of their effects. For we may identify forces acting in a particular situation even where there are balanced forces present. Now Newton certainly intended that his system should deal with balanced forces. It would thus be unjust to read definition 4 in the manner we have been discussing.

The essential point is that there may be good reasons, theoretical and/or experimental, to say of a given situation not exhibiting force-effects that there were nonetheless balanced forces, *rather than no forces*, operative in the situation. (Two attracting, rigid, electrically charge bodies at rest is one example – cf. below.) The Conventionalist, because of his epistemology, is constantly tempted (or willing) to slide (illegitimately) from 'In such circumstances we cannot *know observationally* what forces are present' to 'In these circumstances it is arbitrary (→ conventional, cf. note 1) what is said'. Once this slide is stopped and the power of theory to specify ontology beyond observational access is admitted, then the power of mechanical theories leads us to accept their criteria for the occurrence of forces, i.e. to accept theoretical criteria for the identity and individuation (in the relevant sense) of forces, as for all other unobservables.[6]

But what now of forces in actual experimental situations? How do we there decide that balanced forces are present? The key to answering this question is found in the demand that *physical forces (i) have identifiable physical sources, (ii) act under specifiable conditions, and (iii) in a specifiable manner.* That is, in the case of a physical force, one must be able to identify the force which is acting in a non-arbitrary way from the *physical conditions obtaining in the circumstances.*[7] This identification is, in principle, independent of the effects to which the force gives rise. Let us call these important conditions, *Conditions I.* For example, the sources of the electromagnetic force are charged particles (electrons, protons etc.), the relevant circumstances are determined by the laws of dielectrics and shields and the law of action is the Lorentz force law coupled with the Maxwell field equations.

In the actual practice of science however, identification of forces cannot always be rendered independent of such effects. Normally, we first identify the source, or sources, of a force, its modes of action and so forth by considering a force's *unbalanced* effects. Then we are able to deal with cases involving balanced forces. Of course, in those cases where the forces are directly accessible to experience, this dependence may be broken by the consideration of balanced forces with the human being either directly involved, or else indirectly involved via other forces which have been directly tested in a similar manner. That such circumstances cannot be achieved at all for some forces, and that it can only provide a crude test situation for accessible forces, is logically irrelevant. Such limitations present a purely epistemic barrier to the above procedure, which is in principle universally applicable and precise. Only a rather extreme positivism could erect such epistemic barriers into sources of logical conventionality and such a doctrine may be justly rejected.

In fairness, it must be admitted that there is another way to regard the outcome of balanced forces. Rather than say that neither force is producing its effect here, we may say that both forces are producing their effects – it is just that the effects cancel one another. In this second sense, if both forces were not sufficient for their effects, then something else would have happened other than what in fact happens.

This may seem a more plausible way of looking at the matter, but it also has its difficulties. (It is only a more plausible way, perhaps, given our present situation where we usually cannot directly perceive forces but can directly perceive their effects.) Consider a case where two distinct forces are jointly necessary and sufficient for some visible effect. (For example, a high frequency noise, not by itself sufficient to shatter a particular glass, together with a sharp blow, not by itself sufficient to shatter the glass, are jointly necessary and sufficient for shattering the glass.) The effect of both forces acting in concert is whatever it is that is observed, but what elements of the observed effect shall we assign to either force? Acting alone, each force, let us assume, gives rise to no observable effect. Acting in concert an observable effect ensues. In this case no particular components of the observed effect can be assigned to one force as its effect, rather than to the other. Perhaps it can simply be insisted that there is some deeper level at which to view the matter and on which level individual component effects will be discoverable and assignable. But I can see no *a priori* reasons why this should be so. (Indeed, the modern quantum theory of energy may suggest strong reasons why this particular belief will eventually prove false.) Moreover, if the actual individual effects that would be produced by each force acting singly are absent, in what sense does the force still produce them? The notion of two real, counteracting causes is perhaps intelligible. But can it be intelligibly asserted that there are real, that is, actually existing, effects which nonetheless 'add up' to a total effect in which neither exists as an identifiable component? That latter notion seems scarcely intelligible.

What I am arguing is that the component effects cannot be *physically* present. I am *not* arguing merely that the following claim is false: There are total effects made up of separate effects in which no separate effects are present (this is contradictory and obviously false). I am arguing that two separate effects can give rise to a total effect in which neither is identifiable.

The whole question of the separate existence (or otherwise) of causes and effects in a case of 'conjunct action of causes' (to use Mill's nice phrase) is a tricky one. For example, the components of a machine are still identifiable as individuals within the structure of the machine. But do the so-called components of a force really relate to the net force in the same way? What would it be like if two machine components would 'cancel each other out' so that when both were added to the (as yet uncompleted) machine, the machine, or part thereof, simply disappeared? Looking at that net situation, could we still talk about the components of it? Looking at a wave phenomenon one can, mathematically, analyze that wave into component waves. The nature of these component waves will vary according to the mathematical tools being used. Can one then really say that the components of the wave exist in it in some sense? How much like the case of the counteracting machine parts and the waves is the case of chemical reactions? How wide is the gulf between *identifying joint causes* and *identifying genuine components* of a situation? One might say that the resolution of a resultant force into its components is just a *façon de parler*, that it is no more to be assigned a physical significance than the 'resolution' of a question into sub-questions, that vector 'addition' is only verbally similar to anything like physical 'addition'. But if you push me towards the centre of the door with such and such force and another pushes me at the same time towards the opposite wall with such and such a force, so that I actually do not move, I can sensibly (by means of my senses) distinguish *two* identifiably different causes tending to make me move in different directions with different 'potencies'.

The salient point here is that even if a conventionalist wished to identify forces with 'their' effects (eg. by asserting $F =_{df} \mathrm{d}/\mathrm{d}t\,(mv)$ he would still have to make intelligible that there occur circumstances in which accelerations exist yet cancel one another. Even if a counterfactual analysis were offered to demarcate these situations ('There exist balanced accelerations here because if the circumstances were to be altered thus and so, the accelerations would be thus and so') sense would still have to be made of these situations (accelerations themselves are not dispositions).

Unfortunately, space does not permit the further pursuit of this subtle problem. I shall simply take the foregoing remarks as sufficient defense of the former approach to balanced forces.[8]

These few remarks point the way to a reformulation of the crude form of Definition 4 given above, so that, reformulated, it no longer has the appearance of being inconsistent with the belief characterized in F above. Thus we may read definition 4 in a *conditionalized* manner, as follows: 'Forces are sufficient for producing their effects *if not otherwise counteracted* (by other forces),' where 'effect' is understood to be the accelerations of the bodies they act upon, distortions of such bodies and so forth.[9] Less satisfactorily, we may put the same reformulation in the form: 'forces have a disposition to produce their effects', or, 'forces strive to produce their effects.'

All of this is not without further challenge from the conventionalist, but I leave the challenges to a later context.

6. *The First Law of Motion*

The first law of motion now states of these external forces, as defined in Definition 4, that they are also *necessary* for accelerated motion.

Comment. The form of the first law is, prima facie, as stated above. It reads 'A body continues . . . unless acted upon by an external, impressed force."[10] This statement has the form 'p, unless q' and this is represented as 'q is necessary for $\sim p$'.

There is the old temptation to read 'p, unless q' as 'q materially implies $\sim p$'. But on this reading the first law of motion would say nothing different from what is already said in Definition 4. It is surely unreasonable to believe of Newton that he would repeat himself and fail to notice that he had stated as a law what he had already secured by definition.

This latter remark raises an important point, for the present reading of the first law together with this interpretational assumption as to Newton's logical acuity, further rules out two common views of forces and Definition 4. In the first place, it rules out reading Definition 4 as stating that forces are logically *necessary and* sufficient for accelerated motions. For on this reading of Definition 4 the first law of motion would again repeat part of what Definition 4 had already stated. In the second place, it rules out the view that the first law of motion states a *criterion* for the existence of forces. For a criterion offers at least a *sufficient* condition, and again the first law of motion would, on this reading, repeat part of what Definition 4 has already stated.

There will, of course, be some considerable *pressure* to construe forces in some such way as these if one does not believe that forces are to be taken ontologically seriously. But, as I have said, it is reasonable to assume that Newton took them seriously, and we are here attempting a statement of his mechanics which does so as well. Such an attempt cannot be ruled out *a priori* because this question has been settled, in the negative, in advance. In the present context, and with the present assumptions as regards forces, the first law of motion will therefore state an *empirical* proposition. According to the proposition (F) above, forces are independently identifiable entities and the first law of motion may therefore be *refuted* by producing an acceleration which is not caused by any identifiable force.

To emphasize the empiricalness of the claim which the first law of motion makes, I draw attention to the fact that a *sufficient* condition of the absence of forces between two material bodies is that there be no correlations between the dynamical behaviour of those bodies. Thus, for example, if a body A is moving curvilinearly at a uniform velocity and a body B is made to behave in a dynamically random manner and if *B's* behaviour produces no deviations from uniform curvilinear motion in A then we may conclude that there are no forces between A and B.[11] The question of whether there are correlations between two such material bodies is an empirical one.

It may be objected that the situation envisaged in the last paragraph is never, *in fact*, realized. It may even be added that, given such laws as Newton's law of gravitation, such a situation cannot *in* (physical) *principle* be realized. Both of these statements may be true. But they do not serve to show that the situation spoken of is a *logical* impossibility. For since it is certainly logically possible that all forces should have been truncated at finite distances from their material sources, the situation above is certainly *not* a logically impossible one. Thus the unrealizedness of the contemplated situation constitutes at most an epistemic barrier only to the testing of this proposition. The first law of motion, then, is refuted by finding a body which is independent force-wise of all other material bodies and then determinign whether that body is in uniform rectilinear motion or not.

However, this last remark raises a quite different problem for the first law of motion. For to refute that law, not only must forces be independently recognizable entities, *so also must accelerations be.* I shall later argue that forces are to be retained as non-conventional entities if and only if the first law of motion is ultimately involved in the *specification* of the spatio-temporal metric. Independently of considerations of the status of forces, I believe that the first law of motion is so involved. Thus its status is still open, in this respect, at this stage. Nonetheless, its empirical status is *not* so far in question because of difficulties with the notion of force. Once again, however, the conventionalist will not be satisfied with this situation. And once again I postpone consideration of his objections.

7. *The Second Law of Motion*

The second law of motion specifies, in general, what the precise force-effect is. The law may be stated as "Any body, of mass M say, that is acted upon by a non-zero impressed force, of magnitude F, say, has an acceleration whose magnitude is given by F divided by M and whose direction is the direction of the external impressed force."[12]

Comment. In this form the second law of motion *does not* entail the first law of motion. By explicitly incorporating the clause specifying that the force has *non-zero* magnitude, it precludes itself from having an opinion about what happens in zero magnitude situations. It is the job of the first law of motion to tell us what happens there.

If one accepts this reading of the second law of motion, then it cannot be regarded as definitory of the existence of forces. For it first assumes their existence, and then specifies their effects. To have it otherwise, is to make the first law of motion repeat a tiny fraction of what the second law of motion already says. It is also to have the second law of motion repeat what is already said in Definition 4.

From our discussion of Definition 4 it became clear that we must maintain that forces are identifiable independently of their effects. Their effects, in general, are now specified. Thus we must maintain that forces are identifiable independently of their effects as specified within the second law of motion. To follow this interpretation through consistently we must therefore assert that whilst '*Ma*' (strictly $\mathrm{d}/\mathrm{d}t\,(Mv)$) is a *measure* of the magnitude of a force, it is not definitory of the force's existence. We have already made our first move against this latter view by insisting that the second law of motion does not entail the first.

To see how the remainder of the doctrine is completed, we must notice a central feature of the second law: it cannot be applied in any actual, particular instance *until a specific force is identified.* This is done by meeting conditions I. That is, before the second law of motion can be applied, a force-source must be given, together with a specific dependence of the force's magnitude on the properties of both that source and also all other material bodies present, acting and acted on, and a definite mode of action. When a force is so specified and identified, it becomes an *empirical* matter whether or not the effect is as the second law of motion says it is.

It has often been contended that forces are not to be taken ontologically seriously because of the essential eliminability of forces from the laws of motion. In particular, whenever the second law of motion is applied to the world, we have seen that the expression '*F*' is to be replaced by some particular function of the properties and geometry of the situation in question. In any actual application, it is argued, 'forces' as such simply disappear and are replaced by geometrical and dynamical relations within the situation. But I have been arguing that this is not the correct way to view the matter. In specifying particular forces in particular situations, those forces have not been *eliminated* in any way, rather have they been *described*. Description does not constitue elimination and nor should it be mistaken for such.

Let us consider a parallel example. Suppose that a language contains the names of a number of men, together with a name-determinable of which the names are determinants. Suppose further that I *eliminate* these *names* from the language, by replacing them with definite descriptions of the men concerned. I may say "the name-determinable has no instances in the language. We no longer need it because wherever it holds open a place where a name might have been inserted, something other than a name is always inserted instead." Do we conclude from this that there are also *no men?* Of course not. We conclude only that descriptions *of the very same objects once referred to by determinants of the name determinable* are now used instead of names. Thus also with the determinable '*F*' and the particular descriptions which we use in its place. *The causes of motion remain uneliminated.*[13]

But now a further problem must be considered. We often use the *effects* of a force to *identify* it. For example, we use the accelerations which we observe bodies undergoing to help us locate the sources of a force, its mode of action and magnitude dependencies upon the properties of these bodies. In many cases we can use nothing else than the effects of a force for this purpose. Indeed, let us grant that it will often be empirically impossible to construct any *direct* tests for the presence of forces. Are we not, in these cases, actually using the second law of motion as a *definition* of the presence and magnitude of a force? Does not the second law of motion then become a *non*-empirical statement?

The core of the reply to this objection has already been stated under the discussion of Definition 4. Consider the electro-static force. For macro scopic bodies this force law can be roughly, but *directly*, checked simply by exploring the efforts required to keep two charged bodies apart as a function of their separation distance and, for example, the number of *identical* batteries connected to each.[14] This crude law is then refined and extended by theoretical postulation to all charged bodies. The consequences of this latter act are then tested and the extension and/or refinement modified accordingly. In this last part of the test procedure the second law of motion is taken as *true* and, together with the hypothesis of the force law, provides testable consequences. That the second law of motion, as well as the hypothesized force law, is open to refutation even here can be made clear by consideration of examples. Suppose that at some very high energy the behaviour of charged bodies suddenly changed so that their subsequent behaviour was fully explained if one retained the particular electro-static force law involved at low energies, but changed the second law of motion in some way. If a consistent and fully extended theory could be built around the retention of the electro-static force law at all energies we may well come to judge that the second law of motion in the Newtonian form has only a limited range where it is valid. A not dissimilar situation arose within the Special Theory of Relativity where alternative 'translations' of the second law of motion had to be decided between and *were* decided between by transforming, (i.e. *retaining)*, the particular electro-magnetic force laws and then testing the consequences of assuming the various alternative proposed translations of the second law of motion. (See, for example, Smith [27]; I do not, however, agree with Smith's pragmatic evaluation of the situation.)

What I am claiming then is that there are two ways in which we do in fact identify forces, through the specification of conditions I (a thoroughly theoretically loaded determination) and through force-effects (a more experimentally oriented determination, by cf. my [15]). Further, that *the procedural propriety of the second ultimately rests on the logical validity of the first as the appropriate way to ground the identity of forces.* That by our senses alone we cannot obtain a *completely refined* force law or cannot directly test every force situation is a *purely epistemic* limitation – no matter how physically fundamental its origin. There is no *logical* impossibility about being able to do either of these things. And because we are here concerned solely with sensory (epistemic) limitations, there is nothing inconsistent about claiming that there is no factual absence from the world of features which would uniquely determine such force laws in every situation. The Conventionalist and his Empiricist friends are apt to (willingly) conflate:

(i) Forces cannot in fact be identified except through force-effects

(ii) Forces do not have logically proper identity conditions unless these are given in terms of force effects.

The argument from the epistemic inaccessibility of certain ranges of force situations, even if that inaccessibility itself has a sound physical origin, to the conclusion that there are no empirical features of the world determining uniquely those force situations, is a *non-sequitur*. There is, therefore, nothing inconsistent in the claim that we may, in principle, refine and extend the force laws in the hypothetico-deductive manner and in a wholly non-conventional way.

The above defence has, however, emphasized the necessity and importance of a more or less 'wholist' view of science (cf. [15]). Under this conception of science, the theoretical assumptions which the theories of science employ at a given time are not introduced piecemeal to the science and are not necessarily able to be tested in isolation from one another. Rather, the totality of theoretical assumptions is brought down as a single whole and tested as a single whole. Each such assumption which science brings to the world has an equal empirical status with each other such asumption, in the sense that each assumption is, in principle, open to rejection in the face of the failure of the collective body of such assumptions to make predictions in conformity with our experiences. The unit of empirical significance, and the unit of empirical test, is more or less the whole of science. Thus, for example, whether or not the second law will be rejected will depend ultimately upon a wide variety of theoretical considerations (at least over most of its range of applications). In testing the force law in the first place we already depend upon a prior introduction of a measure of charge (or at least of charge equality, cf. note 14); this charge measure may be determined in part by the recognition of *charge presence* because of force effects. But circularity at this point may be avoided because, for example, our theories of chemistry may lead to reasonable criteria of '*identical batteries*', and we need only know of batteries' general effects via the general sort of forces which are produced when they are present. (Not that chemistry can be developed to this refinement independently of physics.)[15]

The issue involved here is one of the logical propriety of a certain manner of proceeding and it has been argued that there are no logical difficulties in proceeding in the manner outlined above, rather, there are only *epistemological* barriers. All science must spring from 'reasonable assumptions', such as the 'likenesses' appealed to above in the case of batteries, extrapolation to a continuous function of force for springs from the cruder direct testing, and so forth. But this fact does not mean that we need regard these assumptions as not being open to doubt if the consequences of accepting them depart sufficiently from experience. Certainly such laws as the second law of motion need not be rendered conventional in order to rescue us from the 'embarrassment' of proceeding from such assumptions, or of finding that some of them may not be entirely adequate. Much less should it prevent us from taking a thoroughly Realist view of science, for science makes many much bolder conjectures than these in its theory construction. (The entire conceptual edifice of a general ontology for the world is one example, the theoretical conception of a force another – cf. [13] for elaboration.)

But with all of this said, we have not yet done full justice to the force of the objection.[16] Consider the names-definite descriptions analogy again. In this case we are prepared to accept that the objects putatively referred to by the proper names are not eliminated along with the elimination of those names from the language, presumably just because we were sure that the definite descriptions referred to *those same objects*. And how could we be sure of this? Presumably only because we had access to these objects as named, and we had access to these objects as definitely described, or at any rate that it was logically possible to obtain these accesses, *and the one access was not logically dependent upon the other access*. If we wish a similar result to hold in the case of forces, then we must similarly demand that it is logically possible to have access to forces *in a manner logically independent of both their effects and their description in terms of other physical properties*. But is this logically possible?

I have argued an affirmative answer on the grounds that theoretical identification of forces through conditions I is logically independent of determining force-effects. But I have also pointed out that it may in fact be the case that we perceive forces directly when they act on the body, in this case we would have a second access route independent of determining their effects.[17] At the present stage of the debate, it is important to see that the argument from the *prima facie* eliminability of forces cannot prevent a man from consistently holding to a non-Conventionalist position in regard to the status of forces.

This present clarification does, however, focus attention upon the factual question. Though we may yet grant that it is *logically possible* to consistently adopt a non-Conventionalist position with respect to forces, are there *in fact* forces in *re?* Does not the argument for the *prima facie* eliminability of forces after all make it plausible that there are in fact no forces? Does not the argument make it plausible that there are in *re* only systematic relations among sets of properties, the so-called 'force-effects'? This purely factual argument has considerable force to it, but we must postpone further consideration of it until a later point in the debate.

8. *The Third Law of Motion*

The third law of motion states a response condition among sets of force-connected bodies. "To every action there is an equal and opposite reaction." (Newton [26], p. 13.)

Comment. If forces are identifiable independently of their effects, if their magnitudes are empirically measurable and if accelerations can be unambiguously recognized, then the third law of motion is an empirical statement. It is refuted by an instance of a body *A* exerting a force upon a body *B* and there being no corresponding force exerted upon *A*.

Let us consider two examples. Consider bringing our hands together in impact or in steady pressure. In either case, one senses in *each hand* a pressure to move that hand away from the opposite hand. Again, consider the electro-magnetic force law. Since accelerated charges radiate – this being part of the conditions specifying the action of the electro-magnetic force – we can test for the radiation of *both* charges when these are left free of other constraints. (Of course, freedom from other constraints must also be specified in terms of forces, but these forces are then themselves to be specified in terms of their characteristic sources, modes of action and so forth. So that the absence of other constraints is again an empirical question.) Once again, whether or not such tests can be carried out is determined by the physics of the situation in question and thus the carrying out of these tests can present us with at most merely epistemic problems.

But we have not yet dealt with the concept of mass. And if the third law of motion is to remain an empirical proposition, it cannot be used to furnish us with a definition of that concept. On the other hand, both second and third laws of motion employ the concept mass and hence presuppose the introduction of an appropriate and adequate definition. If the second law of motion is to remain an empirical proposition, any such definition must be other than in terms of force-acceleration ratios.

We may introduce an appropriate definition of 'mass' here using a variant of Mach's approach (see [23], cf. [25]). Consider the velocities of rigid spheres moving collinearly before and after collision.[18] It is found *empirically* that the negative differential velocity ratios pertaining to such a collision are constant and positive for any given pair of bodies, independent of the manner of production of the initial velocities, additive for systems of such bodies and so forth.[19] The relative masses of these bodies may then be defined directly in terms of these constants as follows: there is a quantifiable property of the spheres such that they behave under collision as they do; this property, to be called mass, is measured by the negative of the velocity differential ratio. (There is good reason to associate the behaviour with a property of the spheres and with the usual concept of 'quantity of matter' since a sphere twice the size but made of the same material, weighing twice as much etc, will produce a mass ratio, so defined, twice as large when in collision with the original sphere etc. etc.)[20] A definite mass scale may be set up simply by adding to this the choice of a particular body as having unit mass. The concept of mass thus defined is then extended to all pairs of rigid bodies, to all systems of same and ultimately to all physical systems by the usual method of theoretical postulation, in which mass plays the same theoretical role as it does in the simpler situations, i.e. such that momentum and energy are conserved and the usual relations of weight and density are preserved. The questions arising at this point concerning the testing of such extensions are no different, in principle, from those already discussed.[21]

Mach went on to *define* force in terms of mass and acceleration, thus ensuring that both the second and third laws of motion were conventional. But with our insistence on the identifiability of forces independently of their effects, both of these laws now become empirical propositions. However, the third law of motion now has a double aspect to it. That the velocity ratios are as they are is an empirical fact. That the *mass*-velocity products are equal and opposite whenever forces act between pairs of rigid bodies, is not an empirical assertion. But the empirical remarks and the definition of mass do not suffice for obtaining the third law of motion. For the third law of motion is formulated in terms of forces. Under the present scheme of things, however, we do have the following result: The second law of motion, when supplemented by these empirical remarks and the definition of mass, entails the usual third law of motion. Since the second law of motion and the empirical remarks both have empirical content, the third law of motion also has empirical content. That such a definition of mass is possible, is an empirical fact. That when so defined *and generalized* it is linked with the effects of a force in the manner specified in the second and third laws of motion, is also an empirical fact.

Actually, in even developing this current argument in the detailed manner above the non-Conventionalist has been generous with the opposition. An alternative, more ruthless way with the opponents would be this: show by appeal to formal axiomatics that the primitive concepts of Classical Mechanics must include force and mass as well as space and time (cf. [1], [24], [30]); then argue from a general Realist position on physical theories and the empirical adequacy of Classical Mechanics under a hypothetico-deductive approach to scientific method that forces and masses were real entities having the characteristics specified by the theory. Under this approach one would not need to define mass in terms of rigid body collisions at all and then generalize it (nor follow a similar procedure for force laws), one would simply introduce 'mass' as a primitive term, operating in the laws of motion. The reason for not arguing in this manner is that one wishes to meet those who challenge the laws of motion one by one, those who would turn individual laws into definitions, on their own ground so to speak and for the reason earlier adumbrated (Section 1, end).

9. *Review*

At this point we may review the progress of this second non-Conventionalist argument. We have seen that, at least insofar as the roles of forces and masses are concerned, the three laws of motion may all be developed in a non-Conventionalist manner. This development was achieved by a careful development of the laws, paying particular attention to their logical form, together with an emphasis on the identifiability of forces independently of their effects and the making of careful distinctions between logical and epistemological difficulties which may be involved in the empirical testing of such laws. If this much has been shown, then we have been able to show that the non-Conventionalist can at least develop his position in a consistent and adequate manner.[22] As factual support for that preference, the non-Conventionalist can also appeal to actual procedure within science and to his first argument from our perceptual experience. (However, the part of the dispute must be carried several stages furher yet.)

But we have to recognize that our non-Conventionalist formulation of Classical Mechanics is radically incomplete. For up until this point no account has been given of the spatio-temporal terms which occur in each of the three laws of motion. Indeed, the discussion of the spatio-temporal framework within Classical Mechanics raises some peculiarly severe questions for a non-Conventionalist position. The discussion of these questions really forms an entire aspect of the debate in itself and we shall therefore leave the present arguments at this stage in order to concentrate on this one aspect.

IV. THE SPATIO-TEMPORAL METRIC

10. *Introduction*

In order to retain their empirical, non-conventional status the laws of motion require us to be able to assign accelerations in a non-arbitrary, non-conventional manner. (Accelerations of magnitude zero, that is, constant velocities, are but special cases of these assignments.) There are two arguments which bear upon this point and which we are required to consider, that due to Grünbaum ([7], Chapters 1 and 2) which we shall consider in the present section and that due to Ellis [5], which we shall consider in Section V of the essay.

11. *Grünbaum's Metrical Conventionality Thesis*

The present argument has as its point of departure the claim that the spatio-temporal metric is, and must be, conventionally assigned. Grünbaum claims that the conventionality of the spatio-temporal metric follows logically from the nature of space and time alone. That is, given a certain characterization of the space-time manifold, it follows of logical necessity that the spatio-temporal metric must be conventionally assigned. The argument runs as follows. Classical Mechanics treats of space and time as continua made up of individually indistinguishable points. But such continua do not provide, *intrinsically* to themselves, a metric. In the first place, there is no way of *marking* positions within the continuum in order that one might compare 'distance', or separations, within the continua. Furthermore, the equi-polence of all n-dimensional volumes means that no counting procedures for the spatial points of the continuum can establish a unique measure on it. Thus any metric assigned to such a continuum must be assigned by means *external* to that continuum itself. Since there are no intrinsic features of the continuum by which we may assign to it a metric, any metric which is assigned to the continuum is assigned conventionally, in the sense that there are no features of the continuum which determine that assignment. This argument is intended to establish a *logically necessary* feature of a continuum characterized in this manner.[23]

12. *Grünbaum's Thesis and Conventionality in the Foundations of Classical Mechanics*

Off-hand, one might expect the following: Since the metric is conventionally assignable, so also will accelerations be; therefore, forces and the laws of motion will also be conventionally assignable or specifiable. This loose argument will not, of course, do in place of a careful investigation. I begin with Grünbaum's position on the matter and then expound my own view of it.

Grünbaum's view is clear enough: metrical conventionality entails the conventionality, hence fictionality, of forces (= of causally active dynamical agents). To see how he arrives at this conclusion let us follow his reasoning (and expand upon it a little) in Chapter 2 of his *Philosophical Problems of Space and Time*. Consider a *temporal remetrization* within mechanics. If the assignment of a temporal metric is conventional, then such a remetrization is always both possible and legitimate. Moreover, such remetrizations must lead to no change whatever in the empirical content of the theory in question (otherwise we would be able to distinguish empirically among various conventionally assignable metrics.)[24]

In such a remetrization the 'new 'acceleration will not in general vanish when the acceleration based upon the original Euclidean metric does. To see this, let the Euclidean spatio-temporal co-ordinates be r, t and let the new spatio-temporal co-ordinates be r_T, T. Let the temporal remetrization be given by the relation

$$T = f(t).$$

Then we have for the relations between the two accelerations

$$\frac{d^2 r_T}{dT^2} = \frac{1}{[f'(t)]^2}\frac{d^2 r}{dt^2} - \frac{f''(t)}{[f'(t)]^3}\frac{dr}{dt}.$$

Clearly we may so choose the function 'f' that the two accelerations do not vanish together.

In respect of forces, two moves can be made at this point.

(a) Along with the temporal remetrization, the original metric for forces is kept. That is: $F_T = d/dt\,(M\,dr/dt)$.

(b) Along with the temporal remetrization, a new metric for forces is adopted. Thus: $F_T = d/dT\,(M/dr_T/dT)$.

It is now very natural to argue as follows. In the case of (a) we may have $F_T = 0$ and yet have the body undergoing a non-zero acceleration. If this is not to constitute an empirical asymmetry with the original formulation of mechanics then we shall have to drop the requirement that all accelerated motions be explained, that is, we shall have to drop the first law as developed previously. Moreover, we may also have that $F_T \neq 0$ but the body may still be undergoing a constant velocity. Thus alternative (a) seems to involve us in rejecting forces as ontically genuine entities. For now forces are neither necessary, nor sufficient, for any particular sets of effects which we may consider assigning to them. What we have to consider as their effects may always be conventionally changed by a simple temporal remetrization. Any acceleration may be chosen as an 'uncaused' acceleration. But no real entity can have its causal effects conventionally chosen. So on this view it is no longer possible to regard forces as ontically genuine entities – even despite the fact that this alternative allows the class of non-zero force situations to remain invariant.

On the alternative (b), however, we may have that $F_T = 0$ when the original force, F_t, is not 0, and conversely. If these changes are not to constitute an empirical asymmetry between the two formulations of mechanics then the re-assignment of forces must make no empirical difference, that is, that re-assignment must itself be conventional. But if the presence and absence of forces is conventionally assignable, then one cannot maintain that forces are ontically serious entities. Thus on this alternative also a non-Conventionalist approach to the status of forces must be abandoned.

There was a time when I argued in this manner myself, but I now believe that the consequences are a little more complex than this.[25] The above argument clearly presupposes:

(M) A change of metrical description, (for example, from accelerated to unaccelerated motion), is a change of force-effect.

Challenging this assumption alters the situation under alternative (a). Let us fix our attention upon a particular object A in accelerated motion relative to a reference system *O* which employs a particular metric *S*. The following, let us suppose, is a complete, true description of *A*'s behaviour:

(D_1) A has acceleration **a**, as measured on the metric *S*, relative to the system *O*.

Now consider the following description and let it also be true:

(D_2) A has acceleration **a**′, as measured on the metric *S*′, relative to the system *O*.

We all agree, I think, that these are descriptions of *one and the same* sequence of events: *A*'s behaviour during a given interval of time. They are simply *alternative* – *metrically* alternative – descriptions of that behaviour. Therefore, if D_1 is a description of an event which is a force-effect, D_2 is a description of the *same* force-effect, i.e. the same event. *Metrically* distinct descriptions may be descriptions of the same force-effects. (Such alternative descriptions may be generalized so as to include alternative reference frames as well.) This is true irrespective of whether such descriptions are conventional or not, for whether or not there is a true metric for space-time we may choose to ignore it (or be ignorant of it) and choose another metric instead, as a matter of convenience.

Once we reject assumption (M) in this manner it is clear that the argument at the end of the preceding paragraph but one collapses. The apparent shifting around of the extension of the term 'force-effect' can be claimed not to be that at all; rather what occurs is that, while the extension of the term remains fixed (while the *actual events* which are force-effects do not alter), the *metrical descriptions* of those events shift around. For example, an event which was once truly described by a certain acceleration is now truly described by a zero acceleration; but it is the *same* sequence of events that is truly described in each case, hence if the first description was of an event which was a force-effect, so must be the second.

We can now see that it is possible that *both* the metric and the first law of motion be conventional and yet forces be real entities, not (therefore) conventionally assignable. They *need not* be so, but it is *possible* for them to be so.[26] Grünbaum's metrical conventionality thesis does not entail that forces are conventional entities.

The second law of motion played no essential role in this discussion because only alternative (a) was being considered and this is the alternative on which that law retains its original form. Thus if the second law is non-conventional then alternative (a) must prevail and this alternative, we have just argued, is compatible with a non-conventional status for forces. But if the second law is conventional then it must be possible to choose, for example, alternative (b) and on this alternative there is no way to avoid a conventional status for forces as well.[27] On the other hand, the *mere possibility* of remetrizations does not make the second law (or the first) conventional, because both could offer uniquely true descriptions of nature *within the particular metric given*, i.e. there might be *a* pattern to nature of which we can offer metrically alternative descriptions. Differing *(metrical)* forms for the laws does not entail different laws.[28]

The points of the last three paragraphs are enhanced by the following considerations. What suggested the conventionality of forces was the apparent *arbitrariness of force descriptions* under change of metrics. But not only is this suggestion of conventionality not necessary, it is not even true that the only force descriptions accessible to us are of the sort that seem arbitrary in the manner suggested. First of all we have *topological* descriptions as well as *metrical* descriptions. Such statements as 'All spatially closed trajectories are caused by forces' are non-arbitrary, non-conventional empirical claims – or at least, their conventionality, if any, is independent of any metrical conventionality and stems from the conventionality of forces (for which latter independent reasons must be adduced).[29] Secondly, we may relativise the usual force descriptions to obtain non-arbitrary, non-conventional claims of the form D_1 and D_2 above.

What Grünbaum's thesis *does* entail, therefore, is the following:

Either

(i) The second law is conventional (where this is taken to mean that the particular metrical *form* of the law can be conventionally altered) and forces are conventional,

or

(ii) M is true and the second law is not conventional and forces are conventional,

or

(iii) M is false and the second law is not conventional and forces may, or may not, be conventional.[30]

We see that, with the possible exception of alternative (iii), whatever the alternative chosen, metrical conventionality ultimately requires an element of conventionality in the foundations of mechanics as well. Notice however that the existence of metrically non-arbitrary (or non-conventional) statements concerning force strongly suggests that forces themselves might be non-conventional entities and hence that alternative (iii) immediately above is actually the one to consider. I shall, however, now go on also to question the validity of Grünbaum's argument for metrical conventionality.

13. *Critique of Grünbaum's Argument for Metrical Conventionality*

In earlier drafts (1968, 1970) this section opened with the unqualified claim that Grünbaum's argument was a *non-sequitur*. But since he has recently carefully qualified his claims, in [8] (STF) the argument originally presented to some extent only illustrates his new position rather than criticizing it. Nevertheless it still seems worthwhile to present the original argument – chiefly because I believe that Grünbaum still has not quite come to grips with it, but also for the sake of clarity and completeness – and then to indicate how Grünbaum's latest pronouncements bear upon it.

I now argue that Grünbaum's (original) argument is itself a *non-sequitur*. Specifically, I shall now argue that there is a crucial ambiguity in the use of the term 'intrinsic' and that the sentence '*P* (a determinable) is not intrinsic to *A*' does *not necessarily* entail that no statement attributing a determinant of *P* to *A* is non-conventional.

To bring out the ambiguity in the term 'intrinsic' let us suppose that *A* causes *E* if and only if *A* is *R*-related to a_0, where *R* is a two-place relational predicate and a_0 is some other object. Let there also be a property *P*, of *A*, that *A* has as an essential part of its physical nature. Then I shall say in the latter case that it is intrinsic_1 to *A* that it be *P*. And I shall say in the former case that *A*'s causing *E* is an intrinsic_2 property of *A*. But it may not be intrinsic_1 to *A* that it cause *E*. For suppose that it is neither logically nor physically necessary to *A*'s existence *as A* that it be *R*-related to a_0. Then *A* causes *E* only upon the obtaining of some other empirical condition which is not necessary to *A*'s continued existence. Thus *A* may have intrinsic_2 properties which are not intrinsic_1 properties of *A*.

Let me begin the elaboration of the notion of an intrinsic_2 property by reiterating that any alleged *contrast* between intrinsic_1 and intrinsic_2 properties is misleading. Thus we might take it as essential, i.e. intrinsic_1, to the nature of water that it have a certain boiling point *at a certain pressure* but, as the italicised phrase makes clear, this is really an intrinsic_2 property of *the stuff described as water*. That is, it is not intrinsic_1 to the stuff there before me that it stand in those physical relations comprising being under the standard pressure, but it can nonetheless be intrinsic_1 to that *same* stuff *qua water* that it have a certain boiling point under those conditions. I am, therefore, not so much interested in involving myself in the question of essential versus accidental properties as in *distinguishing two uses for a word*, the second use of which (i.e. intrinsic_2) selects a *certain kind of property*, namely *those properties the occurrence of which is physically dependent upon the presence of a physical relation.*[31]

Now let us look for some more examples. We may say of a circle that the ratio of its circumference to its radius is an intrinsic_1 property of the circle. But the circles' bisecting a particular line from the origin forming a chord of the circle in the particular ratio which it does (and suppose it does), is only an intrinsic_2 property of the circle, for that property of the circle depends upon a particular line's standing in a particular relationship to the circle. Surprisingly, examples of intrinsic_1 properties of physical objects are relatively hard to find. For example, if we regard the density, shape, size and colour of physical objects all to be in part dependent upon the physical relationships which they have to their environments, then each of these properties is in fact an intrinsic_2 property of the objects. Another example of an intrinsic_2 property of an object is the particular capacity of a pipe. For the particular determinant of capacity which a pipe has is directly dependent upon the pipe's standing in a particular dynamical relationship to a pressure head of fluid and to the mechanical properties of the fluid itself (viscosity etc.).[32] Perhaps we may be able to locate intrinsic$_1$ properties of physical objects at the quantum level. Thus, for example, the charge on the electron, the spins of the fundamental particles and their other quantum properties may all (possibly) be intrinsic$_1$ properties of those objects. But all of this is rather uncertain. The progress of science has in general been marked by successive demonstrations of the fact that properties of objects which we had hitherto taken to be intrinsic$_1$ to them are in fact intrinsic$_2$ to them. Thus, we find the paradoxical situation where we tend to think of 'intrinsic' as meaning intrinsic$_1$, whereas in fact most properties are intrinsic$_2$.

We can now see that all that Grünbaum's argument proves about the nature of space-time characterized in the manner he gives, is that any metric assigned to it will not be so assigned in virtue of features which are intrinsic$_1$ to it. What Grünbaum's argument shows is that continua thus characterized do not have a metric intrinsically$_1$ to themselves. But this does not exclude the possibility of their having a metric intrinsically$_2$ to themselves.

Despite this criticism, Grünbaum would still be on safe ground if he could argue that intrinsic$_2$ properties were themselves conventional. Mr. Hunt, who originally suggested this move, outlined the following Grunbaumian defense. *X*'s moving with respect ot reference system *F* is surely an adequate basis for '*X*'s moving' to count as an intrinsic$_2$ property of *X*. *A fortiori*, *X*'s moving with velocity *v*, or acceleration *a*, relative to *F* is surely also an adequate basis for counting '*X* moves at *v*' or '*X* accelerates at *a*' as an intrinsic$_2$ property of *X*. If this be so, is not also *X*'s moving with velocity *v*, or acceleration *a*, relative to *F*, *relative to the spatio-temporal metric M* an adequate basis for '*X*'s moving at *v*' or '*X*'s accelerating at *a*' to count as an intrinsic$_2$ property of *X*? But this property is surely convention-laden if the assignment of a metric is.

To this ingenious defense of Grünbaum there are two replies. The first runs as follows. It is necessary to stress that an intrinsic$_2$ property of an object as envisaged here is one had because of the existence of *physical* relations between object and environment. The existence of such relations is then a necessary condition for a property's being intrinsic$_2$ in my sense. Nor is this restriction arbitrary, for if an object is to have a non-conventional intrinsic$_2$ property it should possess it for a substantial physical reason grounded in the physical situation. (One could, therefore, introduce a third term, intrinsic$_3$, to cover all those properties had in the presence of relations which did not have a basis in physical reality. Not all of these would be conventional, as the following examples show.) From this it follows that the statements

A is *D in the English language*,
X moves with acceleration a *relative to the choice of metric M*,

are *not* bases for calling the alleged properties '*A* is D', '*X* moves with acceleration *a*' intrinsic$_2$ properties; though

X moves with acceleration a *relative to reference system O*

is a basis for saying that '*A* moves with acceleration *a*' is an intrinsic$_2$ property of *X*, because of the *physical relations* involved.

On the other hand, possession of the right kind of relation is a necessary, but not sufficient, condition of a property's being intrinsic$_2$. In fact, I hold that such phrases as '*X*'s moving' and even '*X*'s moving at *v*', '*X*'s accelerating at *a*' do not specify properties of *X* at all. They are logically incomplete specifications and need to be completed with expressions of the form '. . . with respect to reference frame F as measured against metric *M*'. (And this remains true even when absolute space and time are admitted.)

The second reply is this: even were this Grünbaumian 'defense' granted it would show only that *some* intrinsic$_2$ properties were conventional (as Hunt himself observed). But from this it does not follow that the spatial metric is one of those. The specific cases need to be argued on their merits.

The present critique will allow an alternative to Grünbaum's conventionalism if, but only if, it can also be claimed that the relevant intrinsic$_2$ properties are themselves non-conventionally assignable. What would a likely intrinsic$_2$ property be that would satisfy this condition and what would the associated relation be? I offer the following suggestion: *The way in which space-time contains material objects is a crucial feature of it. It is possible, therefore, that the containing relations between space-time and material objects, together with the nature of matter, provide a non-conventional intrinsic$_2$ metric for space-time*. I claim that this is the case.[33] Furthermore, I claim that Classical Mechanics offers a unique description of that metric (cf. further Section V below).

It is at this point that the intimate connection between my defense of a non-conventionalist view of the spatio-temporal metric and my erstwhile defense of a Realist view of forces – a connection shortly to be developed in detail – first begins to make itself felt. For consider again a statement such as 'X is moving'. As Hunt points out, whether one chooses to say this or not depends on what one has chosen as a reference system. Once the 'origin' is specified, it becomes a matter of fact whether x is at rest. Certainly 'x is at rest' is an intrinsic$_2$ property, since it is a matter of whether x is moving with respect to the reference system (origin), but the conventionality, the choice, in describing x thus derives from the conventionality of whether a given object or reference frame is *the Origin*. Conventional descriptions can then be a subclass of intrinsic$_2$ descriptions. *Whether the intrinsic$_2$ description is also conventional depends on whether there is a (relatively) intrinsic$_1$ description which is conventional, and on which the former is based.* Thus 'x is the origin' is, putatively, a conventional intrinsic$_1$ description, or more importantly, it can obviously be used in a conventional way. Now I claim that the metrics of space and time are non-conventional intrinsic$_2$ properties of them – where then are my non-conventional intrinsic$_1$ descriptions? It is at this point that forces enter, for I contend that the spatio-temporal metric is specified by Classical Mechanics. That is, I hold *that in the case of force-free bodies to say they are the metrical standard is not a matter of convention.* But before developing this account further let us turn to consider Grünbaum's most recent pronouncements.

This, then, was the situation before the appearance of Grünbaum's STF. In order to get as clear as possible on the nature of my proposal *vis-à-vis* Grünbaum's position, I shall commence by stating my position in the form of five assertions:

(1) The spatial metric is *not* fixed by any intrinsic$_1$ feature of space (i.e. by any non-relational properties of space alone or by any relational properties holding solely among spatial points alone).

(2) The spatial metric is an intrinsic property of space insofar as the manner in which space contains material objects is determined by space alone, but the metric is nonetheless an intrinsic$_2$ property of space.

(3) The *description* of the unique metric which we accept is given by a scientific theory (Classical Mechanics – I am speaking here solely within this framework).

(4) No account has been offered of actually how space – whether

through its $intrinsic_1$ structure or via something else – determines its manner of containing material objects.

(5) The *evidence* for believing 2, despite 4, is the existence of the unique metric as specified in 3 (together with the philosophical account of space actually backing Classical Mechanics).

In STF, assertion 1 above is all that Grünbaum really claims for himself and he concludes that the spatial metric is conventional in the sense that it is not $intrinsic_1$ to space. Insofar then Grünbaum's position has not altered. But he does say a number of other things that bear upon the above claims.

Grünbaum points out, quite correctly, that the truth of 3 is *not* a sufficient condition for the non-conventionality of the metric in his sense, i.e. for the falsity of 1. That a theory normatively specifies preferred metrics – even a unique preferred metric – as Classical Mechanics and General Relativity do – goes no distance whatever in itself to showing that Grünbaum's thesis is *therefore* false, toward showing that the metric is anything more than merely preferred in that version of the theory. (But once one sees the possibility of an alternative such as 2, one also sees that the truth of 3 may lend inductive support to the claim that an $intrinsic_2$ metric actually exists.)

Grünbaum also points out in STF that there exist intrinsic metrics which are describable using *extrinsic* devices, so that the appearance of a metric couched in the language of extrinsic devices does not show that the metric so described is not intrinsic (this holds for either sense of 'intrinsic'). He also allows that we may also be able to set up other metrics, essentially different from the intrinsic ones, using extrinsic devices. *This allows the possibility that a spatial metric may be epistemically conventional but not ontologically conventional.* For we may not in some cases be in a position to know whether any of our extrinsically described metrics are in fact essentially the same as an $intrinsic_1$ metric, though they might be if an $intrinsic_1$ metric exists. This ties in nicely with the position presented here, for I claim that the metric is $intrinsic_2$, though scientific theory describes it in the language of extrinsic devices (rods etc.). Moreover, and stepping outside of Classical Mechanics for a moment, the fact that the scientific specification of that metric has changed over the years, leads me to postulate an epistemological indirectness to its specification – indeed it could yet turn to be epistemically undecidable (if we construct a better theory than General Relativity which does not specify a unique metric), but of course this would not demonstrate the falsity of the claim that the metric was in reality not ontologically conventional, i.e. is consistent with the truth of 2.

But this last development would certainly undermine the most important reason we have to believe in 2. And this brings us to the question of evidence for or against 2. Grünbaum says in STF that "... if no intrinsic [intrinsic$_1$] relational property [property in *intension*] is known of which a given relation-in-extension [specified extrinsically] on a manifold is indeed the extension, this fact creates the presumption that the manifold may well be devoid of such an intrinsic [intrinsic$_1$] property." Indeed it does. This means that, given 4, Grünbaum's position and mine must be judged between by weighing my 5 against this opposite presumption *in the light of all the other relevant information.* I shall return to this question at the end of the paper.

This brings me finally again to the logic of Grünbaum's argument. Grünbaum argues that the spatial metric is not intrinsic to it. From this he concludes that it is conventional. This second move would be acceptable if 'conventional' meant 'not intrinsically$_1$ specifiable'. But Grünbaum wishes to apply his conclusion to *physical* theory and here 'conventional' means 'not uniquely determined by the world'. In this case the possibility of intrinsic$_2$ properties becomes of importance. All that Grünbaum is entitled to conclude, I have argued, is "not intrinsic$_1$, therefore either conventional or intrinsic$_2$." What Grünbaum then needs to show further in order to get out his substantial conclusion concerning conventionality is "if intrinsic$_2$ then conventional (also)." But I have tried to undercut this claim by offering an intrinsic$_2$ property which is not intended to be conventional and by offering grounds for believing that it is not. This returns us to the question of the degrees of support for these positions. To answer this question the Realist, non-Conventionalist position must be developed in more detail.

14. *A Non-Conventionalist Alternative*

The first stage in the construction of an alternative consists in understanding the role of *causes*, (that is, the role of forces) in the defining of *standards* of spatial and temporal intervals.[34]

It is commonly said that the choice of a standard is arbitrary.[35] To see that this is not so, we may suppose that we begin by choosing some fairly crude temporal standard, for example, the human heart. Now on the basis of the human heart as a clock we find that almost every process in the universe is irregular. Yet we also find that many processes in the universe keep step with one another, though none of these keep step with the human heart clock. Equally importantly, we find that the present actions of the owner of the particular heart chosen have instantaneous and often relatively violent *causal* repercussions throughout the universe. (For example, if our heart clock owner plays a game of tennis, then we find that during the game every process in the universe, no matter how spatially distant and how massive the scale of the process, slows down.)

Because of these sorts of considerations and discoveries, we drop the human heart clock as an appropriate temporal standard. In seeing why we do this it is important to distinguish two sorts of reasons. In the first place we drop the human heart clock because we desire to obtain a *descriptively simpler* science. Often, this is represented as being the only consideration involved. And, since the choice of the descriptively simpler theory is a *purely pragmatic* choice, it has been argued that if there is any pattern at all to our choice of standards, that pattern is explained along purely pragmatic, and not along objective empirical lines.

But there is a second sort of reason for which we drop the human heart clock. Our second reason is that we wish to obtain a *credible causal pattern* for the universe. On the human heart clock standard, no unique modes of operation and conditions of operation of forces can be specified, no coherent dynamical pattern can be erected. If we demand for forces the full conditions for their non-arbitrary specification which have been earlier discussed, then we are led to reject the human heart clock on the grounds that it does not allow such conditions to be consistently met. There is another important element to this causal critique. With the crude dynamical generalization concerning the actions of forces which we can raise using the human heart clock as standard, we shall find ourselves forced to conclude that this clock itself decays in time and is therefore not an appropriate choice of standard. (In this particular case the observations needed to arrive at this conclusion are particularly obvious and simple.) But the sorts of considerations involved here in rejecting the human heart clock as standard are *not* pragmatic and are of objective, empirical significance.

Now precisely the same sorts of arguments apply when more refined standard clocks are in question, only now the emphasis shifts away from the pragmatic considerations towards the causal critique. Consider the choice of the earth's rotation as our temporal standard. We know, for example, that although the earth's rotation is very nearly uniform, yet it cannot be completely uniform because of considerations of tidal friction, accretion of stellar dust and so forth. That is, we know from our general dynamical laws, together with observed fact (that there is stellar dust falling on to the surface of the earth and that there are oceans on the surface of the earth), that the rotational frequency of the earth must be very slowly decreasing. On the other hand, the earth as temporal standard was the best temporal standard to which physicists had access for a great many years. On the basis of this temporal standard, they were able to raise the system of dynamics which leads ultimately to the rejection of the earth as an appropriate temporal standard.

In general, the refinement of standards and the refinement of theories goes hand in hand. At each advance, we come upon new causes and more finely delineate the known causes. The subsequent theories which are evolved lead to a critique of the existing standards through which such discovery and refinement of causes was initially facilitated and ultimately to the rejection of those standards because of the perturbing causes acting upon them. We thus have a progression of successive critiques, of a theoretical-causal nature, of the successively refined standards.

Ultimately, we reach the stage where the standard referred to in physical theories is specified totally in terms of a particular causal principle. No body under the action of forces can be used as a completely satisfactory standard. Thus, we refer to *force-free* bodies in rectilinear motion as covering equal distances in equal times. Or we refer to a *force-free* rotating body as rotating through equal angles in equal times. Thus our standards of spatial and temporal intervals are specified *causally* in terms of freedom from physical forces. These standards remain hypothetical standards to which chosen bodies are made to approximate as best we may.[36]

These ultimate standards may then be taken to provide a non-arbitrary metric for space-time. Thus we may take any two of the following three situations: (a) light travelling in vacuo, (b) a force-free rectilinearly moving body, (c) a force-free rotating body. From these we may cull out both a temporal and a spatial metric.[37]

It is now time to tie this line of argument to my previous critique of Grünbaum's argument. At the end of Section 13 I stated that it was possible that the containing relations between space-time and material objects, together with the nature of matter, provide a non-conventional intrinsic$_2$ metric for space-time. It is precisely this claim which I laboured to bring out in the discussion of our choice of standards. According to that discussion, it is indeed the behaviour of material bodies under force-free circumstances which fixes non-arbitrarily an intrinsic$_2$ metric for space-time. It is not being claimed that space-time has a metric assignable to it in virtue of features which it possesses intrinsically$_1$ to itself. Rather, it is being claimed that in virtue of the relations which exist between the space-time structure and material objects, that structure has a non-arbitrary metric which is intrinsic$_2$ to itself.[38]

Notice that the assertion that the cause-free motions, as we have specified them, lead to a *unique* metric is at most *contingently* true. It is also a contingent truth that there is any matter in the universe at all. Thus, it is also contingently true, if true at all, that space-time has a non-arbitrary intrinsic$_2$ metric. But there is nothing particularly objectionable about this, indeed it is precisely what it expected of intrinsic$_2$ properties. Moreover, if we consider a universe empty of matter and attempt to formulate the contrary to fact conditional "if there were to be matter in the universe, then . . ." we find that we can give no answer. For there is in this case no non-trivial universal statement on which to base an answer! Thus the remarks about the intrinsic$_2$ metric made above are limited to what is the case in this world – they do not speak about other possible worlds. Once again, this is what is to be expected for contingent intrinsic$_2$ properties.[39]

15. *Conclusion*

The rejection of Grünbaum's argument removes any threat to retaining a non-conventional status for forces from this quarter. Indeed, forces played a central role in the construction of the counter-claim to that argument, for, as we have seen, the notion of force-freeness is central to our conception of the occupation of space-time by material bodies.

Moreover, the reply to Grünbaum's argument resulted in the specification of a non-arbitrary spatio-temporal metric. Thus our problems over the occurrence of these quantities in the second and third laws of motion are also solved. The empirical content of these laws of motion has now been defended in respect of all quantities appearing in them.

But what of the first law of motion, for it figures in the specification of the spatio-temporal metric? First, notice that even if such specifications of the metric are taken as definitions, at least one of the three laws we may use here (see pp. 158–9) will have an empirical status and this law can be chosen to be the first law of motion. But must these specifications be construed as definitions? To take this line would, it seems, detract from their claim to be non-arbitrary, empirical truths. And it was on the basis of this kind of status for them which the refutation of Grünbaum's argument rested. I shall therefore claim that while such specifications are true and offer necessary and sufficient conditions for the specification of the spatio-temporal metric they are not definitional of it. Instead, I shall regard the metrical terms as primitives for Classical Mechanics. This attitude is in line with my realist, non-conventionalist account of the metric. In this case the first law of motion is not definitory of the metric of space-time, but states the truth about it.

The strength of this latter claim is, however, open to further discussion. Thus suppose that having arrived at such a metric for space-time, we find that there occur *unexplained* irregularities in the motions of bodies. There are really two choices open to us here, either we may reject the first law of motion or we may insist that a force can be found which explains these irregularities. Now there are two distinct modal levels at which the latter claim might be made. It might be asserted that it is not possible for the first law of motion to be false. This turns that law at least into a necessary truth about, and, most plausibly, into a (partial) definition of, the spatio-temporal metric. Alternatively, the possibility of a counter-instance might be admitted but is actually denied. In this latter case the first law of motion remains a straightforwardly true empirical statement. If the existence of counter-examples persist then ultimately we must abandon that law. I shall adopt this view of the status of the first law of motion.[40]

V. MOTION

16. The development of mechanics I have been offering would not be complete without a treatment of absolute and relative motion. To this end, I shall present a development of the subject first given by Professor G. C. Nerlich and contained in a paper which he read before the Sydney Branch of the Australasian Association of Philosophy in mid 1967. We are now in a position to offer a non-arbitrary, non-conventional criterion for an object's being in motion. We have that:

(M_1) If x is acted on at any time t in P by an uncounteracted, motion-producing cause then x is not at rest throughout P.

Here x is any material object and P is an interval of time. As we have seen what is to count as a motion-producing cause (that is, as a force) must be carefully specified. *Whether such causes are present is an empirical question*, namely, the question of whether these conditions are indeed satisfied. Specifically, if C is a cause of this type then:

(i) C has a material body or entity as source
(ii) The conditions under which C acts are specifiable independently of its effects
(iii) The mode of action of C on material entities is specifiable determinately
(iv) The *ceteris paribus* clause in the statement of C's sufficiency for its effects is cashed in terms of other causes.[41]

The case of electro-magnetic forces discussed on page 132 above already provides an example of the non-arbitrary specification of the presence of motion-producing forces. Thus the presence of an electro-magnetic force is determined by what sources are present and motions are determined by what other charged bodies are radiating etc.

What I wish to consider as an example here is the case of a simple mechanical impact. Suppose that two pairs of bodies approach one another, each of them moving uniformly and rectilinearly and all at the same velocity but such that only one pair will collide, as shown below:

————————→ x — — — — — — z ←————————
————————→ y — — — — — — — — — — —
— — — — — — — — — — — — w ←————————

Let all four bodies be in a force-free state initially and let the bodies x and z collide at time t. Then *prior* to time t we have:

(i) x and y, z and w do not separate in a plane perpendicular to their motion.

(ii) The distances, parallel to the motion, between x and z, x and w, y and w, y and z all change with time in the same way.

After the collision we have:

(i) x and y, z and w no longer remain, respectively, in their two vertical planes, they do separate from one another.

(ii) Only the horizontal distance between y and w continues to change with time in the same manner as it had previously done.

Because of the *asymmetry* between the two pairs of bodies we are able to say that there was a motion-producing cause acting on the pair x and z. Moreover this example, if chosen so that after the collision x and z *remain in contact*, is quite independent of a choice of metric – for the contact of x and z and the increasing separation of y and w are purely topological features of the situation and therefore the asymmetry is purely topological. (That there are actually motion-producing causes acting between x and z in such collisions may be verified by placing any force-detecting device, for example, one's hand, between such bodies under these conditions.) We must note, however, that because of the possibility of the operative causes in this situation counteracting existing motions, this example cannot guarantee that all four of the particles are in motion throughout the period of time under discussion. In order to complete the case for the non-arbitrary assignment of motion-producing causes, and hence motions, in this example, we also require that particles should not spontaneously acquire motion. That is, we require the truth of the first law of motion.

To these results may now be added the following:

(M_2) y is cause-free or cause-balanced and x and y do not change their spatial separation if and only if x is also cause-free or cause-balanced.

This emphasises that to choose a rest system one chooses a *universal* system – it is not merely the choice of a system relative to one body only. From this it follows that the choice of a true rest system is a sophisticated empirical problem and not just a trivial pragmatic question of individual convenience.

In general, therefore, rest and motion are distinguishable. For to be at rest is to be cause-free, whereas the presence of a motion-producing cause implies the presence of motion. In Newtonian Classical Mechanics and The Special Theory of Relativity, however, cause-free rest and cause-free *rectilineal* motion cannot, *as an epistemic matter of fact*, be distinguished in this manner, though rest and cause-free rotational motion can be. Certainly, some cause-free rectilinear motions can be distinguished, namely all of those arising from the past action of causes and present cause-freeness, with no counteracting causes intervening. But it is not necessary, *though it is true or false*, that every particle was created at rest. Thus even could we know (and note that this is a purely epistemic problem), whether a particle ever had a motion-producing cause acting on it, we could not know that it was at rest.

Now let us consider the possibility – which is in fact our actual situation: – that *every* body is under the action of uncounteracted motion-producing causes. It follows that *every* body is in motion.[42] It further follows that no body can act as a *frame of rest* (inertial frame), nor even as a cause-free frame of reference. For example, we may describe the situation in the case of charged bodies by saying that every body is radiating. Again, it follows from the non-vanishing presence of the gravitational field, together with the fact that there are no gravitational shields in Classical Mechanics, that every material body is under the action of gravitational forces.

Now since there is no material body which can act as a rest frame and every material body is in motion it follows that that motion must be through a space marked out by reference to other than material rest frames. Such a space is to be identified with the absolute space of Newtonian mechanics.[43] The introduction of absolute time then follows immediately. The argument of this section provides, I believe, a defense of the Newtonian conception of absolute space and time which is essentially similar to, though independent of, Newton's own arguments from rotational motion ([26], Scholium to the definitions). The reader should notice, however, that the absolute space and time here introduced is not precisely that of which Newton had conceived. For in Newton's view space and time surely possessed their metrical, as well as their topological, properties independently of material bodies. Under the present conception of space and time this is not so. In that important sense, therefore, space and time are not absolute. However, (i) there is clearly here a 'container' concept of space and time, for there is something of which we may say that without the existence of bodies it is metrically amorphous and (ii) that it not only comes to possess a metric because of the way in which it contains the material bodies which it does contain, but that this metric is unique. In this latter and excellent sense of 'absolute' there is absolute space and time for classical mechanics.[44]

VI. ELLIS AND THE CONVENTIONALITY OF FORCE-ASSIGNMENTS

17. *Ellis' Argument*

In this section I will discuss, and criticise, a direct argument for the conventionality of forces in Classical Mechanics put forward recently by Professor B. Ellis.[45] Ellis' argument hinges on the assignability of zero-force situations. If zero-force situations are conventionally (and therefore arbitrarily) assignable, then we surely cannot claim forces as part of our physical ontology, for of no real quantity is it true that those situations in which it does not occur (= is assigned zero value) are arbitrarily, or conventionally, chosen. Now in classical physics the (potential) energy scale has an arbitrarily assignable zero. For Ellis, forces are defined in terms of energies and, like energy, Ellis argues that they may also have their zero-force situations conventionally or arbitrarily, assigned to them.[46] Ellis develops his position by way of an argument and an illustration. The argument runs (Ellis [5], p. 31):

P_1 The occurrence of the effects which forces are introduced to explain is logically sufficient, or even necessary and sufficient, for the existence of these forces.

P_2 What are regarded as the effects of forces are conventionally assignable.

P_3 Therefore, the existence of forces is conventionally assignable.

The illustration which Ellis offers is that of a new choice of zero-force situations for Classical Mechanics. Ellis chooses those situations where bodies would, from the Newtonian point of view, be said to be falling freely under gravity to be the new class of force-free situations. He contends that this is an empirically adequate choice and, together with the usual laws of force, provides a theory which is equi-powerful with Classical Mechanics.

It is crucial to see that Ellis' argument claims a conventionality in the assignment of forces which is *over and above* that provided by any metrical conventionality. The two arguments are therefore logically independent of one another and each must be treated on its own merits.

Ellis' premise P_2 actually contains (and conceals) the 'natural motion' approach to dynamical changes. On this approach a system is acted upon by a force or forces if and only if the system persists in an unnatural state or is changing in an unnatural way and a system behaves unnaturally if and only if its behaviour requires a causal explanation. Thus, the natural motions do not have, and do not require, causal explanations. The remaining motions, which are deviations from the natural motions, require causal explanations. Ellis' claim is that what are to be counted as natural motions, natural states and natural changes is a matter of convention. It follows from that claim that motions, and changes in general, not requring causal explanation, that is, situations to which zero magnitude forces are assigned, are conventionally assignable. Ellis' actual choice of a new natural motion enlarges the class of such natural motions, for his class of natural motions contains the Newtonian class. Ellis' conclusion then amounts to the claim that the first law of mechanics is to be taken as definitory of Newtonian force and may be replaced by an alternative definition.

Ellis' mechanics is clearly extensionally equivalent to Newtonian mechanics. For Ellis' laws are:

(I') If $F_E = 0$ then $ma_0 = ma_G = ma_N$.

(I'') If $F_E \neq 0$ then $ma_E = m(a_N - a_G) = F_E$.

Thus $a_0 = a_G + a_E = a_N$.

Here a_G is the acceleration 'due to' gravity, a_N the acceleration in the Newtonian system, a_0 the actually observed acceleration and F_E, a_E the force and dynamically caused acceleration, respectively, in Ellis' mechanics. Clearly, in both cases the observed acceleration in Ellis' system agrees with that in Newtonian mechanics.[47]

18. *Elaboration and Criticism of Ellis' Position*

To begin the examination of Ellis' doctrine, let us consider what would count against his example.[48] In particular, what are we to say about *experience* of forces vis-à-vis Ellis' choice of forcefree situations? It is a striking thing about Ellis' choice of a new natural motion that all of our personal experiences of forces are preserved intact. Clearly enough, this is true of all unbalanced force situations involving forces other than that of gravity. In the gravitational cases, we have only two situations to consider. (i) That of free fall 'under gravity'. In this case no feelings *of dynamical relevance to the motion* are involved. Some displacement of the organs occur, because human beings are not rigid bodies, and these give rise to sensations. But these sensations are not relevant to showing the perception of a gravitational force at work. (ii) That of a balanced gravitational 'force'. Case 1: That of a gravitational 'force' 'balanced' by a non-gravitational force. In Ellis' system we interpret the situation as the presence of a single unbalanced force (the non-gravitational force), preventing the *natural* accelerated ('gravitational') motion from occurring. It is this single force and the tensions which it causes to arise which we feel and not the joint tension of two forces. Case 2: That of two gravitational 'forces' in 'balance'. In this case there are for Ellis no forces whatever present, except the cohesive body-forces preventing the natural motions of the differing parts of the body in the two different directions. It is these cohesive body-forces which we feel.

With this success of Ellis' system in mind, let us consider other choices of force-free situations. Suppose that impact-produced motions were now regarded as natural, that is as force-free motions. Here we find an immediate objection from experience. For in impacts we do unambiguously experience the action of a force and this experience cannot be explained avay in the manner in which it was so where only gravitational contexts were concerned. Again, suppose that the electro-magnetic force-law was used, in the way that Ellis used the gravitational force-law, to provide a new natural motion (that of free acceleration in electro-magnetic contexts). In this case we have the unambiguous presence of the radiation field by which to specify the presence of a motion-producing force. Both of these cases point to the conclusion that one cannot extend arbitrarily the assignment of force-free situations, even where a universal law may be capitalised upon.

Now the example of electro-magnetic theory above illustrates an important point. For in that case we could not, *by personal experience of stresses and strains alone*, tell for or against an Ellis-type move capitalising on the electro-magnetic force-laws. The decision is made solely on the basis of the presence or absence of the radiation field. This strongly suggests that Ellis is really exploiting a *purely epistemic* latitude in Classical Mechanics in order to make it *appear* that forces are conventionally assignable. (That is, it may be suspected that what Ellis shows is that forces are *epistemically* conventionally assignable, which is quite compatible with their being real entities, and not *ontologically* conventionally assignable, which is not.) We might begin to suspect that it is an unfortunate fact that we cannot distinguish *experimentally* Ellis' choice of force-free situations from those of the usual Newtonian mechanics, but that this does not entail that there is no difference between them.[49]

But the battle with Ellis is far from over, for the above replies to Ellis rest on a conception of force which he does not share. On Ellis' view forces are ontologically peculiar, seeming to be created only in order to explain their effects. This viewpoint suggests that we might take the further step of eliminating force-effects altogether. Thus Ellis may claim that charged bodies *naturally radiate* as they *naturally accelerate* toward one another according to certain laws derivable within electro-magnetic theory. It is but a small extension of this view to go on to claim, via the observation that our sensations are produced by physical bodily states, that in certain situations these bodily states also *naturally arise* and thus the sensations so experienced are *natural consequences* of these situations. The general upshot of this move is that there are no forces whatever and our sensory experiences and other such criteria as we might have (for example, radiation), are powerless to show that there are. Every law of mechanics, indeed, of science in general, states a law of *natural* succession.

This view is indeed a natural consequence of the role which Ellis assigns to forces. Forces are intended to explain the motions of bodies, that is, to explain force-effects. But if forces are nothing over and above these same effects, then it is difficult to see how they can be appealed to to explain them. Ellis admits as much, but does not go on to reject the explanatory value of forces. But on the present generalization of the Ellisian position there would be no force explanations whatever. Individual events would be explained, by subsumption under general laws, but no dynamical causal explanations would be offered. Indeed, Ellis himself suggests this move – without actually taking it up seriously (Ellis, *op. cit.*, p. 47).

Moreover, *so long as one is willing to do sufficient violence to the Classical Mechanical concept of force*, one can even *enlarge* the range of non-zero force situations to include the Newtonian class of force-free situations as a sub-class. For suppose that Ellis were to attempt a choice of force-free situations under which, by convention, rectilinear motion at uniform velocity (or uniform rotation) was a force-effect. We then must conjure up the existence of forces throughout the most distant regions of empty space, forces whose mode of action on a body is independent of all of the body's physical properties (since all objects undergo rectilinear, uniform motion in force-free situations according to Classical Mechanics). Moreover, the 'strength' of these forces must be object-specific and depend on the prior history of the object involved. For consider two objects ejected into inter-galactic space by two different physical processes along identical trajectories, but following one another. They will, let us assume, be travelling at widely different velocities. Then these mysterious forces must be assigned different strengths *over exactly the same spatial locations* and these strengths are determined by the ejection processes which the two particles have undergone. And, of course, that we do not sense the action of such forces is because our bodies are so constructed as to only respond to accelerations. This then gives a theory which is again empirically equivalent to Classical Mechanics, but only at the expense of violating our normal conception of force (see above, p. 161) and introducing an utterly bizarre conception of force in its place.

On these grounds one could argue for an element of conventionality in *every* theoretical term. For example, one could describe as electron-effects what are now regarded as the effects of neutrinos, as well as the effects of electrons. A neutrino is now to be regarded as an electron moving at c to whom the usual laws of physics do not apply (they apply to electrons only when they are producing electron-effects). The two theories are empirically equivalent (both have all and only electron – and neutrino – effects as their observable consequences) yet not theoretically equivalent. What has happened here is that we have exploited the *epistemic* fact that theoretical terms are known only through their observable effects, to alter the extensions of those terms.

Does this show a deep-seated, physically interesting conventionality in these terms? Of course not. The reasons why physicists separate out electrons from neutrinos are not idle, they are buried deep in our meta-scientific (metaphysical) conception of what the world is really like. Thus part of our fundamental conception of what it would be to have an atomistic world is that there would be a unique set of laws of nature which each natural kind would obey and these laws could not be arbitrarily altered. Since we take this as a *deep truth* about this kind of world we believe that if we can successfully develop a science within the atomistic framework then it will show us the kinds of individuals (atoms) there actually are.[50] Of course, if you violate these deep conceptual canons concerning the nature of an atomistic world, you can – *exploiting also our epistemic limitations* – alter the extensions of our fundamental theoretical terms. But then there is no longer any reason to believe that what you finish with bears any relation to the real world.

Such at bottom, would be an Ellisian programme for enlarging the non-zero force situations to include the Newtonian force-free situations. The programme would violate the best canons we have for handling forces. In the bizarre situation we would find ourselves in if we violated these there would no longer be reason to believe we had any contact with the real world. Moreover, the programme is in any case only made possible – at least in the case of electrons – by exploiting our epistemic limitations, so there is again no reason why any ontological significance should be attached to it. Similarly, why should we not take the same attitude to this Ellisian programme?

This leaves us with the other half of an Ellisian programme, namely extending the range of force-free situations (universally if necessary). This move also has a similar trivialization for all theoretical terms. Thus what counts as an electron-effect can not only be widened at will (cf. above) it can also be narrowed at will. Indeed, *again exploiting our epistemic limitations*, I can say "Electrons don't exist, only electron-effects do". This Instrumentalistic position is very old – Ellis' move is of a piece with it. The move has long since failed to impress Realists with any more than the fact that if you espouse a theory of what there is you run a risk of being wrong. That it is *logically possible* to retreat to the so-called observational level goes no distance whatever toward showing that it is *plausible* or *reasonable* to do so, or that we are not on to the right thing in taking our theories seriously.[51] The logical possibility of this move is not in itself sufficient reason to claim that a fundamental ontological discovery has been made (there are no electrons, 'electron' is a conventional term), rather than the mere exploitation of our *epistemic* limitations. And so it is with Ellis' move over forces.

Except for this additional point: Forces are a subspecies of causes – at least so far as their *dynamical role* in Classical Mechanics is concerned – and we have Hume's celebrated attack on any ontologically substantial notion of cause. The Ellisian move to remove forces entirely in favour of natural successions can be seen as part of a wider Humean programme to institute a regularity theory of cause. Seen thus, the Ellisian programme may have more philosophical appeal than did its conterpart in the case of electrons.

What are we to say in reply to these moves? An effective reply must be two-pronged. On the one hand it must attempt to criticise Ellis' argument and Ellis' conception of force. On the other hand it must be able to present a coherent alternative conception of force and construction of Classical Mechanics.

19. *Critique of Ellis' Argument*

The critique of Ellis' conception of force (i.e. the premise P_1 of Section 17) has been developed in some detail by Hunt and Suchting [17], and I shall not recapitulate it here. From the point of view of the present essay the essential point is this: Ellis' treatment of force suggests that the term 'force' is nothing more than a short-hand device for 'force-effect', a convenient way of drawing attention to the systematic connections among the appearances of various force-effects but nothing more. As I argued in Section 18 above, and as Hunt and Suchting argue, this sort of argument can hardly fail, for if Ellis does adopt this semantics for 'force' he is correct in his claim that no forces can occur in the absence of force-effects, but he will be *vacuously* correct and his argument for their conventionality will be a *petitio*, simply question begging. Moreover, the burden of the earlier part of the essay, and of much of Hunt and Suchting's argumentation, is that this simple view of forces does not do adequate justice to the actual situation in mechanics – that forces are related to their effects in much more complex ways than this treatment allows.

Consider the claim that the occurrence of forces is equivalent to the occurrence of force-effects. I have argued above, on pp. 131–3 and 142 that this claim is false by examining cases of balanced forces. In these cases, I suggested, the forces were not sufficient for their effects. Nor is it easy to see how Ellis can avoid admitting this. Here is another illustration: Imagine a universe, or otherwise physically isolated situation, which contains only two rigid, non-conducting bars, both of which are electrically charged and charged with positive and negative electicity respectively. The bars are in contact. According to the electro-magnetic theory there is an attractive force between the bars. And yet there are no force-effects.[52]

A further illustration will serve to re-emphasise the importance of this conception of force. When several forces act the effect produced is the effect of the *resultant* force. If force-effects were logically necessary and sufficient for the existence of forces, then we should be able to infer only the existence of the resultant force from these situations. We should not, it would seem, be able to make any clear cut decisions about the existence of the various components of the resultant force. In fact, however, what component forces are present in a given situation which displays the action of a certain resultant force is not an arbitrary question but a genuine empirical issue.

Conclusion: the concept of a force is, as we found earlier, bound up with physical situations which are characterised by references to physical objects as sources of force, by the modes of interaction of the forces and so forth and *not* merely with force-effects. The upshot of this is that Ellis' premise P_1 is rejected, at least in its '. . . logically necessary and sufficient . . .' reading. For forces are sufficient for their effects only in the absence of countervailing forces – cf. the discussion of Definition 4 above. Correspondingly, it is in the spirit of this same Realist view of forces to reject even the weaker version of P_1 and claim that forces are also at most empirically necessary for their effects – this is what the Realist reading of the first law of motion teaches us to say. And later on, when we have identified forces with an aspect of the physical ontology of mechanics this is exactly what it will seem natural to say (cf. p. 175 below).

Now let us consider Ellis' premise P_2: This premise divides itself into two claims.

(P_2') In any formulation of dynamics physical, motions (dynamical behaviour in general) are always divisible into two distinct classes, the natural and the unnatural motions. The former of these remains unexplained and unexplainable (in that formulation) whilst the latter, the force-effects, are explained by forces.

(P_2'') The division between the two classes of motion is conventional.[53]

As against P_2' it does not seem to me either necessary or self-evident that some motions remain inexplicable. There may be theories on which *every* motion is explained. But in these cases the division spoken of in P_2'' would be quite non-conventional so long as the choice of that theory was itself non-conventional. (That is, the theory was not a mere conventionally chosen reformulation of a theory under which there was indeed a non-empty class of natural motions.) By way of analogy, consider the following example. Suppose that it can be taken as an 'external force-free' law of organic populations that they evolve in a certain manner. This law functions at the 'molecular' (= whole population) level of population theory. Yet this law of natural succession can be fully explained, let us agree, in terms of a deeper bio-chemical theory of the individuals which make up the population. Similarly, we may suppose that the so-called natural mechanical motion of a body is explicable in terms of a more fundamental dynamical theory. Suppose that the realisation of a certain configuration of quantum properties always issues in spontaneous motion. Suppose further that these 'dynamical powers' of material structures are found to be but a part of a complete quantum theoretical specification of those properties, so that the quantum theory entails that rectilinear motion at a certain velocity always ensues on attaining these particular configurations, whether or not there are external forces present. The so-called 'natural motions' of Classical Mechanics would then be completely explained in terms of these more fundamental quantum properties of material structures. Moreover they would be explained by a theory, as I imagine the case, which itself had no natural *dynamical* states (and possibly no *natural* states of any kind, though it will have fundamental laws, for example, which are unexplained in the theory). In this important sense every motion would have an explaining *cause*.[54]

The other half of the premise, P_2'', has also been refuted by this same example, because in this situation the division into natural and unnatural motions is not at all conventional. But I would wish to deny the truth of P_2'' irrespective of the truth or falsity of P_2'. P_2'' is false even within Classical Mechanics. For there what motions are cause-free and hence natural is, as I have argued in earlier sections of this essay, an empirical question. It is a question of the form "Which motions are such that we cannot identify forces, that is, specify their modes of action, locate appropriate material sources for them etc., which will explain these motions?", and this is an empirical question.

I have rejected the premises of Ellis' argument and therefore I am able to reject his conclusion. I am at liberty, therefore, to adopt the view which I hinted at early in this section, namely that Ellis' counter-example presents at most an *epistemic* undecidedness or conventionality in the extension of the term 'force-free situation'. But from this situation there follow no conclusions about the *ontological* conventionality of forces.

VII. TWO NON-CONVENTIONALIST POSITIONS

20. We may distinguish two non-conventionalist positions concerning the status of forces within Classical Mechanics and one will belong to the one or the other according to whether or not one wishes to retain forces as a serious element in the mechanical ontology. The position which takes forces ontologically seriously I shall call the 'strong' position, and the other the 'weak' position. The weak non-conventionalist position is only semantically distinct from Ellis' position. On this position there is agreement with the ontology of conventionalists, that is, that there are no forces in re. But there is on this view disagreement with the semantics of conventionalists for it is claimed that talk of forces in Classical Mechanics, that is, the logico-conceptual structure of Classical Mechanics, is non-arbitrary and non-conventional. The weak non-conventionalist position essentially boils down to the claim that what functional relations of a fairly simple order we can find connecting the physical properties and dynamical behaviour of bodies is not conventional and that therefore the conceptual and syntactical structure of the theory which we use to

describe such situations is not conventional. Within this theory the term 'force' is a unifying concept, which may be cashed out in terms of talk of force-effects, but which brings a conceptual coherence and economy to the theory. On these grounds Ellis' conventionalism can be rejected precisely because a non-conventional descriptive difference can be drawn between all other motions and inertial motions. But there no longer remains any *ontological* dispute between Ellis and the weak non-conventionalist.

The strong non-conventionalist position, on the other hand, takes the ontological status of forces seriously. And we must now face the fact that he has something of a *factual* case to answer. This case stems from the fact that, whilst it has been shown that the argument from the *prima facie* eliminability of forces to a conclusion asserting their ontological vacuity has been shown to be invalid, it was pointed out at the same time that one may nevertheless feel that forces are, as a matter of fact (though not of necessity), eliminable from one's ontology (see pp. 137–42). Why not, it may be asked, reduce our mechanical ontology to the minimum by insisting that the particular functional expressions which are put in place of 'F' are not, after all, definite descriptions of members of our ontology, namely forces, but simply express the fact that there are certain regularities present in the dynamical behaviour of mechanical systems which can be characterised by certain physical properties? This move apparently has the advantage of greater ontological simplicity without loss of empirical content. The argument from our sensations, or experience, of forces would be rejected under an Ellis-type move on which such experiences were merely part of those same sequences of regular events, but no more than this. Finally, the argument from the ability to construct a mechanics in which forces play a significant role would then be held to be inconclusive as well.

There is another aspect of forces which might also be thought to militate against the strong non-conventionalist position. Forces may be felt to be strictly theoretical entities within Classical Mechanics with little role to play outside it and having at most a doubtful contact with the observational level. On this latter point, it may seem, for example, that there is something puzzling about the notion of actually observing the force itself as we would, say, a particle. And yet the strong non-conventionalist appears to be driving us in this direction by wishing to speak of forces as part of our ontology. It is all too easy and natural to say that we do not observe the force itself but only its effect. On the former point, it may be said that we usually require of theoretical entities that they be fecund, that is that they turn up in a richer variety of contexts than that in which they were originally introduced. And on this score forces may seem not to match up to our requirements for taking them seriously. Gravitational forces, for example, will not appear elsewhere than in gravitational situations. And forces in general may seem to have little role in modern field theories or even, indeed, in such formulations of Classical Mechanics as the Hamiltonian formulation.

There is, however, a further argument for his position which the strong non-conventionalist can raise and which runs as follows. Forces are to be identified with energy gradients. Now the concept of energy is an extremely ubiquitous and fecund fundamental theoretical concept. It plays a crucial role in every fundamental physical theory. Indeed, the Hamiltonian formulation of Classical Mechanics centres around that concept as the most fundamental of all. Indeed, it is not unreasonable to maintain that science has been moving in the direction of establishing a view of the physical world in which energy is the sole substance of which all physical things consist. The strong non-conventionalist has, therefore, a way of understanding forces which leave them as structures in our fundamental (energy) ontology and a way of understanding physical theories involving forces which is surely as penetrating an understanding as any found in science. He can then go on to maintain that in our experience of stresses and strains we are directly perceiving the occurrence of energy gradients and the transfer of energy.

Now of energy, *qua* fundamental substance whose measures are mc^2, $\frac{1}{2}\varepsilon_0 E^2$ etc., it is true that it is at most an empirical truth that energy gradients play the dynamical causal role which Classical Mechanics ascribes to forces. Thus it is at most empirically true that energy gradients, *qua* gradients in the fundamental substance whose measures are mc^2 etc., are, *ceteris paribus*, causally necessary and sufficient conditions for the presence of force-effects. This view supports and completes the view of Newton's laws developed earlier and the rejection of Ellis' argument. Before I leave this argument, there are two important things to be said about energy which affect its role in the argument. The first is this: The distinctive characteristic of energy, namely its place in so many distinct physical domains and their theories, can give rise to an interesting argument (concerning energy), but this argument can also stimulate a corresponding objection. The argument runs as follows. No one property appears in all of the various expressions for the measures of energy. For example, no one property is common to Gm/r, Q^2/r, JH, and so forth. Yet since all energy scales agree (where they overlap) and since energy can be transformed from one form to another, there must be something in common between all of these varying situations. That which is in common is energy itself, the various forms which it manifests being determined by the particular properties, or collections of these, which appear in the various expressions above. The appropriate combination of values of such physical properties are *measures* of energy.

The related objection is, of course, just the question why we should not simply say that there are indeed found in nature regular connections amongst various sets of physical properties *but no more than this*. Why, that is, should we not be prepared to remove energy from our physical ontology and treat it purely as a 'metrical convenience'. The major difficulty with this objection seems to me to be that it simply does not take modern science seriously. For modern science certainly does treat energy as a substance and it does seem to be assigning it an increasingly fundamental role in its theories. At all events, the Realist attitude on this issue provides a possible and coherent view of science.

On the present conception, energy is not a property of bodies. To *have* a certain energy is certainly a property of physical systems. And the amount of energy which a given physical system has can be given a *measure* in terms of the values of other physical properties. But energy itself is to be construed as a substance, in just the way that material bodies and physical fields are substances.

The second important feature of energy is that, unlike other theoretical terms which we do wish to take ontologically seriously (mass, charge etc.), the scale of energy apparently has an arbitrarily assignable zero. And if the situations of zero energy are indeed arbitrarily assignable then it would appear difficult to claim that energy is itself an ontologically significant quantity, for surely of no real quantity is it true that those situations in which it does not occur (= are assigned zero value) are arbitrarily chosen. Indeed, this feature of energy seems to make it very difficult to construe it as a substance.

I shall now argue that it is reasonable to accept that the free assignment of the zero of energy is a *purely epistemic* matter, reflecting an experimental inaccessibility of the true zero of energy to human beings and no more. On this view, there would be a true zero of energy, but our theories and experiments would allow us access only to relative energies. Reasons for thinking in this way are the following. It is certainly implausible to believe that in a universe completely devoid of matter and physical fields, there should be a non-zero energy present. And it is no more plausible to believe that the interactional energy is non-zero in a universe where each material object was indefinitely greatly separated from each other material object and the interactional forces decrease with increasing separation of the objects. Here then are at least two non-arbitrary zeros of energy – though we find no practical use for them in our theories. Again, we must note that the expressions for the kinetic energy of material objects and the electro-magnetic energy of objects and the electro-magnetic field etc., are quite precise. Indeed, on reflection it becomes clear that the only energy term which creates any difficulties in this connection is the potential energy term. And since even this we have just seen to have at least one non-arbitrary zero, it is not unreasonable to believe that the potential energy of the universe also has a non-arbitrary zero, but one which is not empirically significant for us. It is not impossible, however, that we should at some future time find a way of assigning a non-arbitrary zero to potential energy. For example, at one time no true zero of temperature was discoverable, but it would have been wrong to conclude from this that there was no such true zero. Indeed, in the case of temperature we are now even able to give a theoretical account of this true zero. It may eventually be possible to give, for example, a quantum mechanical account of a true energy zero (though this does not seem likely at the present time). These two features of energy, therefore, do not prevent the strong non-conventionalist from making use of the argument from the ontological centrality of energy which was outlined above.

VIII. PHILOSOPHICAL POSITIONS

21. Most broad philosophical positions find themselves with competitors which are equally broad and which diverge from one another and from it at such a basic level that no real cross-criticism, based on even more

fundamental agreements, is possible. Rather, the positions must simply be laid out and their advantages and disadvantages left to each man's assessment. The present strong non-conventionalist doctrine of Classical Mechanics is no exception. The position I have sketched rests crucially upon the claims made about the nature of physical forces. Unless these can be argued to be non-arbitrarily specifiable, independently of their effects, then the whole position is in danger of collapsing. Let us consider the overall position put forward in this essay *vis-à-vis* that of Grünbaum and Ellis.

(1) Grünbaum. To arrive at my conclusions concerning the uniqueness of cause-free motion I require that the nature of causes be taken seriously in the manner in which I have previously outlined.

Since I do not want to claim that there must *necessarily* be a unique cause-free motion in our universe, it is not sufficient merely to demonstrate that Grünbaum's argument for the conventionality of forces is invalid. Rather, I must actually succeed in showing the conclusion to be false in our particular world. I attempted to do this by offering a particular counter-example, that concerning the nature of our standards (pp. 156–9). But Grünbaum can, it seems to me, consistently adopt a reinterpretation of that counter-example which is consistent with the claim that there are no non-conventionally assignable causes. From Grünbaum's point of view the claim concerning standards becomes merely a claim about certain pragmatic *descriptive features* of our world having the characteristic of maximum descriptive simplicity.[55] Under these conditions I cannot show Grünbaum's position to be false – though I can show his own arguments for the *necessity* of that position to be invalid – and hence I cannot finally demolish Grünbaum's alternative to the strong non-conventionalist position.

On the other hand, however, Grünbaum cannot show that the strong non-conventionalist position is either inconsistent or false. For his own reply to my treatment of standards, a reply in terms of descriptive simplicity rather than empirical fact, already assumes that forces can be discounted as conventionally assignable. Thus this reply would beg the question against the strong non-conventionalist position, just as any attempt to show that Grünbaum's position was false would succeed only in begging the question against him. Thus these two opposed, but consistent, positions on Classical Mechanics part company at the outset, motivated as they are by differing philosophical traditions and by differing attitudes towards the fundamental concepts.

(2) Ellis. Essentially the same remarks apply to Ellis' position *vis-à-vis* that of the strong non-conventionalist. There is no stopping Ellis from *legislating* about the use of the term 'force'. And his system of mechanics is certainly a consistent one. One can therefore only propose the alternative construction of mechanics and attempt to assess their relative merits.

Let us now turn, therefore, to the relative merits of the strong non-conventionalist position. I wish to claim for that position three basic advantages over and above the alternative conventionalist positions (for example, those of Ellis and Grünbaum).

(i) Both Ellis and Grünbaum admit that they must do some violence to the common conception of force in science. And both of their positions require that rather counter-intuitive results be accepted.[56] Both, for example, find themselves forced to admit that certain central features of scientific practice (for example, looking for sources of force, attempting to specify modes of action of forces, treatment of standards, etc.), are purely pragmatic and that certain *prima facie* possible situations are not after all possible (for example, the explanation of natural motions, the causal explanation of changes of standard objects and so forth).[57] The strong non-conventionalist conception of force, I would claim, does more justice to the scientific conception of force, to the scientific practices based on it and to the range of explanatory situations which we think are open to us.

(ii) Any conventionalist account of Classical Mechanics emasculates the putative empirical content of that theory, for some laws are construed as definitory of some of the concepts involved. Such laws then lose their empirical status. On the strong non-conventionalist position all laws have empirical status and thus the empirical content of Classical Mechanics is maximized.[58] Moreover, there is a certain justice done to the historically developed theories of mechanics which the conventionalist position is not able to provide.

(iii) Because forces are identical with energy gradients, no conventionalist account of forces can afford to admit energy to the fundamental ontology of science. Yet in the history of science, and in modern science especially, energy seems to have an excellent claim to be treated as the basic substance of the universe. The strong non-conventionalist position, on the other hand, not only is incompatible with the giving of a place of primary importance to energy, but actually requires that energy, or at least energy gradients, be counted as part of the fundamental ontology of science. Moreover, a straightforward account of our sensory experiences in dynamical situations is then forthcoming.

In previous sections of this essay I have attempted to argue that it is possible to consistently adopt a non-conventionalist position on the ontological status of forces. I have attempted to rebut arguments to the contrary. I have now argued that there are several distinctive advantages to adopting such a position. I hope that this has served to show, not only that it is not unreasonable to adopt a non-conventional doctrine of Classical Mechanics, but that it is in fact one of the most reasonable positions to adopt on that subject.

University of Western Ontario

NOTES

* This paper grew out of a seminar on the foundations of Classical Mechanics held at Sydney University during 1967. Other members of the seminar were Dr. W. A. Suchting (Director), Mr. K. Campbell and Mr. I. Hunt. Clearly it is to be expected that the development of the paper owes a great deal to these people. At a general level, the development of the foundations of Classical Mechanics in part III owes much to ideas put forward by Suchting in conversation whilst the role of arguments surrounding energy in the conventionalist controversy (part VII) also benefited from his criticisms. Mr. Hunt has since offered detailed criticism of earlier drafts of this paper and I am especially indebted to him for his critical analysis of the relevance of Grünbaum's metrical thesis to the issue of conventionality (though Grünbaum's own, recent, analysis – see [8] – has also helped greatly in this respect). Since that time the material has been presented to a graduate seminar at the University of Western Ontario whose members were Messrs. P. English, B. Fargen, M. Moffa, R. Slack and the arguments of the earlier sections especially owe much to their careful criticism. Where further acknowledgement is appropriate and can be more specific, it will be given.

1 There is a close connection between conventional assignment and arbitrary assignment. If the assignment of an extension to a term from among a range of possible extensions is conventional in the sense here to be used – cf. below, then that assignment may also be said to be arbitrary. It is arbitrary just in the sense that there are no empirical considerations which can determine the assignment towards one alternative rather than another. It may, of course, be a non-arbitrary assignment from a pragmatic point of view.

Notice that it is an arbitrary assignment only among some limited range of *possible* alternatives. Conventionality does not imply that literally anything can be done with the term or law thereafter. In general the restriction is to those statements which yield theories empirically equivalent to some original theory.

[2] Mach is actually a somewhat more complex thinker than this caricature suggests, but it captures that dimension of his philosophy of relevance here. For material on Mach see eg. the relevant essays of volumes 5, 6 of this series, and [21].

[3] Professor Adolf Grünbaum [7] calls this kind of conventionality 'Trivial Semantic Conventionality'. Grünbaum also discusses the kinds of conventionality with which we are concerned in this essay.

[4] In particular, I am aware from the writings of Ellis [5], and before him Dijksterhuis [4], that Newton did not formulate his theory in terms of forces and accelerations, but in terms of impulses and velocities. I am also aware that the modern mathematical formulation of Newton's second law of motion entails the first law of motion, thereby rendering that law redundant. I shall therefore find myself at once both discussing Newton's mechanics in a more modern garb than that in which he presented it (for ease of presentation), and also developing the system in a manner more closely akin to Newton's and other early developments than the modern axiomatic mathematical treatments in order to bring out certain features of its logico-conceptual structure. Later on I will comment on alternative formulations of Classical Mechanics.

[5] Actually this remark is a *non-sequiter*, even with the 'merely' stressed. The concept of a gene may be the concept of something merely sufficient for the normal transmission of hereditary characteristics, but we can (now) identify genes independently of their function (as parts of DNA molecules). What *is*, perhaps, a little more plausible is the claim that since we normally do not perceive forces (action on and by ourselves excluded? – cf. discussion of Ellis below) but perceive only force-effects (accelerations, etc.), if the concept of force is merely the concept of something sufficient for its effects then here is no reason to hold that forces are anything over and above their effects. But even this argument does not show that forces are not distinct from their effects. Much of the initial plausibility of doing away with forces seems to stem from the kind of move that I have just criticized. It was a remark of Suchting's which stimulated these comments.

[6] Actually, we may have *prima facie* sensory evidence for the occurrence of balanced forces. Consider the case of the balanced pressure between two hands pressing together. In this case we *perceive* the presence of the balanced forces! Moreover, if there had been other pairs of balanced forces and these had been sufficiently powerful to rise above the physiological threshold so as to be noticeable by us, then we should certainly also perceive their presence from the *other* sources of pressure on the hands. Nor can we say that there are no forces present because no effects are observable (externally to us), that is, because no *external* observable *dynamical* effects occur. (It is not really of relevance to claim here that there are in fact *inner* dynamical adjustments by which we detect the presence of these forces. It is logically possible that our bodies be perfectly rigid bodies and yet still inform us of pressures upon their parts.)

[7] Cf. also the discussion on pp. 137f., 161f. I am indebted to Professor G. C. Nerlich for making me see the relevance and importance of these conditions.

[8] The remarks above were again stimulated by Suchting, some of whose comments made in an exchange of letters on the subject I have incorporated into the above.

[9] Since the presence or otherwise of counteracting forces is not a trivial question, this formulation of the definition is not the tautologous 'forces are sufficient, except when they are not'; for *when they are not* can be spelled out.

[10] 1st Law: Every body continues in a state of rest, or of uniform motion in a right

line, unless it is compelled to change that state by forces impressed upon it. Newton [26], p. 13.

[11] Of course it is logically possible that there is a force between *A* and *B* but that it is permanently balanced by other forces on *A* and *B* (separately). But one can then easily so arrange the motions of *A* and *B* that given the arrangement of the surroundings, no appropriate sources and/or laws of action for these forces could be found. Thus they would be rejected as violating conditions I for forces.

[12] 2nd Law: The change of motion is proportional to the motive force impressed; and is made in the direction of the right line in which that force is impressed. Newton [26], p. 13. This is Newton's form of the 2nd law, for its interpretation and relation to the modern version see the references of note 3 above.

[13] More precisely, what replaces F is a description which specifies the magnitude of the force; this of course alters nothing, it is like replacing the names of men by descriptions that specify their heights in various circumstances.

[14] Identity of batteries can be reduced to an identity assumption concerning non-batteries (hence amply attested to by other theories) and an assumption of symmetry for charge flows, as follows: connect two identical metallic spheres to a circuit in parallel, charge the spheres by completing the circuit, break the circuit and also separate the spheres by opening a switch between them, they are now identically charged. Fractions of this charge may be prepared by connecting these spheres to numbers of identical uncharged spheres. (That this supplies a constant fraction can be tested by comparing spheres prepared in this way independently from the two original identical spheres, and among one another from a single such sphere.)

[15] In this respect, gravitation would be a cleaner, but similar, example, if only the force were stronger. For then we could directly determine its action in a rough way and then refine it à la Cavendish, and finally we could extend the refined law by postulation to all material bodies. Again, we would need a prior concept of *mass* – however in this case we do indeed have independent access to this particular quantity. (See below p. 143. But cf. the qualifications which may be necessary in the light of the discussions of note 21 and Section 9 and the references contained therein.)

[16] For the further development of this objection I am indebted to Suchting.

[17] This latter is not, however, a very strong argument. As I noted when first discussing it in Section 3, if one really thought of forces as very peculiar and rather dubious theoretical entities, then it would be very plausible to say that what we are experiencing in such circumstances are, not forces, but only *force-effects*. The claim would be then that one never had direct access to forces, but only to force-effects. The argument would then conclude that, although the *formulation* of mechanics in terms of sources of forces, modes of action and so forth is *verbally* different from that in terms of accelerations, *materially or in fact*, the two formulations are tested in one and only one way, via what is called in one formulation (only) force-effects (in the alternative formulation the description of certain types of physical effects as 'force-effects' would be regarded as reflecting only a vestigial remnant of a discredited theoretical ontology).

This dispute about the existence of forces to which this discussion leads is a real one but at this stage it is crucially important to see that the argument *is not the right logical type to constitute a rebuttal of this non-Conventionalist defense*. The situation is this. The Conventionalist has put up an argument, based upon the character of the second law of motion, for the eliminability of forces. In reply, the non-Conventionalist has constructed a parallel argument which is designed to show that the Conventionalist argument is *invalid*. What is required on the part of the Conventionalist is an attack upon the soundness of this latter argument. But that part of the argument which is attacked here is merely the claim that it is possible to perceive forces directly. Clearly, what the Conventionalist must show in order to break down this defence in this way is that it is *not* possible to directly perceive forces. But the above argument succeeds, at most, in showing that the situation is ambiguous and that experience might be so construed that we do not as a matter of fact perceive forces directly. It is obvious that this claim, even if granted, will not be enough to rescue the Conventionalist's original argument from rebuttal.

[18] We can begin, as usual, with a fairly rough notion of rigidity. Those who are sensitive to questions about the refinement of the notion of rigidity, especially in the light of the more recent problems concerning non-Euclidean geometries and General Relativity, would invoke from us at this stage but another long digression into the more or less 'wholist' concept of science, and its advances.

[19] I.e. if these masses M_1 M_2, approaching collinearly, respectively having initial velocities U_1 U_2, and velocities V_1, V_2 after collision then we obtain, by conservation of momentum,

$$M_1U_1 + M_2U_2 = M_1V_1 + M_2V_2$$

hence

$$M_1/M_2 = -(U_2 - V_2) / (U_1 - V_1).$$

[20] Anticipating a later argument of Ellis' in connection with forces we observe that I have defined *mass* in terms of *mass-effects*; if these latter are only conventionally determined then so will the former be. One might wish to say, e.g., that objects naturally rebounded in certain ways and that no reference to the properties of objects was called for. In reply I point out first that the actual 'definition' of mass is much more complex than this and that the simple collision situations only serve to introduce the concept of a property of objects in virtue of which they displayed certain dynamical behaviour and obeyed certain dynamical principles. Second, some such property is needed to explain the varying behaviours of different objects and the relations between these behaviours and other phenomena, e.g. weight and density. One could only eschew such properties by eschewing theoretical explanation altogether, a position my opponents, but certainly not I, might espouse.

[21] The emphasis here on the hypothetical or *theoretical* generalization of the notion of mass is quite important, for it is certainly not the case that all mechanical systems can have the masses of each of their constituents determined by such simple measurements. See, for example, Pendse [27]; Simon [28]; McKinsey *et al.* [24]. But the fact that the term 'mass' is *not eliminable* from Classical Mechanics in favour of spatio-temporal observables does not mean that it must be introduced as a *conventional* term to the theory. The generalization of 'mass' spoken of in the text leads to perfectly empirical, testable general laws. In particular, it leads to many ways to determine masses alternative to the one mentioned in the text. But cf. also the ultimate paragraph of this section.

Notice that I do not define *mass* to be the velocity ratios, even initially, but rather a property of objects which gives rise to these ratios. Mach used mutual gravitational acceleration ratios and seems to have defined a mass ratio to be such an acceleration ratio. But this obscures the issue of gravitational versus inertial mass. Kleiner [21] has given a detailed discussion of the role of Mach's conception of mass in Classical Mechanics.

[22] The Conventionalist might here try the manoeuvre of arguing that which development of Classical Mechanics one chose was itself conventional (no empirical differences are involved). That argument would be a special case of the general argument that the dispute between Realism and Conventionalism can only be conventionally settled since I concede that, via Craig's Theorem, one can always select out just the observational content of any theory and do what you like with the rest, preserving it. There is no space here to reply to this issue, except to remark that I do not believe that philosophical dispute can be trivialized in this way – cf. my [11, 15].

[23] Questions concerning the intrinsicness of metrics are very complex and admit of a wide variety of interesting answers, depending on the type of manifold in question. In the above I have given only the crudest account of what is a very sophisticated position. In particular I have neglected to point out that 'conventionally assignable' is to be read as 'convention-laden' in Grünbaum's sense. For a definitive treatment of the topic see Grünbaum's recent [8].

[24] In a comment on this paper Professor Grünbaum drew my attention to the fact that the foregoing wording may possibly give rise to an unnecessarily superficial understanding of his position. The reader may gain the impression from my phraseology that there are no facts of physical life which are pertinent to the choice of a metric – but this would simply be incorrect. Even if the spatio-temporal metric cannot be assigned on the basis of *intrinsic* facts about space-time, the fact that practically 'rigid' rods preserve their coincidence relations under spatio-temporal transport is an operational fact pertinent to *setting* a spatio-temporal metric. And, as Reichenbach and Grünbaum have emphasized, there will in general be many more such facts pertinent to any decision, in science, even when it is conventional in Grünbaum's sense. In other words, we must distinguish between conventionality à la Grünbaum – which is the claim that there is no *directly relevant* factual basis for determining the outcome of a decision (applied to our present case it is the claim that there is no intrinsic factual basis for determining metric equalities and inequalities) – and the claim, in general false, that there are no facts whatever pertinent to the decision (applied to our particular case, that there are no facts whatever pertinent to the ascription of metric equalities and inequalities). 'Direct relevance' as used here is a technical term and what it amounts to must be spelled out in each case – in the present situation it means '*intrinsic*' to space-time'.

This distinction does not, however, alter the argument of the paper. For it remains true that if a decision is conventional in Grünbaum's sense then all theories which differ only by employing differing outcomes for the decision must be empirically equivalent with respect to the directly relevant factual basis concerned, for they disagree about none of the features of reality in that basis. In order to clarify the fate of the pertinent, but not directly relevant, facts in such a change, consider again the (assumed) fact of rigid rod coincidence under transport. Consider a theory T_1 under which space is Euclidean and practically rigid rods remain length-invariant under transport, and another theory T_2 under which space is metrically non-Euclidean (but topologically Euclidean). What can T_2 now do about the *pertinent* fact that practically rigid objects have invariant coincidence relations among themselves under transport? First, it must retain this feature because it *is* a *fact*. (Note, however, that this particular fact may not be preserved verbatim as such, it may reappear as consequences of other statements of the theory, for example about travel times of light.) Second, it can incorporate this fact by denying that these objects remain length-invariant under transport. This latter assertion in T_1 was, of course, conventional (at least if Grünbaum is correct) and so does not represent a fact that must be retained. In this way, by similarly adjusting the set of conventional assertions (metric, change under transport etc.) both theories preserve exactly the same pertinent facts.

So long as the reader reads the remarks of Section 2 of the paper and this section of the text in the light of this note – reading 'directly relevant empirical features' for 'empirical features' – nothing need be changed.

[25] I am grateful to Mr. I. Hunt who first convinced me of the insufficiency of this argument; much of the remainder of Section 3 originates with his remarks.

[26] Forces would still be conventional entities if, for example, we found it possible to assign a spatio-temporal metric *and* a first law of motion such that the extension of the term 'zero-force situation' was *not* preserved but this without any alteration in empirical content. This is what Ellis claims to do and his claim is, therefore, over and above Grünbaum's conventionality claims.

[27] Because the first law must not conflict with, but 'butt on to', the second law there is no possibility of removing this consequence by a suitable conventional choice of the first law. Thus on alternative (b), $F_\mu = 0$ when $a = (f''(t)\, \mathrm{d}r/\mathrm{d}t) / f'(t)$ which conflicts with the original extension of 'force-free situation' and a new first law framed to make this a genuine force-free case would conflict with the old first law (this *not* being a case of alternative description).

[28] This point is *in addition* to the formal point – made by Grünbaum in [8] that mere alternative remetrizability does not entail the conventionality of the metric.

[29] Notice that I am not arguing here that these statements are non-conventional *because* they are topological but only that many topological descriptions may be, and are, non-conventional. On this distinction see again Grünbaum [8].

[30] This is an appropriate place to stress that the ascription of a conventional status to forces, or to the laws of motion does not imply, in this context, the ability to write down entirely arbitrary expressions for the laws of mechanics or to define 'force' in an entirely arbitrary fashion. The permissable changes are those produced by transforming Newtonian mechanics under remetrizations of space and/or time – for only then can it be claimed that empirically equivalent theories result. (All this assuming Grünbaum correct of course.) Cf. also note 1.

[31] I am indebted to the persistent questioning of Suchting for a great deal of whatever clarity this distinction possesses.

[32] Notice that the pipe's capacity in given circumstances is a *relative*, but *not a relational*, property of it. It is a relative property *because* of the physical relations present, but is not itself a relational property. The same is true of the other intrinsic$_1$ properties mentioned in the text above. On the other hand, the property '. . . is a parent of . . .' is a relational, intrinsic$_1$ property of all members of species that generate their kind sexually, for such beings must stand in a sexual relationship to another member of the population before they can stand in the relation '. . . is a parent of . . .' to get a further member of the population.

[33] Miss C. Whitbeck [31], has the following, interesting argument against Grünbaum's position. The ability to impose a metric on, or assign a metric to, space-time follows from Grünbaum's metrical conventionality thesis; (otherwise we could not obtain a space-time with a metric at all). But

'. . . the properties of a distance function or simultaneity relation require that the respective assignments of distance and time coordinate values be systematic. Since the manifolds of space and space-time are nondenumerable point sets, one cannot check the assignments individually to see that each is compatible with those remaining. Thus. a reference system of metrical relations is required in order that another be defined It follows that the topological relations are not prior to the metrical relations of these manifolds, but merely invariant under certain transformations of the metrical structure.

There remains the Whiteheadian view that nature exhibits a system of uniform relations, metrical as well as topological, which constitute space and time.'

[31], pp. 339–340. Here I shall not, however, attempt to evaluate the argument or Miss Whitbeck's other remarks, but content myself with bringing them to the reader's attention.

[34] My elaboration of this doctrine of standards now to follow is heavily indebted to the ideas of Professor G. C. Nerlich of Sydney University with whom I had many conversations on these and related matters and whose lectures and public addresses I was fortunate to be able to attend. But on this topic cf. also Broad [2] and Lenzen [22], among others.

[35] At the outset, let me undercut one source of support for the view that the choice of a standard is arbitrary by distinguishing between a *standard* and a *unit*. The unit provides only a *particular scaling* of the measurable quantities involved and is always arbitrarily specified. It is the standard which determines the actual ordering of the set of entities in question (in this case spatial and temporal intervals), the unit determining only a particular mapping of that ordered set on to the real number system. But often the standard and the unit are one and the same body. (Thus, for example, the platinum bar now kept in Paris is regarded by physicists as both a standard for spatial intervals and also as the unit of length.) That this is so, often leads people to the mistaken belief that the choice of a standard and the choice of a unit are one and the same. But the choice of a standard is a quite different matter from the choice of the unit.

[36] This view of standards is not, of course, uncontroversial. For an entrance into the controversy the reader could consult Grünbaum [9], [10].

[37] The question arises here whether such bodies can provide a complete specification of a spatio-temporal metric. Thus the cases of rectilinear velocity, (a) and (b), provide *ratios* of spatial intervals to temporal intervals, but cannot provide any absolute scaling of both, either singly or together. Nonetheless, this is not a serious problem – it amounts to choosing a *unit* for either space or time. When rotational motion is introduced, however, we seem to be able to deal with one metric at a time, for rotational motion seems to provide a temporal metric alone. This situation arises because an entire rotation is a *topological* and not a *metrical* affair. If this is so, we need only a time unit and then either of the other rectilinear motions can provide a spatial metric. Of course, the rotational constancy of a rigid body depends upon it not changing its *size* and thus in turn involves the spatial metric, but in an unusual way.

Thus consider a sudden doubling in size of bodies in the universe. In the case of rectilinear motions all times between the 'same' marks on bodies will alter in the same ratios and hence not show up in the alleged doubling. So also will times for the rotation of, say, spheres of differing sizes. But now consider two *differently shaped* bodies rotating so as originally to exactly pace one another. Such pacing is a purely topological affair. Now let us suppose that the laws of nature do not change during this alleged doubling. (This makes clear the *ad hominem* character of my argument, for this is the position that Grünbaum would defend because he maintains the empirical vacuity of such an alleged doubling.) In particular, suppose that the law of conservation of angular momentum holds good throughout the change. Under these circumstances a purely topological difference will be noticed, the two bodies will no longer pace one another. Moreover, from which body now lags behind which, the direction of the spatial change can actually be inferred (that is, whether a doubling or a halving).

But we must still face the charge that an entire rotation is not a purely topological affair after all, for it must be with reference to a particular and fixed direction. But since we have specified no such fixed direction, the introduction of force-free rotating bodies leaves us no better off than the cases of rectilinear motion. But now let us consider the presence of two bodies which exert no forces upon one another (such bodies are of course hypothetical), the one rotating and the other moving rectilinearly. The rectilinearly moving body defines a constant direction in space for us against which we may determine entire rotations of the rotating body. Only the axis of spin of the rotating body can move, and since both bodies are force-free, the axis of spin can only move rectilinearly as well. So after all this situation allows us to obtain a pure temporal metric. We need then only introduce a temporal unit and then we may use the rectilinearly moving body to provide ourselves with a complete spatial metric.

[38] Note that what Grünbaum's argument shows is that space, characterized in a certain way, is metrically amorphous. The characterization includes the requirement of continuity. That requirement has nothing whatever to do with the imagined properties of matter which might be contained in space. Moreover, that space may have, as an intrinsic$_1$ property, a disposition (whether in virtue of its topological characterization or not I do not know), to contain material objects in a certain fashion and hence to possess a certain kind of metrical structure as an intrinsic$_1$ property of itself, but this disposition obviously cannot itself be based upon properties of space which in their own right are sufficient for assigning an intrinsic$_2$ metric to space.

[39] It is, perhaps, also worth remarking that the intimate connection between matter, force and metric here envisaged may provide the basis for a better understanding of such theories as the General Theory of Relativity where the metric of space-time depends upon the distribution of matter. On metrics in General Relativity, see Grünbaum [8].

[40] Notice that the truth of the first law of motion and Definition 4 together imply a special case of the principle of determinism, namely:

(D) Every *dynamical* change has a cause (where inertial motion is not an instance of dynamical change). Conversely, (D) allied with the assertion that forces are the only dynamical causes implies the first law of motion. Thus support for the truth (or the necessity, if desired) of the first law of motion is provided by whatever appeal the principle (D) and the remarks about forces have.

[41] Note that (iv) implies the principle of the uniformity of action of a cause only if at least some causes act uniformly. Since this principle is required to prevent a 'normally' motion-producing cause from acting but not producing motion, we may add it to our required properties of causes. It could also be added as a necessary truth, rather than an empirical truth, concerning causes if it is thought so to be.

The four conditions listed in the text have already been discussed on pages 131 and 161 above. These conditions, I argued there, also suffice to give an empirical status to the question of whether or not balanced forces are present.

[42] Doubtless there are or have been one or two out of all the atoms in the universe which at some instant or other in its few $\times 10^9$ years have been momentarily under the action of perfectly counteracted forces. (The reason why this is not the general rule is simply that most forces are finite at all finite distances from their sources, so that

force-freeness is impossible and perfectly balanced forces require accidental configurations of the entire matter of the universe.)

[43] The argument here hinges on the claim:

(C) If any body is in motion then there is some frame of reference relative to which it is in motion.

Without C's existential import, the existence of an absolute space does not follow. Hunt has suggested that C could be denied in favour of the claim that some accelerations simply are not relative. He points out, for example, that one can construct a measure of an object's acceleration from *intrinsic* co-ordinates to the set of material point events making up the space-time track of the object. But this latter move seems to me to be employing those events as a co-ordinate frame. For myself, I find the notion of an acceleration in itself, relative to nothing, more difficult to grasp than I do the acceptance of an absolute space-time framework. But for more detail concerning Hunt's position, see [16].

[44] This argument forms part of a wider examination of the status of the concepts of absolute space and time. For an examination and rejection of the alternative Relational account see my [14]. And for an examination of Newton's arguments see my 'On the Traditional Dynamical Arguments for Absolute Space' (with R. Slack), unpublished manuscript.

[45] See [4]. The argument of this section grew out of a paper originally presented by Hunt to the seminar mentioned at the outset. Hunt and Suchting have since produced their own examination of Ellis' argument, see [17].

[46] An assignment, *qua conventional*, is certainly arbitrary. This does not mean, however, that there may not be other factors *drawn from outside of the specific aspects of the situation involved*, which determine us in favour of one conventional assignment and against another. This fact does not, however, affect my equating conventionality and arbitrariness of assignment in this context.

[47] This last sentence states all of what is intended in the claim that Ellis' mechanics is 'extensionally equivalent' to Newtonian mechanics. It does not imply, and it is not true, that they have the same ontology, or that their corresponding statements have the same empirical status (Ellis' first law is a definition, the Newtonian first law, I argued, is an empirical claim). Cf. note 58 below.

[48] There is one criticism of Ellis which is worth mentioning in passing and that is that his claims are suited only to the Newtonian formulation of mechanics. Once moved to the Hamiltonian formulation of mechanics, for example, and inertial motion loses its significance as being different from any other kind of motion. In this case Ellis' natural motion construal of mechanics has no place. Fine [6] adopts this line of criticism.

But this objection is surely misplaced. In the first place, the mathematical form of the second law of motion as usually adopted implies the first law of motion, so that that law certainly does figure in the formulation of Hamiltonian mechanics. But it also figures in the Hamiltonian formulation in a special way directly analogous to its function in Newtonian mechanics, for the Hamiltonian functions are explicitly built on the criterion of deviation from inertial motion. Thus, for example, the Hamiltonian formulation gives the usual inertial motion if and only if the interaction potential of the Hamiltonian is zero. Thus inertial motion retains its special status even within the Hamiltonian formulation of mechanics.

[49] On this score it is significant to note that if the Einsteinian field theory of gravitation

turns out to be the correct one, then gravitational forces can be expected to be in much the same situation as electro-magnetic forces now are. For we should be able to detect the presence of a gravitational force by the presence of 'gravitational radiation' emitted by a gravitationally accelerated body.

[50] For more on the role of meta-science or metaphysics in science see also my [12] and [13].

[51] On this assessment of the debate see further the closing section of my [11].

[52] It is true that there are *possible* force-effects. These can be formulated in terms of contrary-to-fact conditionals (for example, 'If you were to attempt to separate the bars . . .'). But possible force-effects are not force-effects.

It is possible, however, that Ellis may wish to reply to this argument, not in terms of possible force-effects, but in terms of *dispositions* to produce force-effects which are present. The claim would be that the occurrences of these dispositions are themselves force-effects. But this would simply be another version of the same move which was criticised in the text above. For forces are fundamentally dispositional – cf. discussion of definition 4 above. Thus, this move is another move to legislate that the terms 'force' and 'force-effect' shall be logically equivalent.

In any case the defense of Ellis contemplated in the first paragraph of this footnote would be a weak one, for plainly *we* cannot use dispositions to identify forces and Ellis clearly intends to refer only to force-effects which are in principle observable. His own philosophical attitude would surely prevent him from treating these unobservable, inoperative-except-for-the-situation-they-were-invented-to-explain dispositions seriously.

There is another characteristic of the two-bar example which might be picked on in order to defend Ellis and that is the electro-magnetic field patterns. The electro-magnetic field connecting two rigid bodies in consequence of these bodies being charged will be quite different from that existing if these two bodies had been electrically neutral and merely resting against one another. In this case it seems reasonable to claim that the differences in the field patterns here are force-effects (of the balanced forces when the bodies are charged). But it is surely not *logically necessary* that the electro-magnetic force be mediated by an electro-magnetic field. If it had been an action-at-a-distance type of force (as gravity is in Classical Mechanics) then there would have been no such field patterns present and hence no force-effects of this type. But Ellis must argue that forces are *logically* related to force-effects and what the above logical possibility shows is precisely that they are not.

[53] P_1' and P_2'' may seem to conflict unless one remembers that in P''_2 the phrase 'remained unexplained and unexplainable' *refers only to the current theoretical context*, i.e. those states which a particular version of Classical Mechanics counts as natural are, *within that theory*, unexplained and unexplainable; but there may be another formulation in which they are not counted as natural and here they will be both explainable and explained. Cf. in this respect the natural motions of Newton's and Ellis' formulations of mechanics.

[54] This is really only a modernised version of the old Aristotelian idea, an idea possibly also held by Newton, that bodies in 'natural motions' were so in virtue of an inner causal principle.

It is possible that Ellis might claim that P_1' is too restrictive in specifying *motions* as that which is unexplained; for example, it might be claimed that certain states of strain or other statical states might be made to be the natural states (at least under some conditions). But I cannot see why the same kind of reply is not possible to any such changes: in each case it is surely *logically* possible that these states be causally explained by some deeper theory within which there may be no natural states of affairs, (though there will be fundamental laws and facts which are not explained by that theory).

[55] This is, in fact, the move he makes against the argument from standards. See Grünbaum [7], pp. 71–4; and also [10].

[56] Some of Grünbaum's comments on this latter score are to be found in his article cited in note 55.

[57] Suchting has argued (private communication) that the position of the conventionalist is not as drastic as I make it out to be. Thus he can understand the command to look for sources of force as a command to look for bearers of properties functionally related to the dynamical behaviour of objects, for a determinate classification of causal chains and the formulation of determinate physical laws (= generalizations).

There is perhaps some truth to Suchting's charge, but his game can be played all too readily. We can always strip off from our description of science everything except the generalities covering our ways of building physical theories – we can always retreat to the purely linguistic level if we insist. The question is, how plausible is this retreat? Can it offer a deep understanding of why forces (or electrons, or masses . . .) were ever introduced to mechanics, of what we mean by a *credible causal account*, (cf. pp. 157–8 above, = why some possible functional relations are acceptable and not others), of our experiences and the practices of our sciences? The next sentence in the text proper is my own response. (Cf. also note 58).

[58] One might ask (as Suchting did) how, if Ellis' mechanics, for example, is equivalent to Newtonian mechanics, the two can have differing empirical contents. Now the sense of equivalence involved was spelled out in note 47 above. They are equivalent only in the sense that both yield the same *observable dynamical behaviour*. They are *not* equivalent, however, in the sense of both saying the same things about the same physical ontology. What the two theories share in common is a certain, we might say 'superficial', description of the world. What they do not share in common is a deeper lying description of the structure responsible for the truth of the superficial description. Thus the conventionalist theory describes the conditions under which bodies move, the non-conventionalist theory does both this *and* also explains why they move under these conditions (viz. the presence of forces of certain sorts). The conventionalist development, as it were, capitalizes on the regularities which arise because of the deeper lying structure of the world to wipe off the (unobservable) deeper lying structure and retain simply the surface regularities. In this respect it is closely akin to Craigian moves to eliminate theoretical terms and to a behaviourist psychology.

BIBLIOGRAPHY

[1] A. Bressan, *A General Interpreted Modal Calculus*, Yale University Press, New Haven, Conn., 1972.

[2] C. D. Broad, *Scientific Thought*, Harcourt, Brace and Co., N.Y., 1923.

[3] R. Colodny (ed.), *Beyond the Edge of Certainty*, Prentice-Hall, N.J., 1965.

[4] E. J. Dijksterhuis, *The Mechanization of the World Picture*, Oxford University Press, 1961.

[5] B. Ellis, 'The Origin and Nature of Newton's Laws of Motion', in [3].

[6] A. Fine, 'Explaining the Behaviour of Entities', *Philosophical Review* **75** (1966) 496–509.

[7] A. Grünbaum, *Philosophical Problems of Space and Time*, Knopf, N.Y., 1963, Chapter 1. See 2nd edition, Boston Studies, Vol. XII, D. Reidel Publ. Co., Dordrecht-Holland, 1973.
[8] A. Grünbaum, 'Space, Time and Falsifiability', *Philosophy of Science* **37** (1970) 469–588.
[9] A. Grünbaum, *Geometry and Chronometry in Philosophical Perspective*, University of Minnesota Press, Minnesota, 1968.
[10] A. Grünbaum, 'The Denial of Absolute Space and the Hypothesis of a Universal Nocturnal Expansion', *Australian Journal of Philosophy* **45** (1967) 61–91.
[11] C. A. Hooker, 'Craigian Transcriptionism', *American Philosophical Quarterly* **5** (July 1968) 152–163.
[12] C. A. Hooker, 'The Nature of Quantum Mechanical Reality: Einstein versus Bohr', in R. Colodny (ed.), *Paradox and Paradigm, Pittsburgh Studies in the Philosophy of Science*, Vol. V, 1972.
[13] C. A. Hooker, 'Physics and Metaphysics: A Prologema for the Riddles of Quantum Theory', in C. A. Hooker (ed.), *Contemporary Research in the Foundations and Philosophy of Quantum Theory*, Dordrecht, Reidel, 1973.
[14] C. A. Hooker, 'The Relational Doctrines of Space and Time', *The British Journal for the Philosophy of Science* **22** (1971) 97–130.
[15] C. A. Hooker, 'Systematic Realism', to appear in *Synthese* (1974).
[16] I. E. Hunt, 'On Absolute Space in Newtonian Physics', not yet published.
[17] I. E. Hunt and W. A. Suchting, 'Force and Natural Motion', *Philosophy of Science* **36** (1969) 233–250.
[18] M. Jammer, *Concepts of Force*, Harper and Row, N.Y., 1957.
[19] M. Jammer, *Concepts of Space*, Harper and Row, N.Y., 1960.
[20] M. Jammer, *Concepts of Mass*, Harper and Row, N.Y., 1964.
[21] S. Kleiner, 'A New Look at Mach's Foundations of Classical Mechanics', Unpublished.
[22] V. F. Lenzen, *The Nature of Physical Theory*, J. Wiley and Sons, N.Y., 1931.
[23] E. Mach, *The Science of Mechanics*, La Salle, Illinois, 1942, p. 304.
[24] J. C. C. McKinsey, A. C. Sugar and P. Suppes, 'Axiomatic Foundations of Classical Mechanics', *Journal of Relational Mechanics and Analysis* **II** (1953).
[25] E. Nagel, *The Structure of Science*, Routledge and Kegan Paul, London, 1961.
[26] I. Newton, *Principia* (translated by Motte, revised by Cajori), University of California Press, 1966.
[27] C. G. Pendse, 'A note on the Definition and Determination of mass in Newtonian Mechanics', *Philosophical Magazine*, Ser. 7, **XXIV** and **XXVII**, 1939; addendum, **XXIX**, 1940.
[28] H. A. Simon, 'The Axioms of Newtonian Mechanics', *Philosophical Magazine*, Ser. 7, **XXXIII**, 1947.
[29] J. H. Smith, *Introduction of Special Relativity*, Benjamin, N.Y., 1965, pp. 201–2.
[30] J. D. Sneed, *The Logical Structure of Mathematical Physics*, Humanities Press, N.Y., 1961.
[31] C. Whitbeck, 'Simultaneity and Distance', *Journal of Philosophy* **LXVI** (1961), No. 11.
[32] G. J. Whitrow, 'On the Foundations of Dynamics', *British Journal for the Philosophy of Science* **1** (1951) 92–107.

LASZLO TISZA

COMMENTS ON C. A. HOOKER: SYSTEMATIC REALISM

During the first half of the century, positivism had a firm hold on the great majority of physicists. It appeared, however, in recent years that this method of analysis has lost its early creativity.

Professor Hooker is among the increasing number of philosophers of physics who are concerned about this situation. His determination to redirect philosophical study to significant human problems is most refreshing. Unfortunately, I am afraid that his technique does not measure up to the difficulty of this task. He polarizes the issue by formulating two comprehensive and mutually incompatible world views and then proceeds to destroy positivism and vindicate realism. Since he cannot discern any merit in positivism, it is hard to avoid the implication that Einstein, Bohr, and Heisenberg were utterly misguided when they used positivist arguments to challenge classical dogmas.

Instead of analyzing his paper point by point, I prefer to discuss the problem from my point of view. Let me start by clarifying first how I intend to use the terms 'real' and 'reality'. Curiously, the commonsense usage serves as a good point of departure. We usually know what 'reality' means in contrast to 'dream', 'illusion', 'deception' or 'error'. We know how to compare a photograph or a map with reality.

There are situations in physics which are quite similar. We distinguish a real object from a hologram, and we know how to tell a 'real' effect from an experimental error, at least in the long run.

There are instances, however, in which the problem is more delicate. According to commonsense reality the Earth is at rest, but astronomical observations, *interpreted in terms of mechanical theory*, induce us to attribute a deeper reality to a mobile Earth.

What is noteworthy in this example is that we are occasionally willing to subordinate the testimony of our senses to theoretical constructions. This increased reliance on theory, in the formation of our conception of reality, becomes more pronounced and also more tenuous as we deal with entities quite removed from direct observation, such as atoms and molecules.

Therefore, my next step is to examine briefly the conception of reality imposed upon us when we accept a particular theory, say, classical mechanics (CM) as 'fundamental'.

According to CM, force is proportional to acceleration, and if we observe a dissipative force proportional to velocity, this must be an 'appearance' behind which there is the 'reality' of Newtonian conservative forces on the microscopic level. This view imposes an obligation to 'reduce' appearances to 'reality'. The traditional programs of reducing thermodynamics and electrodynamics to CM have indeed received a great deal of attention, an indication that physicists do accept the implication of their ontological committment to the 'reality' implicit in CM.

Although the reduction program was not completely successful (think of the attempts of Maxwell to reduce electromagnetic field theory to mechanical models!), until the close of the last century, there seemed to be no compelling reason to question the 'deeper reality' of CM. In fact, this claim became so firmly anchored in tradition that it became a roadblock to further progress.

It is at this juncture that positivism exerted a liberating influence by reasserting the priority of experience, or phenomenology in confrontation with a theory which had turned dogmatic. However, even while we acknowledge that positivism earned its place in the history of physics, we should realize that its reaction to the crisis of classical physics was an emergency solution.

Although the crisis is still not entirely over – remember the unsatisfactory state of the theory of elementary particles – the emergency has certainly abated, and it is high time to reformulate the positivist critique in accordance with our present knowledge.

I have been concerned with this problem for a long time and have presented some of my ideas in the early days of this Colloquium.[1]

I propose to summarize and rephrase these ideas succinctly. The core of my suggestion is that theoretical physics ought to be considered as a cluster of basic theories, or 'languages', organized as more or less closely knit deductive systems. The use of a multiplicity of languages reconciles the precision of deductive discourse, insisted upon by the classical physicist, with the flexibility needed to deal with a wide range of experience emphasized by the positivist. However, the joint use of different deductive systems raises novel problems involving the identification of conflicts and the enforcement of mutual consistency.

These problems have to be analyzed in terms of a *metalanguage* in which the deductive systems are the elements of our discourse. We ensure consistency by various devices, such as assigning logical priorities and ranges of validity to the various languages.

Note that this use of deductive systems is very different from various traditional uses. The classical physicist attached overly great importance to a selected system: CM. For him this was the entire universe of discourse. In contrast, the positivist is willing to consider any number of hypothetico-deductive systems without taking them very seriously, and without accepting the responsibility of making them consistent and building out of their joint use a richer reality.

One of the difficulties of using our metalanguage is that such theories as classical mechanics (CM) and classical electrodynamics (CED) have undergone considerable evolution in the course of time, and the precise identification of a theory requires the specification of its postulational basis as explained l.c. Here I confine myself to a very sketchy discussion of the evolution of the mutual relation of CM and CED.

CM had reached relative maturity at a time when the theories of electricity and magnetism were still in their infancy, and were framed closely after the gravitational action at a distance theory. At this stage these theories could be considered simply as special cases of CM. Beginning with Faraday, Maxwell and Hertz, CED was established as a field theory, the autonomy of which emerged almost unintentionally because of the failure of persistent efforts to reduce the theory to a mechanical model. By the turn of the century, the coexistence of the autonomous systems turned into conflict. This was removed by Einstein's special theory of relativity.

The establishment of this greatly extended range of consistency has a price: we have to live with apparently paradoxical notions according to which the simultaneity of spacially separated events, and the mass of an object are dependent on the velocity of the observer. I would conjecture, however, that these notions may appear paradoxical mainly because they contradict the ontology of CM.

The positivist skepticism toward theory provided the justification to challenge the classical dogma. However, raised to a general principle, this skepticism is unwarranted and counterproductive.

Theorizing enables us to come to terms with the unmanageable wealth of experience: we find ways to simplify and to predict; this involves risk. The positivist warns: do not generalize by framing abstract concepts, you risk an agonizing reappraisal. Yet the only way to avoid an agonizing reappraisal is to reappraise, without agonizing, as a routine watchdog operation. I believe that the metalanguage of l.c. is eminently suited for these purposes. It proposes to make a general paradigm out of the procedure that led from a contingent CM to the deeper reality of the special theory of relativity.

Dept. of Physics, MIT

NOTES

[1] L. Tisza, 'The Logical Structure of Physics', in *Boston Studies in the Philosophy of Science*, Vol. I, D. Reidel Publ. Co., Dordrecht-Holland, 1963. Also 'The Conceptual Structure of Physics', *Rev. Mod. Phys.* **35**. (1963), 151. Both papers are reprinted in L. Tisza: *Generalized Thermodynamics*, MIT Press, Cambridge, Mass., 1966.

BAS C. VAN FRAASSEN

THE FORMAL REPRESENTATION OF PHYSICAL QUANTITIES*

In earlier papers I have attempted to develop a certain perspective on the structure of physical theories, and to provide correlatively a critical account of the logic of quantum mechanics.[1] After a brief outline of the basic framework, the present paper aims to distinguish various non-statistical and statistical aspects of the representation of physical quantities. I shall attempt to elucidate recent work in quantum logic in the following way: when certain mathematical entities are used to represent physical quantities (in a manner successful relative to certain purposes), then logical relations among statements about physical quantities can be defined in terms of these mathematical entities. The most important question about these logical relations is then: to what extent can we expect them in the context of physical theories in general, and to what extent do they reflect the peculiar features of quantum theory?

I

There are many different kinds of physical theories, and I would not pretend to say anything non-vacuous applying to all of them. However, I expect that everyone is acquainted with the sort of physical theory that addresses itself to a certain kind of physical system and aims to describe the behavior of such systems through time. In such a theory, a system of the requisite kind is regarded as capable of a certain set of *states*, and these states are represented by vectors, or functions, or some other variety of mathematical entities. The family of the representing entities I shall call the *state-space*, and my two main examples will be the use of Euclidean *n*-spaces in classical mechanics, and of Hilbert spaces in quantum mechanics, for this purpose.

The theory will secondly specify a family of transformations that govern the evolution of the state of such a system – that is, its trajectory through the state-space – with time. This aspect, the dynamical aspect, of these theories I shall ignore entirely in this paper. The third aspect of these theories, an aspect which will form our main concern here, is the specification of certain physical magnitudes characterizing systems of the requisite kind, as having certain values in certain states.

These physical magnitudes are also represented in the mathematical model in some way or other. Before discussing this in detail, I wish to show my philosophical bias by introducing some linguistic considerations. A statement U expressing the proposition that a certain physical magnitude has a certain value will be called an *elementary statement*. Whether U is true or not, with regard to a specific system, will depend on the state of that system. So regardless of how physical magnitudes are represented, we must have a function h assigning to each elementary statement U the set $h(U)$ of states that satisfy U. This function we shall call the *satisfaction function*. Semantic relations among elementary statements can be defined in terms of this function; for example U semantically entails U' exactly if $h(U) \subseteq h(U')$. The properties of this entailment relation, and hence the principles of inference applicable to the elementary statements, are determined by relations among physical quantities and states in the theory, and hence cannot be known entirely without examining the theory.

II

The first question that we may ask concerning a physical quantity is what values it may have. These values are always real numbers, a circumstance not without its philosophical puzzles, but one that we shall take for granted here. The next obvious question is which values the quantity has in various states. This is the first point at which contemporary physics has placed new demands on our tolerance of ambiguity. For in classical mechanics each quantity pertinent to the system has a definite value in each state, but not so in quantum mechanics.[2] For in quantum mechanics, the quantity m is represented by an operator M such that m has value r in state φ exactly if $M\varphi = r\varphi$. Since this may fail to hold for any value r, the quantity does not have a value in every state. Because of this we introduce the terminology of *eigenstates*: φ is an eigenstate of quantity m exactly if m has a definite value in φ.

The question may now arise whether it is restrictive to discuss only single values. Is not the statement, for example, that the momentum is within an interval $(p-d, p+d)$, of as much importance as the statement that the momentum is exactly p? The answer would seem to me to be affirmative, so we must consider statements of the form

The value of m is in set X

where X is a set of real numbers. In the foundational discussions with which I am acquainted it is universally accepted, however, that limitations must be placed on the kind of set X can be. And the restriction most commonly accepted is that X should be a Borel set. The reason for this I suppose to be as follows: we wish to be able to speak of the probability that the value of m shall be found to be in X. This requires that a probability measure be definable on these sets X; and the restriction to Borel sets ensures this.

One objection has been raised: Birkhoff and von Neumann argued that if X has measure zero, there is no way of empirically verifying the statement.[3] As example they give the statement that the angular velocity of moon around the earth, in radians per second, and at a given time, is a rational number. I am afraid that the pernicious influence of operationism or early logical empiricism befuddled the issue here. For in classical mechanics it is clearly entailed that the value of that quantity is a specific real number – and they would deprive us of the means of formulating that consequence.

In any case, given the restriction to Borel sets or not, we have in classical mechanics the following circumstance: the value of m is within a certain set X iff the value of a certain other quantity m' is a certain real number. We may define m' by:

m' has value 1 if m has a value in X, and m' has value 0 if m has a value in $R-X$

where R is the set of real numbers. But this cannot simply be transposed to quantum mechanics. For example, if $X=R$, then m' would have to have the value 1 in state φ exactly if φ is an eigenstate of m. But a Hermitean operator M' representing such a quantity m' can generally not be found.[4] Yet as we shall see below, something *similar* can be done in quantum mechanics for Borel sets of values.

The upshot of this discussion is that physical quantities can at least partially, and in the classical case wholly, be represented by functions m_e of subsets of the state-space into the set of real numbers, where

$m_e(\varphi)$ exists and equals exactly if φ is an eigenstate of m, and m has value r in φ. We shall call m_e the *eigenstate function* of m, and we shall designate as $U(m, r)$ any elementary statement U such that $h(U)=\{\varphi:m_e(\varphi)=r\}$.

III

However, this cannot be the end of the story. For when φ is not an eigenstate of m, the representation so far establishes no relation whatever between φ and m. This point is not germane to the classical case, in which all states are eigenstates of all quantities, but is very pertinent to quantum mechanics. And there we find the following principle, which we shall take as a clue: suppose quantity m is represented by an operator M, and we have a base $\{|r\rangle\}$ of eigenvectors of M for the space. Then φ can be expressed as a sum of these eigenvectors.

$$\varphi=\sum c_r|r\rangle$$

and $c_r^*c_r$ is the probability that measurement of m will find value r in state φ. To put it quite generally, if φ is not an eigenstate of m, we still have a probability measure on the real line, $P(\varphi, m)$, such that for any Borel set X, $P(\varphi, m)(X)$ is the probability that a measurement on a system in state φ will show a value in X. The eigenstate function clearly does not give such information: the quantity m is represented in addition *by a family of probability measures on the real line, one for each state.*

At this point we can take care of a problem we had to leave in the preceding section. Is there an elementary statement that says that the value of m lies in a set X? The answer for the classical case was *yes*, and at least in one sense, the answer for quantum mechanics was *no*. However, we find there that the analogous statement

> The probability that the value of m will be found in a set X, upon measurement, equals 1

is an elementary statement, given that X is a Borel set. That is, if quantity m is represented by operator M, as above, we can find an operator P such that $P\varphi=\varphi$ iff $\varphi=\sum c_r|r\rangle$ with $\sum_{r\in X} c_r^*c_r=1$. So the statement that the quantity represented by P has value 1 is the requisite elementary statement.[5] For every quantity m we can therefore introduce a function m_b from Borel sets into elementary statements, such that $m_b(X)$ is the statement that the value of m would certainly be found in X, upon measurement.[6] I shall call m_b the *range function* of m.

Now this is one use of probabilities in the representation of physical quantities, and we must make a clear distinction between this use of probabilities, found in quantum mechanics, and another use of probabilities common to quantum and classical mechanics. Let us use the term *distribution* to refer to a probability measure on the state-space. Such a distribution may be regarded as representing an indefinite kind of state, or a lack of information with respect to the actual state.[7] Suppose now that the distribution is such that we assign a probability of p_1 to the set of states φ such that $m_e(\varphi)=r$ and a probability p_2 to the set of states φ such that $m_e'(\varphi)=r'$. Then it follows specifically that there is also a definite probability p_3 assigned to the intersection of these two sets.[8] So if we can speak of the probability of m_e equalling r and of the probability of m_e' equalling r', we can also speak of the probability of (m_e equalling r and m_e' equalling r'). So much should be unsurprising. But it is very different with respect to the first discussed use of probabilities.

Suppose for example that we expand a state-vector φ in quantum mechanics first in terms of eigenvectors of the position operator and then in terms of eigenvectors of the momentum operator. In this way we can find the probabilities, for the various Borel sets, that the values of position and momentum would be found, upon measurement, to lie in these sets. That is, we can find these probabilities for each quantity – but we cannot find joint probabilities. For we do not have an operator that represents the 'conjunction' of position and momentum.

This disparity in the two uses of probabilities seems to have led to some confusion. At least if I do not misread him, the simple fact of this disparity led Patrick Suppes to advocate the development of a quantum probability theory based on a non-classical quantum logic.[9] To refute this argument for quantum logic. Arthur Fine produced a reformulation of the axioms of quantum theory.[10] But I truly fail to see how the disparity in question has any surprise in it. For in the first case we have a single probability measure which, if defined on two sets, is clearly also defined on their intersection. In the other case we have *two* probability measures, $P(\varphi, m)$ and $P(\varphi, m')$; and there is simply nothing in classical probability theory that could lead us to expect the calculation of a measure P on R^2 such that $P(Y \times R) = P(\varphi, m)(Y)$ and $P(R \times Y) = P(\varphi, m')(Y)$, which is what seems to be wanted.[11]

There is still another way in which the desire for a non-classical probability theory may raise its ugly head. Recall that for a quantity m and Borel set X, there is an elementary statement U which expresses, in a certain sense, the proposition that the value of m lies in X. We have specifically a function m_b defined on the Borel sets such that $m_b(X)$ is a statement U of that sort. Then instead of directly introducing the functions $P(\varphi, m)$ to help represent m, we can introduce for each state φ a function φ^* on the elementary statements such that

$$P(\varphi, m)(X) = \varphi^*(m_b(X))$$

Clearly φ^* and m_b give the same information as $P(\varphi, m)$, and so P is not really needed; it can be introduced by definition. And if we use φ^* as basic, it is easy to see that φ^* gives us the probability for each statement U that it will be verified by an appropriate measurement in state φ. So φ^* is then said to assign probabilities.[12] But since φ^* is not necessarily defined on a Boolean σ-algebra – the elementary statements forming a structure that we have not yet determined, but that is in fact not very ordinary in the quantum case – there is again a temptation to speak of 'non-classical probabilities'. For the probability assignments φ^* are very different from the measures discussed in classical probability theory.

Again there seems to me to be a confusion. For the values of φ^* are called probabilities only because they are the values for corresponding arguments of $P(\varphi, m)$ – and the latter is a probability measure. It is only the detour via propositions, effected by the mapping m_b, that makes the assignment of probabilities seem so unusual.

IV

We have connected elementary statements of a theory with the model the theory provides of its subject matter in two ways: by eigenstate functions and by range functions. These connections we shall now use to explore semantic relations among the elementary statements. Such exploration is the typical concern of logic, although the present case is quite different from those usually encountered in logic.

For in logic, the set of statements is usually characterized by syntactic structure, and this structure plays a main role in logical studies of those statements. For example, if I ask for a conjunction of the sentences 'John is a male' and 'John is a sibling of Mary', you are most likely to give me a sentence formed by placing the word 'and' between these sentences (taken in either possible order). But from the semantic point of view this is not at all the only possible kind of answer: what is necessary is that you give me a sentence which is true exactly if both the given sentences are true. From that point of view you could have answered me satisfactorily by displaying the sentence 'John is a brother of Mary'.

In the case of elementary statements, the syntactic structure is not very important. If I had to describe them syntactically, I would give them all the same structure, 'quantity *m* has value *r*' for example. So if we ask for conjunctions, disjunctions, negations, or whatever, of elementary statements, the question must be understood semantically. These concepts must each have a semantic characterization, which here will have the general form:

> U is a Γ of $U_1, \ldots, U_n$ if and only if $h(U)$ stands in relation Γ^* to $h(U_1), \ldots, h(U_n)$.

And if, as is convenient from the semantic point of view, we identify U and U' when $h(U) = h(U')$, then the logical connectives are represented by operations:

$$U = 0(U_1, \ldots, U_n) \text{ if and only if } h(U) = 0^*(h(U_1), \ldots, h(U_n)).$$

Given an operation on the subsets of the state-space, however, there is *a priori* no certainty that the family of sets $h(U)$ will be closed under that operation. Hence the set of elementary statements need not be closed under a given logical connective.

From this perspective we can evaluate to some extent the very divergent approaches that have been taken in quantum logic in the past four decades. In an earlier paper, I have contrasted Reichenbach's approach to that of Birkhoff and von Neumann;[13] in the present paper I shall contrast the more recent work of Kochen and Specker, and of Jauch and Varadarajan.

V

We turn now to the relations among elementary statements that appear when we consider the representation of physical quantities by means of probability measures. Recall that for each quantity m and each state φ we have a probability measure $P(\varphi, m)$ on the real line. We then defined the range function of quantity m to be a mapping m_b of the Borel sets into elementary statements that such $h(m_b(X)) = \{\varphi : P(\varphi, m)(X) = 1\}$. The range of the range function of m (that is, the set of statements U such that $U = m_b(X)$ for some Borel set X) we shall designate as $[m]$, and we can assume that every elementary statement belongs to some such range.

The Borel sets themselves form a Boolean σ-algebra with $\subseteq, \cap, \cup, -$. Let us define similar operations among the elementary statements in $[m]$:

$$-m_b(X) = m_b(R - X)$$
$$\wedge_i \{m_b(X_i)\} = m_b(\cap \{X_i\})$$
$$\vee_i \{m_b(X_i)\} = m_b(\cup \{X_i\})$$

where the subscripts range over countable index sets. In addition the statements U are partially ordered by inclusion among their satisfaction sets $h(U)$. It can be seen then that m_b is a homomorphism of the Borel algebra onto the system of statements $[m]$ with the indicated order and operations. For specifically, if $X \subseteq Y$ then $p(X) \leqslant p(Y)$ for probability measure p, so if $X \subseteq Y$ then $h(m_b(X)) \subseteq h(m_b(Y))$. Thus the range $[m]$ is itself a Boolean σ-algebra, with maximal element $\mathbf{I} = m_b(R)$ and minimal element $\mathbf{0} = m_b(\Lambda)$. We do not assert an isomorphism of course;[14] only that if the polynomial $\varphi(X_i) = R$ holds in the Borel sets then the corresponding polynomial $\varphi^*(m_b(X_i))\ m_b(\varphi(X_i)) = m_b(R) = \mathbf{I}$ holds in the elementary statements in $[m]$. This is one point generally made in quantum logic as practiced by the heirs of Birkhoff and von Neumann (such as Jauch and Varadarajan), and seems to be independent of the specific physical theory under study.[15]

Since there is no isomorphism, it may be well to inquire just how deceptive the 'classical' look of $[m]$ is. Recall that U is true in a given state (of a system in that state) exactly if the state is in $h(U)$. Now we have, for any probability measure, that $p(X \cap Y) + p(X \cup Y) = p(X) + p(Y)$. Since all probabilities are in [0, 1], it follows that $p(X \cap Y) = 1$ if and only if $p(X) = p(Y) = 1$. Therefore $h(m_b(X \cap Y)) = h(m_b(X)) \cap h(m_b(Y))$; that is, the defined conjunction is true if and only if both conjuncts are true. But it does *not* follow similarly from the probability principles that $p(X \cup Y) = 1$ if and only if $p(X) = 1$ or $p(Y) = 1$. Specifically

$U \vee -U$ is always true by our present definitions, but it does not follow that either U is true or $-U$ is true in any given case. So that the logic is classical when all elementary statements are 'compatible', and similar assertions, must not be taken too seriously: logical calculations with statements in $[m]$ follow classical patterns, but the underlying semantics is far from classical.

The family of all elementary statements is the union of the ranges $[m]$, and so it is natural to look for a structure in that family derived from the structures discerned in these ranges. To begin we may continue to identify statements when they are satisfied by the same states; then the whole family of elementary statements is partially ordered by inclusion among the sets $h(U)$. In addition we have then that $m_b(R)=m'_b(R)$ and $m_b(\Lambda)=m'_b(\Lambda)$, so that the ranges $[m]$, $[m']$ have their unit and zero elements in common.

Kochen and Specker take the most direct approach.[16] Roughly following them, we proceed: if there is an observable m such that U and U' are in $[m]$, then U and U' are compatible (U⫰U'). Polynomials in $U_1,\ldots, U_n$ are defined only for statements $U_1,\ldots, U_n$ of which any two are compatible; such a polynomial *holds* (is valid) exactly if it equals H whenever it is defined. So construed, the set of elementary statements forms a partial Boolean algebra.

Is this result general? The main problem is that, although ⫰ is symmetric and reflexive, it is not transitive. I can put this another way: the main problem is that, for a given U, there may be distinct maximal observables m and m' such that U is in $[m]$ and also in $[m']$. If this were assumed not to be so, we could deduce that the elementary statements form a partial Boolean algebra, but of a special kind: one in which ⫰ is an equivalence relation. If we do not make the assumption that each elementary statement belongs to a unique range m_b, then ⫰ is not transitive, but then the following principle is not deducible:

> If any two of $U_1,\ldots, U_n$ are compatible, then there is an observable m such that $\{U_1,\ldots, U_n\} \subseteq [m]$.

Does this principle hold in quantum mechanics? Well, it does if we identify U_1 and U_2 when $h(U_1)=h(U_2)$. The fact that this principle holds, given this identification, must be counted as a basic feature of the logic of quantum mechanics, as opposed to the logic of physical theories *überhaupt*.

Could the operations, specifically the join or sum, be extended to pairs of incompatible elementary statements? In 'orthodox' quantum logic, which takes its cue from Birkhoff and von Neumann, the answer is yes, for the elementary statements are asserted to form a lattice (the join exists for any two statements). How general is that result?

To begin, let us make two simplifying assumptions, to be scrutinized later: that if $h(U) \subseteq h(V)$ then there is an m such that $U, V \in [m]$; and that if $[m] \neq [m']$ then $[m] \cap [m'] = \{\mathbf{I}, \mathbf{0}\}$. Under those conditions, we can define:

$$\wedge_i \{U_i\} = \mathbf{0},\ V_i \{U_i\} = \mathbf{I}$$
if there is no m such that $\{U_i\} \subseteq [m]$.

With this extended definition, the elementary statements form an orthocomplemented lattice. (That is, the lattice laws for $\subseteq$, $\wedge$, $\vee$ hold, and in addition, $-U = U$, if $U \subseteq V$ then $-V \subseteq -U$, and $U \vee -U = \mathbf{I}$.) We find that *modularity*[17] will not hold; suppose $U \subseteq V$ with $U, V \in [m]$ and $W \in [m'] \neq [m]$. Then

$$V \wedge (W \vee U) \subseteq (V \wedge W) \vee U$$

amounts to

$$V \wedge \mathbf{I} \subseteq \mathbf{0} \vee U$$

hence

$$V \subseteq U$$

which will not hold for all U, V in $[m]$. But *weak modularity* holds: If $U \subseteq -V$ and $V \subseteq W$ then U, V, W all belong to the same range $[m]$ and hence $(U \vee V) \wedge W \subseteq (U \wedge W) \vee V$.

The result that the family of elementary statements forms a weakly modular orthocomplemented lattice is a main result of quantum logic. We have obtained the same result without specific reference to the structure of quantum mechanics. However, we made two simplifying assumptions which may be attacked. In fact we can use quantum mechanics to show that they amount to an oversimplification. Let us consider a three-dimensional vector space, and let vectors $\mathbf{r}_1$, $\mathbf{r}_2$, $\mathbf{r}_3$ form an orthonormal base, and also be the unit eigenvectors of an operator M, corresponding to eigenvalues a_1, a_2, a_3. Now we can find two new unit vectors r_2^1, r_3^1 orthogonal to each other and to r_1 but not to r_2 or r_3. So r_1, r_2', r_3' form a new base and are the unit eigenvectors of a new operator M', corresponding to eigenvectors b_1, b_2, b_3. If M and M' represent physical quantities m and m', we shall have $U(m, a_1)$ is true if and only if $U(m', b_1)$ is true. But $U(m, a_2)$ is not thus equivalent to any sentence $U(m', b_i)$; when m has the value a_2, the probability that m' has value b_1 is zero, but not conversely. So we see that if we identify the statement U by the sets $h(U)$, then $[m]$ and $[m']$ have a common element, other than **I** or **0**, though they are distinct ranges. This shows the implausibility of our assumption that if $[m]\neq[m']$ then $[m]\cap[m']=\{\mathbf{I}, \mathbf{0}\}$. Our other simplifying assumption cannot be similarly refuted,[18] but neither can it be justified in the present context. So we must consider what happens when these simplifying assumptions are dropped.

We may characterize our problem as follows: our sentences U are all elementary statements, but they are semantically characterized from several points of view. As long as we considered physical quantities only from the the point of view of the eigenstate functions, it was natural to consider sentences U and V indistinguishable unless $h(U)\neq h(V)$. But that m and m' may be distinct physical quantities, while the sentences $U(m, r)$ and $U(m', r')$ or the sentences $m_b(X)$ and $m'_b(Y)$ are indistinguishable. Another way to put this, perhaps somewhat metaphysically or metaphorically, is: the propositions expressed by the sentences U are represented by the sets $h(U)$, but this representation may ignore important features of those propositions.

What assumptions are needed to deduce that the set of elementary statements forms a lattice I shall discuss in another paper.[19] For now I wish only to note this: that the difficulties begin, in our attempt to find familiar logical structures, exactly when we identify elementary statements U and V when $h(U)=h(V)$. This was true in our discussion of Kochen and Specker, and again in this discussion of the more orthodox approach of Jauch and Varadarajan. This identification has many equivalent forms: it began with von Neumann's assumption that there is a one-to-one correspondence between physical quantities and a certain family of operators on Hilbert space. This yielded in turn the result that $m=m'$ when, for every state φ, $P(\varphi, m)=P(\varphi, m')$ – recall the general discussion of probabilities in Section III.

This assumption of von Neumann's played a central role in his proof that there can be no 'hidden variables'. In a later paper I hope to discuss the role such an identification plays in the Kochen and Specker proof of a similar theorem.[19] In any case, the results of quantum logic are surprising not in the sense of displaying logical structures we could not expect to find in the case of other theories, but only in the sense that such logical structures would not normally survive an identification of statements which are about distinct observables.

V

It is time, I think, to sum up. What I have attempted to do is to present a general picture of the representation of physical quantities, not in some specific theory of physics, but in any theory concerned to describe the behaviour of physical systems. When I did refer to classical or to quantum mechanics, it was mainly to check that my discussion was abstract enough so that these prime cases of physical theories did not provide counterexamples. I then used this general representation of physical quantities to explore semantic relations among elementary statements.

With respect to quantum logic, I may remark here that I cannot help but find it somewhat surprising (though gratifying) that this has seen much emphasis in foundational work in physics in recent years. For the typical results in quantum logic turn out to have such generality that it is hard to see how they can throw much light on the foundations of one particular theory.

But in philosophy of physics, such generality is exactly what is desired. A philosopher coming to quantum mechanics is at first struck, rather unpleasantly perhaps, by divergencies from the classical case. Indeed, philosophers have spent many hours of quiet desperation, worrying about how to adapt our world-picture to the exigencies of the quantum theory. Our first need is here for a general account of the structure of theories concerning the behavior of physical systems in general, and to see how quantum mechanics fits this general structure. My present effort is meant as part of such a general account; even if this part is acceptable, much remains to be done.

University of Toronto

NOTES

* The research for this paper was supported by Canada Council Grant 69-0650.

[1] 'On the Extension of Beth's Semantics of Physical Theories', *Philosophy of Science* **37** (1970), 325–339; and 'The Labyrinth of Quantum Logics', this volume. For the semantic approach adopted in these papers, see my *Formal Semantics and Logic*, Macmillan, New York, 1971.

[2] For didactic purposes, our illustrations from quantum mechanics will be in terms of pure states, represented by normalized non-zero vectors in Hilbert space.

[3] G. Birkhoff and J. von Neumann, 'The Logic of Quantum Mechanics', *Annals of Mathematics* **37** (1936), 823–843; especially p. 825.

[4] Suppose that $M'\varphi=\varphi$ exactly if φ is an eigenvector of M. Then if φ and ψ are eigenvectors of M, we have $M'(\varphi+\psi)=M'\varphi+M'\psi=\varphi+\psi$; but $M(\varphi+\psi)=M\varphi+M\psi$ $=r\varphi+s\psi$ which is in general not a multiple of $\varphi+\psi$.

[5] P is a projection operator, projecting on the subspace spanned by the vectors $|r\rangle$, $r\in X$. We then have $P\varphi=\varphi$ if φ is in that subspace and $P\varphi$ is the zero vector if φ is orthogonal to that subspace, i.e. if the probability that a measurement of m will find a value in X equals zero.

[6] Cf. J. Jauch, *Foundations of Quantum Mechanics*, Reading, Mass. Addison-Wesley, 1968, pp. 98–99; and V. S. Varadarajan, *Geometry of Quantum Theory*, vol. 1, Van Nostrand, Princeton: 1968, pp. 108–111. Note that in the present context, $h(m_b(\{r\}))=$ $=\{\varphi: m_e(\varphi)=r\}$ i.e. the probability that m will be found to have value r in state φ equals 1 if and only if φ is an eigenstate of m corresponding to value r. There are reasons for not wishing to generalize upon this fact; see D. L. Reisler, *The Einstein Podolsky Rosen Paradox*, unpublished dissertation, Yale University, 1967, especially pp. 162–164.

[7] In the case of quantum mechanics, the term used is 'mixture', and we do not here wish to enter into the question whether the use of mixtures represents ignorance, or whether mixtures represent a distinct kind of state.

[8] For a probability measure is defined on a family of subjects closed under intersection (a Boolean σ-algebra).

[9] P. Suppes, 'The Probabilistic Argument for a Non-Classical Logic in Quantum Mechanics', *Philosophy of Science* **33** (1966), 14–21.

[10] A. Fine, 'Logic, Probability, and Quantum Mechanics', *Philosophy of Science* **35** (1968), 101–111.

[11] We do expect that there will *exist* such a measure on R^2, but not a unique such measure, and then again we have no reason *a priori* to expect this measure to be the function $P(\varphi, m'')$ for some physical quantity m'' represented by some Hermitean operator M''.

[12] Cf. Jauch, *op. cit.*, pp. 93–95.

[13] 'The Labyrinth of Quantum Logics', this volume.

[14] In quantum mechanics, $\mathbf{I}=m_b(R)$ and $\mathbf{O}=m_b(\Lambda)$, for any quantity m; $h(\mathbf{I})$ comprises all state-vectors and $h(\mathbf{O})$ only the zero-vector.

[15] Jauch, *op. cit.*, p. 100.

[16] S. Kochen and E. P. Specker, 'Logical Structures Arising in Quantum Theory', in J. W. Addison *et al.* (eds.), *The Theory of Models*, North-Holland Publ Co., Amsterdam, Holland, 1965, pp. 177–189; especially pp. 183–184.

[17] The lattice is *distributive* if (D) $a\wedge(b\vee c)\leqslant(a\wedge b)\vee c$ holds, *modular* if (D) holds when $c\leqslant a$. Jauch defines *weak modularity* as: if $a\leqslant b$ then $\{a, b\}$ generates a Boolean

sublattice; Varadarajan defines it as: if $a \leqslant -b$ and $b \leqslant c$ then $(a \wedge b) \wedge c \leqslant (a \wedge c) \vee b$. See Jauch, *op. cit.*, p. 86, and Varadarajan, *op. cit.*, pp. 107–108. The credit for noticing the connection between lattices encountered in quantum logic and unions of Boolean lattices must apparently go to *S*. Watanabe, 'A Model of Mind-body Relation in Terms of Modular Logic', *Synthese* **13** (1961), 261–302; for a treatment of this topic with greater generality and rigor see P. D. Finch, 'On the Structure of Quantum Logic', *Journal of Symbolic Logic* **34** (1969), 275–282.

[18] Cf. Jauch, *op. cit.*, p. 100.

[19] See my 'Semantic Analysis of Quantum Logic' in C. A. Hooker (ed.), *Contemporary Research in the Foundations of Quantum Theory*, D. Reidel Publ. Co., Dordrecht-Holland, 1973.

E. J. POST

COMMENTS ON 'THE FORMAL REPRESENTATION OF PHYSICAL QUANTITIES'

The title of this talk might suggest a bird's eye view of the formal mathematical representation of physical quantities. Professor van Fraassen, instead, selected a very specific, but most fundamental problem, namely that of the statistical aspects of quantum mechanics and its underlying systems of logic.

The speaker's choice reflects a prevailing philosophy that all of physics is in some way an asymptotic case of quantum physics. Hence, any discussion of formal fundamentals is best served by starting with what we presently believe to be the most basic discipline in physics, i.e., quantum mechanics.

This rationale for starting with quantum mechanics would be completely justifiable if our thinking about quantum physics could be sufficiently divorced from the vast body of knowledge that is normally referred to as classical physics. However, even today, correspondence type arguments are still essential as guiding principles to ascertain consistency of our overall picture of physics. Therefore, the tenor of my commentary will be one of de-emphasizing the conceptual separation between classical and nonclassical physics to minimize its use as an escape from essential difficulties.

Let me first mention what I have missed most in the speaker's survey: A discussion of physical quantities should cover the important question of units and physical dimensions. By virtue of man's perceptual limitations in taking cognizance of nature, most units in physics never transcend the restriction of a 'local' definition. The 'global' use of 'locally' defined units depends on our notions of space-time structure. If the definition of physical units depends on space-time structure, the formal mathematical representation of physical quantities will certainly depend on space-time structure. In fact we have here a choice preoccupation for epistemology to weed out preconceived notions from contemporary physics.

Secondly, where the major content of the talk centered around the probability interpretation of quantum mechanics I believe I might do well to recall some of the history relevant to the subject. In fact, the speaker's principal conclusion of de-emphasizing the need for introducing a so-called 'nonclassical' quantum logic is at least indirectly but well supported by a historical analysis of the arguments that led to the probability interpretation of quantum mechanics.

One may indeed argue that the introduction of new principles in logic ought to stand on its own. Quantum mechanics may at most serve as an inductor for introducing new logical principles. Moreover, the physical weaknesses of that part of quantum mechanics that is normally considered for an axiomatic analysis (non relativistic quantum mechanics) hardly justifies an overinvestment from a point of view of philosophy. Let me, therefore, attempt to delineate some of these physical weaknesses by stepping back in history.

The probability interpretation of quantum mechanics which emerged in the late twenties, stands and falls with the possibility of at least denoting a single characteristic point for a particle about whose size, shape and structure one commits oneselve in a minimal fashion.

The assumption of a characteristic point for a particle is not yet equivalent to the conept of a point-particle. It contains the possibility of using the point-particle model as a conceptual expedient, it does not yet imply the actual physical existence of point particles.

It is the natural limitation of two finite (nonzero sized) particles which interferes with the idea of an equal apriori probability for all points of space, including the neighborhood of a characteristic point of a particle. If one wants to include the behavior of particles in close proximity as a basic feature of the theory, one either assumes with Schrödinger that the field is the particle or one accepts with the Copenhagen school that the point particle is a physical reality. In the latter case the field is used to derive a probability for the expectation of finding the point particle in a given domain of space.

An early and major objection against the actual physical existence of point masses and point charges is an ensuing infinity of energy interaction. The energy-mass equivalence in addition leads to an inertial mass infinity for a finite point charge. Conversely, a finite mass does not permit a nonvanishing point charge.

In modern theory, these infinities are formally eliminated by renormalization procedures that leave for many of us a wake of conceptual difficulties which are sometimes too conveniently relegated to the garbage pail of the classical-nonclassical dichotomy. The common appeal to accept the absence of a classical counterpart cannot be justified for the point particle abstraction, because the latter is of purely classical origin. The only alternative to the point particle abstraction was Schrödinger's idea of indentifying particle and field. It is not surprising that he witnessed with dismay the rejection of his idea on the grounds that it led to problems for multiple particle systems.

At this point, it is useful to recall some further details about the classical point-particle abstraction. The so-called characteristic point (gravity or electric center) of a mass or charge distribution of known finite geometry and density is used as the basis point of a multipole expansion for its outside field. This multipole expansion is a mathematically permissible point simulation of a non-zero sized source; it is only valid for the outside domain. The actual field and its multipole expansion vastly differ for the domain inside the source; the expansion has singularities that are absent in the actual field.

The multipole expansion has two properties that particularly invite the simple point particle abstraction of the monopole type. First, the expansion breaks off at the monopole term for a source of spherical symmetry and secondly, the monopole contribution prevails at distances large with respect to the size of the body. For the far-field, all sources appear as monopoles. To describe the near-field, more 'structure' has to be given to the characteristic point of the source by adding multipoles with their orientation and magnitude.

In the micro-world of quantum mechanics, one is in no position to start out with detailed statements about the geometry and physical structure of the particle. In fact, one is forced to follow a reversed procedure. The structure of the particle is now inferred from its behavior in fields. To explain phenomena, such as the anomolous Zeeman effect, it became necessary to endow particles with dipole features reminiscent of a multipole expansion in the classical sense.

A nonzero sized source distribution can lead to a unique multipole expansion but conversely a given multipole expansion does not uniquely determine a source distribution. The Copenhagen school capitalizes on this ambiguity, because a characteristic basis point for the source is essential for the probability interpretation. The ensuing field singularities at the basis point are dealt with as physically unessential. In an asymptotic sense, they are situated inside the (point) source. It is the field in the outside domain that matters physically. The success of renormalization supports this point of view as permissible. Yet the scattering work in high energy physics pursues a course of action in which the size and distributional structure of particles is being probed. These observations and conclusions tacitly undermine a fundamental aspect of the Copenhagen interpretation.

I hardly have to go further depicting for you these known developments in order to share a common feeling of discomfort about the physical basis of the probability interpretation of quantum mechanics. However, no theory is perfect. Its lack of perfection is part of its fascination, and certainly no excuse to refrain from making an analysis of its logical structure.

When beautiful results beckon, the need may arise for making a seemingly illogical step. Illogical in that context may just mean a temporary lack of logical perspective, rather than an implication of false logic. The probability interpretation of quantum mechanics conveys to me the impression of a situation in which we have a workable recipe, yet a model oriented insight is missing.

In summary, I suggest that a wider area of (quantum) physics be included for a more encompassing analysis. The restriction to the subdiscipline of nonrelativistic quantum mechanics, because of its logical appeal, could well lead to a divorce form physical reality. I concur with the speaker's conclusion of toning down the somewhat overemphasized contrast between classical and nonclassical physics and their associated logics; that is to say, to the extent where this contrast is unduly disruptive for the unity of physical experience.

University of Houston

JOHN STACHEL

COMMENTS ON 'THE FORMAL REPRESENTATION OF PHYSICAL QUANTITIES'*

In this paper, Prof. van Fraassen proposes a formal model for a class of physical theories which generalizes the formalisms used in classical and quantum mechanics. He shows how semantic analysis of his model, inspired by the work of Beth, leads to a study of the logic of the theory, here defined as the study of the relations of validity and entailment between the propositions, or elementary statements, of the theory and their logical compositions. He shows that this logic, or rather these logics, share many features with the quantum logics which have been the subject of much recent discussion. He is thus led to raise the question whether there is anything very useful about the quantum logic approach to quantum theory, since many of its features are not specific to that theory.

Several interesting questions are raised by his paper, which I shall discuss in my comments:

(1) Is the formal model unique, for a given physical theory; or if not, is his model always the most fruitful?

(2) Is it disturbing that many features of the logic of quantum theory, as he defines it, are applicable to a much wider class of theories?

(3) Can one get any interesting results about quantum theory using the quantum logic approach that might be harder to get in other ways; or that have actually been derived first via this approach?

(4) Could the quantum logic approach lead to possible fruitful generalizations of quantum mechanics?

(5) Does van Fraassen's semantical analysis of various quantum logics help us to understand them better?

On the last question, the semantic approach of van Fraassen, leading him to distinguish between various ways in which negation and the connectives between elementary propositions may be introduced, continues the useful work he started in his paper on *The Labyrinth of Quantum Logics* (which he was kind enough to show me before its publication in this volume). It has shown itself to be one fruitful way of distinguishing between various approaches to quantum logic, enabling him to bring some order into a confusing field, in which a number of workers seem to start afresh without attempting to relate their work to that already done in the field. He shows how, once appropriate conventions are adopted in the various versions of quantum logic, the various algebraic structures used in each quantum logic arise, and how they relate to each other.[1]

I must say I am a bit puzzled by van Fraassen's slighting reference to 'non-classical probability theory', as applied to certain attempts to formalize the use of probability concepts in quantum theory. The subject, unfortunately, abounds in misnomers, and has from the beginning. The wave function doesn't refer to a wave, the uncertainty principle can yield quite certain results about the dispersions of complementary variables, 'observables' aren't necessarily observable – you continue the list with your favorites. Prof. van Fraassen is willing to accept the use of the term 'logic' to refer to the "system of axioms and/or rules which characterize the set of valid sentences and the set of valid arguments for a certain language pertaining to quantum mechanics. We shall take this language to be the language of elementary statements (and perhaps complex sentences built up out of these)." (*The Labyrinth of Quantum Logics*, p. 75.) He is quite ready to accept, on the basis of this definition, and indeed to discuss quite ably, the existence of various possible quantum logics. Yet this is certainly not the traditional use of the word 'logic' – so that we might add to our budget of paradoxes that quantum logic is not a logic, but rather an elucidation of certain algebraic[2] structures inherent in the theory.

Given this background, I fail to see what he finds so disturbing about "the desire for a non-classical probability theory... rais(ing) its ugly head," (p. 63). Varadarajan and others who have used the term 'non-classical probability theory' to describe their attempts to axiomatize probability theory so that it may be applicable to systems of propositions embedded in a non-Boolean lattice are doing no more or less violence to the language than the quantum logicians. Van Fraassen points out that the traditional concepts of probability measure, as axiomatized by Kolmogoroff, for example, may still be applied in quantum theory, if we do not choose to apply the concept of probability to the elementary statements (propositions) directly. But what harm is there if we do so choose? No more harm, it seems to me, than in applying the alternate logics of quantum mechanics to the elementary statements as he defines them. Each of the quantum logics, in terms of its own conventions, builds on some aspects of the structure of the theory, and so does non-classical probability theory.[3]

Now let me turn to some of the other questions raised by the paper. Van Fraassen sets out to "present a general picture of the representation of physical quantities," using classical and quantum mechanics as guides; and then uses "this general representation of physical quantities to explore semantic relations among the elementary statements." (p. 69). About the exploration of semantic relations, given that one accepts his starting point, I have nothing to add to my admiration expressed above for the care with which he proceeds in the elucidation of these formal relations.

But one may raise the question, is the starting point too special for the proposed general theory of physical quantities; and, indeed, even for the study of classical and quantum mechanics in particular? Van Fraassen starts from the description of the physical system in terms of its possible states in a state space, with some preassigned mathematical structure. Phase space for classical mechanics, and Hilbert space for quantum mechanics are his examples. Incidentally, it is implied in this paper, and stated in another paper by Prof. van Fraassen, that the structure of phase space is that of a $2n$-dimensional Euclidean space; actually, the structure of phase space is that of a symplectic space constructed as the cotangent bundle over an n-dimensional configuration space. The latter may be taken as Euclidean for many applications, although it is not for such simple systems as a particle constrained to move on a non-flat surface. This is a minor technical detail here, but is of some importance when we later discuss the question of the relationship of the state space to space-time. Since the appearance of von Neumann's book on quantum mechanics, Hilbert space has been the hallowed starting point for the rigorous treatment of the quantum-mechanical formalism in most texts. Curiously enough, von Neumann was also responsible for initiating two rather different approaches to the quantum mechanical formalism: the algebraic approach, via the algebra of the operators representing the observables (horrid word!); and the quantum logic approach used as a way of postulating the structure of the theory.

The algebraic approach was originally developed by von Neumann,

Wigner and Jordan, and more recently is associated with work by Segal, Haag and Kastler among others.[4] The basic difference between the Hilbert space and algebraic approaches is this: in the former we start with the states, represented as rays of a Hilbert space, and then define the mathematical representation of observables in terms of linear, bounded Hermitean operators on the Hilbert space; in the latter we start with a Jordan or C* algebra of the observables, and then define states in terms of positive linear forms on the algebra. States in the algebraic formulation can be physically specified, for example, by giving the expectation values of all observables in a given state. Now, for systems with a finite number of degrees of freedom, the two approaches can be proven to be equivalent. But for systems with an infinite number of degrees of freedom, notably statistical mechanical systems and relativistic field theories, they are not. There are reasons to hope that the algebraic approach may eliminate some problems of the quantum theory of fields in the usual formulations. But that is not the important point here. The point I want to make is that such an approach provides just as 'natural' a starting point for a 'general picture of the representation of physical quantities' as the more familiar state space approach that van Fraassen uses. Indeed, classical physics can also be treated from this point of view; here, the algebra of observables will be commutative.[5] Possibly van Fraasen might be able to apply some variant of the semantic analysis approach to such algebraic formulations, and get interesting results. But his starting point in this paper is really more closely tied to one conventional approach to analyzing classical and quantum mechanics than he seems to realize. The same theory may have widely different formalizations, which may lend themselves to rather different types of analysis, interpretation, and generalization. One need only think of the Newtonian, Lagrangian, Hamiltonian and Hamilton-Jacobi approaches to classical mechanics, and the various ways that these approaches have proved suggestive in the development of physics, to see my point.

The quantum logic approach has itself been much developed in recent years as a possible third formalism for quantum mechanics. Here, neither the Hilbert space nor the algebra of observables is taken as the starting point from which to deduce the logic of the quantum-mechanical propositions. Rather, the aim is to formulate enough postulates about the structure of the lattice of propositions (or whatever substructure of the full lattice the particular approach uses) to reconstruct the full quantum theory, Hilbert space and all. For some the motivation for this attempt is just the possibility of an alternate axiomatization of quantum mechanics, not introducing the Hilbert space initially. For others, the desire to avoid initial introduction of the Hilbert space is itself motivated by the 'non-intuitiveness' of postulating the Hilbert space (or the algebra of the observables, for that matter). While the view may be expressed with more or less subtlety, a number of people are still under the impression that the foundations of a physical theory can be drawn practically directly from 'experience'.[6] I have not the time, nor is it necessary to go into the critique of such a viewpoint before this audience; but I should point out that the illusions of such investigators do not necessarily detract from the usefulness of their researches in giving us an alternate formalism from which to build up the quantum theory. It is also true that the use of the *Gedankenexperiment* approach as a heuristic device has actually been important in talented-enough hands in suggesting how to proceed with the development of a new theory (Einstein in relativity, and Heisenberg and Bohr in quantum theory come to mind at once).

In any case, the quantum logic approach has been more or less successful in recent years; various sets of postulates for the lattice of propositions have been proposed (notably by Piron) which lead more or less to an isomorphism of the lattice with the lattice of subspaces of a Hilbert space.[7] What we need for conventional quantum mechanics, roughly speaking, is a complete orthocomplemented atomic lattice with weak semi-modularity (I omit the definitions of technical terms, which would not be useful here to those unfamiliar with them[8]). Now this is rather more than van Fraassen is able to deduce as properties of the lattice of propositions of his general class of physical theories. I fail to see what van Fraassen finds surprising or paradoxical about his conclusion that "typical results in quantum logic turn out to have such generality that it is hard to see how they throw much light on the foundations of one particular theory." (p. 69). It all depends on what is meant by 'typical results'. If one looks at results that the quantum lattice has in common with the weaker lattice structure van Fraassen deduces from his approach, naturally the results could apply to a much wider class of theories than quantum mechanics. But it is well known that if one weakens the quantum lattice postulates they are no longer categorical, in the sense that they do not lead uniquely to an isomorphism with the subspaces of a Hilbert space. Until one has a categorical set of postulates, or more generally a categorical characterization of a system, naturally one can deduce results which hold for the system as well as for any other model satisfying the postulates or other characterization given. Would one say, speaking of ordinary vector analysis, that he failed to see the importance of the fact that vectors formed a linear vector space, since the antisymmetry of the vector produce cannot be deduced from that fact?

The point is, that in attempting to axiomatize some subject, one must look for some set of assumptions which will limit the class of models satisfying these assumptions to the class one had in mind from the beginning. One can then also experiment with weakening (or adding) assumptions to see what has been lost (or gained) thereby. For example, one may find he has been too restrictive initially in his assumptions for what he really wanted to accomplish.

In these respects, the work on quantum logic is certainly valuable, and I shall return later to one aspect of the possible fruitfulness of such work.

But one can make an even better defense of work in quantum logic as a tool of research in quantum theory, by pointing out that this approach has already led to important new results. The work of Kochen and Specker comes to mind at once. They start from the Hilbert space point of view, and use the partial Boolean algebra of propositions to study the hidden variable problem. They put a few simple requirements on the hidden variables, more or less amounting to assuming that they are local in character (i.e., serve to characterize only the microsystem); and are then able to show that a necessary and sufficient condition for the existence of the hidden variables is that the partial Boolean algebra be embeddable in Boolean algebra. The work of Gleason showed that this cannot be done in the quantum mechanical case if the Hilbert space is more than two dimensional. Thus, no local hidden variable theory can reproduce all the results of quantum mechanics.

One might also credit the quantum logic approach with giving rise to more precise formulations of the notion of complementarity, in the work of Emch and Jauch and Heelan, for example.[9]

On the question of whether the quantum logic approach might lead to possible fruitful generalizations of quantum theory, one cannot be dogmatic, of course, in legislating for the future. Just as an example of how such generalizations may be suggested by another formal representation of a theory, let me mention some recent work by Mielnik. In the attempt to give intuitive physical content to the quantum logical propositions, they have often been identified with *filters*, idealized experimental devices of a yes-no, or pass-no pass type. Filters having been identified with the elementary propositions, an attempt was made to correlate other filters with negations or combinations of these propositions representing the logical connectives, so that an order structure of the filters could be identified with some order structure of subspaces of Hilbert space. So far, this is only a variant of the quantum logic approach, attempting (with more or less success) to tie it more closely to idealized measurements. But Mielnik has recently shown that it is quite possible to generalize the concept of filters in such a way that they could still be regarded as leading to a generalization of quantum theory; but one that is not characterised by the usual Hilbert space structure. It is either directly assumed in the quantum logic approach, or deduced as a result of other assumptions, that filters can be represented mathematically by projection operators acting on the Hilbert space. As postulated by Mielnik, the properties of the filters are such that they cannot be represented by such projection operators. This leads to theories in which quantum states cannot always be represented by vectors in a Hilbert space. "The physical reality can be too complex in order to fit in any Hilbert space," as Mielnik says.[10]

Thus, the representation of physical quantities by methods quite different from those suggested by van Fraassen has led to an interesting proposed generalization of quantum theory – which has yet to prove its physical fruitfulness, I hasten to add. But this seems to me to point up the moral that it is not only, or even not so much the 'many different kinds of physical theories' that can be formally represented in one framework that one should be concerned with; but rather with the many different kinds of formalization of one theory that often prove important in leading to progress towards the next stage of development of a science. It seems to me that philosophy of science must investigate such questions much more fully.

Let me finish by raising one last question. Van Fraassen, in his paper, does not discuss the question of the relation of the state space formalism to the evolution of the physical system in space and time, a common failing among quantum logicians. In the early (Newtonian) formulations of classical physics, the question of the relation of the mathematical structure used to build up the theory to the space and time structures hardly arose, because the embedding of the system in space and time formed a direct part of the mathematical formalism. This is no longer the case, even in the more abstract formulations of classical theory, for example in the phase space formulation of classical mechanics. It is possible to develop large parts of classical mechanics as a formal theory of Hamiltonian structures in a symplectic space. However, these mathematical results have no direct physical interpretation unless one can interpret the symplectic space as the cotangent bundle over some configuration space, which represent the actual physical space of the theory. The invariance group of the abstract phase space theory, the symplectic group of canonical transformations, for example, is vastly wider than the class of space and time symmetry transformations which are of direct physical significance (their generators serve to define vital physical concepts such as energy, linear and angular momentum). At any rate, the rules for relating the abstract theory to the evolution of a physical system in space and time are by no means trivial. This is even more the case in quantum theory. Without going into any details, the structure of the space-time manifold is taken as given in both classical and quantum mechanics and field theories, even in their special-relativistic generalizations, and it is always related in some fashion to the mathematical formalism of the theory in question, to the extent that it is not interwoven into that structure from the beginning. In the theory with which I spend most of my time wrestling, general relativity, there is an additional complication: the space-time structure, in its affine and metrical aspects, is not given *a priori*, but must be built up along with the behavior of all other physical processes. In all of these cases if one wants to formalize the theory in the usual way, one cannot even formulate the vital semantical correspondence rules which single out certain mathematical elements of the theory and correlate them with physical quantities which are presumed to be meaningful and therefore measurable until one has related the mathematical structures of the theory to the space-time structure. One can talk about 'positions' and 'momenta' abstractly as elements of a mathematical formalism, but without correlation of these elements with the space-time structure, these labels have no physical content.[11] It seems to me that even the most abstract formulation of physical theory, if it does not want to abstract away the physical content, must discuss the relation of the space-time structure to the formalism. The lack of such a discussion thus constitutes a serious omission in van Fraassen's approach.

Boston University

NOTES

* My comments are set down here more or less as they were delivered at the time of his talk: I have added a few footnotes and references to bring some ideas expressed into better accord with my current views. I plan to publish a more extended discussion of quantum logic elsewhere.

[1] I have become convinced that the conventional elements involved in the choice of a quantum logic are of such a nature that the adoption of such a logic adds nothing to the physical interpretation of the theory, which can be carried out using standard logic. Also, the use of the phrase 'algebraic structures', although common in the literature on quantum logic, is not accurate. If the word structure is used in its technical sense, and the distinction made, following Bourbaki, between algebraic, topological and order structures; then each quantum logic relates to a structure of order within the mathematical apparatus of quantum theory.

[2] See previous note. Michael Gardner was kind enough to report my comments on misnomers in quantum theory in his excellent paper on quantum logic: *Philosophy of Science* **38** (1971), 508.

[3]. I should have added that one can get along with the traditional concepts of logic if one does not choose to apply the concept of elementary statement or proposition to such word groupings as "the position of the electron is *x*," as Van Fraassen, following the quantum logicians, does. They may be regarded as incomplete phrases, in somewhat the same sense as "the viscosity of water is *v*" is incomplete without a specification of conditions of pressure, temperature, etc. of a definite sample of water. As Bohr always emphasized, the attribution of properties such as position to microsystems must always be in the context of a complete specification of the classically described arrangement in which these properties take on their meaning. For a full discussion of Bohr's viewpoint, with a critique of the quantum logic approach, see the important paper by C. A. Hooker, 'The Nature of Quantum Mechanical Reality: Einstein Versus Bohr', in R. G. Colodny (ed.), *Paradigms and Paradoxes*, U. of Pittsburgh Press, 1972.

[4] At last, one can recommend a textbook on this approach which has recently appeared: G. Emch, *Algebraic Methods in Statistical Mechanics and Quantum Field Theory*, Wiley-Interscience, New York, 1972.

[5] J. M. Cook in an article on 'Banach Algebras and Asymptotic Mechanics' outlines the algebraic approach to classical and quantum mechanics, as well as classical and quantum statistical mechanics. *Application of Mathematics to Problems in Theoretical Physics* (ed. by F. Lurçat), Gordon and Breach, New York, 1967.

[6] Perhaps I exaggerate slightly; but Jauch in his Varenna lectures, says "When a theory is to be generalized or modified a postulational formulation is particularly useful since the empirical justification can be explicitly identified in a few well defined places and possible modifications can be more easily studied and surveyed" (p. 23)

If only it were so! Later he says of the *elementary* propositions, interpreted as *yes-no* experiments, "These concepts are relatively easily interpreted in physical terms and therefore a good motivation for the axiomatics is available" (p. 24). J. M. Jauch, 'Foundations of Quantum Theory', in B. d'Espagnat (ed.), *Foundations of Quantum Mechanics*, Academic Press, New York, 1971.

[7] Actually, this situation is a bit more complicated technically. For a review of this work, see Jauch's book, J. M. Jauch, *Foundations of Quantum Mechanics*, Addison-Wesley.

[8] For a survey of lattice theory, the article by Abbott is most useful: *Trends in Lattice Theory* (ed. by J. C. Abbott), Van Nostrand Reinhold, New York, 1970.

[9] It should be pointed out that these authors do not regard the presence of a quantum logical structure within the formalism of quantum theory as showing the need for using a non-classical logic in the physical interpretation of the theory, a viewpoint which I share.

[10] B. Mielnik, 'Geometry of Quantum States', *Commun. Math. Phys.* **9** (1968), 55.

[11] Stein has recently emphasized this point, following the work of Weyl, in the context of the interpretation of the quantum mechanical formalism: H. Stein, 'On the Conceptual Structure of Quantum Mechanics', in R. Colodny (ed.), *Paradigms and Paradoxes*, Univ. of Pittsburgh Press, 1972.

BAS C. VAN FRAASSEN

THE LABYRINTH OF QUANTUM LOGICS*

1. The development of quantum logic

The conceptual structure of the new quantum theory is in some respects so different from that of classical physics that it has from the very beginning suggested radical departures in philosophy and logic. Specifically, a number of writers have considered non-standard systems of logic in connection with quantum mechanics (see bibliography). Two main directions may be discerned, initiated by Reichenbach, and by Birkhoff and von Neumann. The aim of the present paper is *first* to present a unified exposition of the main systems found in the literature, and *second* to discuss and evaluate the main logical and philosophical theses and arguments which have concerned the subject of quantum logic.

The starting point for our exposition is Beth's semantic analysis of physical theories. This will make possible a semantic analysis of each of the systems of quantum logic to be discussed. Since the original presentation of these logical systems was in most cases rather imprecise, it would perhaps be better to speak of a *reconstruction* than of an analysis. But we do not think that our treatment does more violence to the original intent of the authors than does, say, the current semantic analysis of modal logic to the intent of the originators of the standard modal systems.

2. The formal structure of a physical theory

We begin by outlining a model for a certain kind of physical theory, derived from Beth's ideas concerning the application of formal semantic concepts to the study of scientific theories. In the next section we shall apply this to the specific case of the elementary quantum theory, and show how this leads to the conception of a 'logic of quantum mechanics'.

Beth addressed himself specifically to non-relativistic theories which use a mathematical model to represent the behavior of a certain kind of

Boston Studies in the Philosophy of Science, XIII.

physical system [1], [3], [4], [5]. A physical system is capable of a certain set of *states*, and these states are represented by the elements of a certain mathematical space, the *state-space*. Specific examples are the use of Euclidean n-space in classical mechanics and Hilbert space in quantum mechanics.[1]

Besides the state-space, the theory uses a certain set of *measurable physical magnitudes* to characterize the physical system. This yields the set of *elementary statements* of the theory: each elementary statement U formulates a proposition to the effect that a certain such physical magnitude m has a certain value r at a certain time t. (Thus we write $U=U(m, r, t)$; or $U=U(m, r)$ when abstracting from variation with time.) Whether or not the magnitude m has the value r depends on the state of the system; in some states m has the value r and in others it does not. This relation between states on the one hand and the values of physical magnitudes on the other may also be expressed as a relation between the state-space and the elementary statements. Thus, there corresponds to an elementary statement $U=U(m, r)$ a certain subset $h(U)$ of the state-space H: m has the value r if and only if the state of the system is represented by an element of H which belongs to $h(U)$. (We also say that $h(U)$ is the set of elements of H which *satisfy* U.) The mapping h is the third characteristic feature of the theory; it connects the state-space with the elementary statements, and hence, the mathematical model of the theory with the measurement-results.[2]

This mapping h induces the semantic relations among the elementary statements:

(1) U is *true* if and only if the state of the system is represented by an element of $h(U)$.

(2) U is *valid* if and only if $h(U)=H$.

(3) U is *semantically entailed* by V if and only if $h(V)\subseteq h(U)$.[3]

That this is as yet only a rough and preliminary characterization of these semantic notions will be clear especially from the informal nature of (1). But one can already see how this may lead to a consideration of various logical systems: it has traditionally been the task of logic to give a systematic account of validity and entailment.

3. The Elementary Statements of Quantum Mechanics

In the case of quantum mechanics, the states of a system are represented by the elements (*state-vectors*) of a Hilbert space.[4] For each measurable physical magnitude ('observable') m there is a Hermitean operator M on Hilbert space, with the following significance: m has the value r in state x if and only if $Mx = rx$. The following terminology is used here: when $Mx = rx$, then r is called an *eigenvalue* of M, and x an *eigenvector* of M corresponding to the eigenvalue r. So the vectors which satisfy the elementary statement $U = U(m, r)$ are given by:

(4) $h(U) = \{x\colon Mx = rx\}$.

This abstracts from the possibility of variation with time. Quantum mechanics was originally developed in two forms: Heisenberg's matrix mechanics, in which the operator M is a function of the time, and Schrödinger's wave mechanics, in which the state-vector is a function of the time.[5] Thus, depending on which of these specific formulations we choose, we have either

(4a) $h(U(m, r, t)) = \{x\colon M_t x = rx\}$, or
(4b) $h(U(m, r, t)) = \{x(t)\colon Mx(t) = rx(t)\}$.

But for our purpose the shallower analysis given by (4) will do in most cases.

An operator transforms a vector into another vector, and in general the new vector is not merely a scalar multiple of the first one. So if the state of the system is represented by the state-vector x, we have three possibilities:

(5a) $Mx = rx$,
(5b) $Mx = r'x$ for some $r' \neq r$,
(5c) $Mx \neq r'x$ for any value r'.

If the third possibility obtains, then the magnitude m does *not have any* of its possible values $r, r', r'', \ldots$. This sounds strange to the classical ear; the Bohr-Heisenberg explanation is that a measurable magnitude has a (sharp) value only in a certain kind of experimental situation.[6] But this feature is quite independent of the Copenhagen interpretation; for example, Feyerabend sees a central difference between classical and quantum theory exactly in the breakdown of the principle that "each entity possesses *always* one property out of each category" [13], p. 51, and pp. 52–53.

Of special interest here is the question of *compatibility or incompatibility* of two physical magnitudes m and m'. This corresponds to the question whether or not two operators M and M' commute, i.e. whether $MM' = M'M$. In particular, it is found that

$$QPx - PQx = i\hbar x$$

where Q and P are the operators corresponding to the X-coordinate of position and momentum respectively. Thus Q and P do not commute. It means specifically that if x is an eigenvector of the one, it is necessarily not an eigenvector of the other; for if it were, say corresponding to the eigenvalues r and r', we would have

$$QPx = Qr'x = r'Qx = r'rx = rr'x = rPx = Prx = PQx\,.$$

in which case the difference between QPx and PQx would not be $i\hbar x$ but the zero-vector (since x cannot be the zero-vector, which represents no possible physical state). Therefore when the X-coordinate of position has a value r, then the X-coordinate of momentum does not have any of its possible values r'.

4. Alternatives in Quantum Logic

A *logic* is a system of axioms and/or rules which characterizes the set of valid sentences and the set of valid arguments for a certain language. Thus a logic of quantum mechanics must do this for a certain language pertaining to quantum mechanics. We shall take this language to be the language of the elementary statements (and perhaps complex sentences built up out of these).

At this point we must present a schematic formalization of this language. This formalization is schematic in two ways. First, we intend to remain on the level of abstraction assumed by the writers whose work we discuss below. (There can be no doubt that quantum logic represents, in each case, a relatively shallow analysis of the language of quantum theory. But then, so does standard logic with respect to mathematical discourse; the term 'logic' is traditionally associated with such a high level of abstraction.) Second, the various writers (implicitly) differ on certain points in this construction, and we mean to accommodate all their approaches. The general structure of this language L is then given by:

(6)(a) *Syntax of L*: The set of sentences of L comprises (at least) a set of elementary statements $U(m, r)$.

(b) *Semantics of L*:

(i) Associated with L is a *state-space* H.

(ii) Associated with each elementary statement $U(m, r)$ is a Hermitean operator M on H; r denotes an eigenvalue of M.

(iii) A *model* for L is a couple $K=\langle X, f\rangle$ where X is an element (the system) to which the function f assigns a location $f(X)$ in H. (Here $f(X)$ is the state-vector for the system X.)

(iv) The elementary statement $U(m, r)$ is *true* in the model $K=\langle X, f\rangle$ if and only if the vector $x=f(X)$ is such that $Mx=rx$.

The definitions of validity and semantic entailment can be given in the form

(7) U is a *valid* sentence of L if and only if U is true in every model for L;

(8) U is *semantically entailed by* V in L if and only if U is true in every model in which V is true;

which are equivalent to those given by (1)–(4) for elementary statements U.[7]

We may note about L, *first* that it is what we have elsewhere called a *semi-interpreted language* [32], and *second* that its syntax and semantics have been left under-specified in a number of respects. Specifically, in the syntax we have left open the possibility that there are sentences of L which are not elementary statements. About the elementary statements, we have said when they are true in a model, but not when they are false. For both points there are two main alternatives, and various writers have chosen differently among these alternatives.

Let us first consider the second question, the question when an elementary statement is false. We have already noted that $U(m, r)$ may fail to be true for either of two reasons: because for some values r' other than r, $U(m, r')$ is true, or because for no value r' whatsoever $U(m, r')$ is true. The latter possibility represents the divergence from the classical case.

One possible reaction is to say that $U(m, r)$ is false whenever it is not true – for whatever reason. However, this is not the only possible alternative; the history of logic provides ample precedent for distinguishing *false* from *not true*. This distinction is usually made via the principle:

(9) A sentence is *false* if and only if its denial is true.

If we construe the denial in such a way that it is true whenever the original sentence fails to be true, then this reduces to the previous case. When denial is construed in this way, we speak of *exclusion negation*.[8] But, as has been pointed out many times, a denial is usually understood in the context of a definite set of alternatives. And then the denial is construed to *assert* that one of the *other* alternatives obtains. In this case we speak of *choice negation*. Specifically, we may take the set of eigenvalues of the operator M as the set of alternatives providing the context for $U(m, r)$; its choice negation would then assert that the statevector x is such that $Mx=r'x$ for a value $r'\neq r$. If we now uphold principle (9). for choice negation rather than exclusion negation, *false* no longer coincides with *not true*. This yields the first set of basic alternatives for quantum logic:

(IA) $U(m, r)$ is *false* in the model $K=\langle X, f\rangle$ if and only if $Mx\neq rx$ for $x=f(x)$.

(IB) $U(m, r)$ is *false* in the model $K=\langle X, f\rangle$ if and only if $Mx=r'x$ for $x=f(X)$ and some value $r'\neq r$.

In the first case, the principle of bivalence holds for elementary statements: $U(m, r)$ is always either true or false. In the second case, the truth-or-falsity of $U(m, r)$ presupposes that the system is in an eigenstate of the operator M.[9]

The second question, whether L comprises sentences which are not elementary statements, must be understood in the context of our intention that L be a language of elementary statements. That means that any sentence of L must either be an elementary statement, or have been constructed out of elementary statements. We shall call an n-ary function f of sentences into sentences an *n-ary connective* provided the truth-value (or lack of truth-value) of $f(U_1, \dots, U_n)$ in a model K depends entirely on $K, f, U_1, \dots, U_n$. So for any sentence V of L there must be elementary statements $U_1, \dots, U_n$ of L such that $V=f(U_1, \dots, U_n)$ for a certain connective f.

Connectives are usually associated with certain symbols of the language; for example, the conjunction of two atomic sentences in the language of the propositional calculus is a non-atomic sentence produced by infixing a dot or ampersand. If we follow this course of action then connectives will map elementary statements into sentences which are not elementary statements. It might be thought that the question of the logic of L will become trivial if we do not introduce non-elementary statements by means of such connective symbols. But this is not so, due to the existence of various semantic relations among the elementary statements (in the language of the standard propositional logic, the atomic sentences are semantically independent of each other). So one might ask the question: is there an elementary statement V which qualifies as the exclusion negation of $U(m, r)$, or as the choice negation of $U(m, r)$, or as the conjunction of $U(m, r)$ and $U(m', r')$ – and so on. This yields our second set of basic alternatives for quantum logic:

(IIA) The language L has a set of connectives, each of an integral degree $n<0$ and defined for a certain class of n-tuples of elementary statements; the values of the connectives are again elementary statements.

(IIB) The language L has a set of connectives, each of an integral degree $n<0$ and defined for a certain class of n-tuples of sentences of L; the values of the connectives are not elementary statements.

Only when the structure of L has been further specified, by a choice among these alternatives and a characterization of the set of connectives, can the question of the logic adequate with respect to L be broached.

In the following sections we shall present some main approaches to quantum logic, explicating their intent to be that of formulating a logic adequate with respect to L, given a certain choice from the pairs of alternatives (I) and (II). That a certain amount of extrapolation will be involved here, is unavoidable. Our purpose will be served if under our explication, each of these approaches to quantum logic is intelligible, in the sense of having a well-defined objective and well-defined criteria for success. The criteria for success are those which are standard in metalogical appraisal: the system must be at least sound and preferably complete with respect to validity and semantic entailment. (In the present paper, questions of completeness will mostly be disregarded; most of the writers discussed did not raise these questions.)

5. Reichenbach's Approach to Quantum Logic

Reichenbach chose alternatives (IB) and (IIB); this choice is what we shall mean by 'Reichenbach's approach'. (His discussions in [26], Sections 30–34, are always in terms of measurement, but reference to his theoretical discussion of this subject in Section 21 shows how to represent his choice in terms of the mathematical model of operators and eigenvectors.) We shall first discuss Reichenbach's use of *three-valued matrices*, to implement this choice, then Lambert's use of *supervaluations* for the same purpose, and finally, consider the limitations of this approach.

5.1. *Reichenbach's Use of Matrices*

Reichenbach introduced his choice of alternative (IB) as follows:

> Ordinary logic is two-valued; it is constructed in terms of the truth-values *truth* and *falsehood*. It is possible to introduce an intermediate truth-value which may be called *indeterminacy*, and to coordinate this truth-value to the group of statements which in the Bohr-Heisenberg interpretation are called *meaningless*. Several reasons can be adduced for such an interpretation. If an entity which can be measured under certain conditions cannot be measured under other conditions, it appears natural to consider its value under the latter conditions as indeterminate. It is not necessary to cross out statements about this entity from the domain of meaningful statements; all we need is a direction that such statements can be dealt with neither as true nor as false statements. This is achieved with the introduction of a third truth-value of indeterminacy. ([26], p. 145)

Much philosophical puzzlement has been occasioned by the use of the term 'third truth value'. This term is unfortunate; the case can be stated much more perspicuously by saying that certain statement are neither true nor false, rather than that they have a third truth value, the value *indeterminacy*. A value assignment may assign T, F, or I to a sentence; or perhaps 1, 0, or $\frac{1}{2}$; but this assignment is only a marker for the corresponding class of sentences.

The connectives introduced by Reichenbach are all defined by three-valued truth-tables. These connectives are given by syncategorematic expressions in the language L; thus we see that Reichenbach chose alternative (IIB). It must be noted that Reichenbach lists only a few of the connectives definable by means of a three-valued matrix, and does not use even all of these in his subsequent discussions. It is not clear whether he meant all such connectives to appear in L, or only those which he lists.

The main use made of this machinery is to express the statement

(10) The elementary statements U and V correspond to incompatible magnitudes (are incompatible)

in L by means of the sentence

(11) $(U \vee \sim U) \rightarrow \sim\sim V$

where the connectives $\vee$, $\sim$, $\rightarrow$ are given by the tables

$\vee$	T	I	F	$\rightarrow$	T	I	F	$\sim$
T	T	T	T		T	F	F	I
I	T	I	I		T	T	T	F
F	T	I	F		T	T	T	T

It is easy to see that (11) was constructed to satisfy the criterion that it be true if and only if U and V are such that if U is true, or U is false, then V is indeterminate. This criterion is also satisfied by any connective ꝺ defined by placing T or F or I in the blank spaces of the following table

ꝺ	T	F	I
T	F	F	T
F	F	F	T
I	–	–	–

There are 27 such connectives, and the one given implicitly by (11) is one defined by placing a T in each blank space. In any case, a single connective could be used to express (10) in L. Certainly this connective is one which is not an obvious generalization of any two-valued truth-functional connective; but neither is Reichenbach's $\sim$, which does not correspond to either choice negation or exclusion negation.

From another point of view, however, no such connective can be used to express the notion of incompatibility adequately. For when m and m′ are incompatible magnitudes, then (on choice (IB)) $U(m', r')$ fails to be true or false in *any* model in which $U(m, r)$ is true or false. But of course it could happen that U is true in a model, and V does not have a truth-value in that model, although U and V do not correspond to incompatible magnitudes. For example, let Q and Q' be the operators corresponding to the X and Y coordinates of position respectively and P the operator corresponding to the X coordinate of momentum. Then suppose that in

$K = \langle X, f \rangle$, $x = f(X)$ is such that $Qx = rx$ and $Q'x = r'x$. The former entails that x is not an eigenvector of P. Hence $U = U(q', r')$ is true in U and $V = U(p, r'')$ is indeterminate in K; by the above table we see that $U ⫰ V$ is true in K. Yet U and V do not correspond to incompatible magnitudes.

In this sense, the semantic relation of incompatibility is not adequately expressed by any three-valued truth-functional connective, but at most by a three-valued *modal* connective. Reichenbach's sentence (11) stands to (10) approximately as does material implication to semantic entailment. Instead of choosing an object-language equivalent for (10), however, Reichenbach might have treated incompatibility as a relation among statements (to which one could appeal in arguments) and this relation he might have defined by:

(12) $U \vee \sim U$ semantically entails $\sim \sim V$

The logic adequate to Reichenbach's formulation of L must rely both on the systems of three-valued logic developed by Post, Lucasiewicz, and Tarski (to which he refers in [26], p. 147) and special rules concerning this relation of incompatibility. An example of such a rule would be one validating the following argument form:

(13) $A \vee B$
C (C incompatible with B)
Therefore, A

– a generalized form of the disjunctive syllogism. We may understand Reichenbach's discussion of (11) as implicitly conveying the conjecture that a complete system need only have the special rule:

(14) If U is incompatible with V, infer (11)[10]

The inferable sentences of form (11) are not three-valued tautologies of course; they are rather what Carnap would have called *meaning postulates*.

We may point out here that if it were decided not to use a connective corresponding to incompatibility, then one could also choose the three-valued connectives for the language L in such a way that the classical propositional logic would govern them. One way to do this is to define all the usual connectives in terms of disjunction and negation in the usual manner, and then to let these two basic connectives be *exclusion negation* and *exclusion disjunction* defined by:

+	T	I	F	–
T	T	T	T	F
I	T	F	F	T
F	T	F	F	T

The result would be that bivalence holds for complex sentences, classical propositional logic is sound, and for example, the conjunction of two incompatible sentences is an always-false sentence. In that case alternatives (IB) and (IIB) would be implemented by a procedure which only adds certain special rules to the standard logic (and which would require no reschooling of our logical intuitions).

Two objections might be raised against this by Reichenbach: first, he appears to feel that (10) ought to have an object-language counterpart, and second, the non-standard connectives play a role in the 'suppression of causal anomalies' in [26], Section 33. We have already suggested a sense in which no three-valued truth-functional connective can be an adequate object-language counterpart to the semantic relations of incompatibility. But the further alternative of using a three-valued modal connective for this has its drawbacks also: it still leaves us with the tasks of giving a sense to the iteration of this connective, in such sentences as 'That U and V are incompatible, is incompatible with W'. This use of 'incompatible' is so far from its original intent that it seems to me to raise serious doubts concerning the introduction of any such connective.

Secondly, if an anomaly can be shown to be avoidable by an analysis of its reformulation in L, this must be so due to the semantic structure of L. We can make this point clear by showing that the mere choice of alternative (IB) makes Reichenbach's argument possible in one of his main examples. The following passage contains Reichenbach's use of the properties of his three-valued disjunction in his analysis of the well known n-slit interference experiment.

> Let B_1 be the statement: 'The particle passes through slit B_i'. After a particle has been observed on the screen we know that... if the particle did not go through $n-1$ of the slits, it went through the nth slit, and that if it went through one of the slits, it did not go through the others.... But since from these relations the disjunction $[B_1 \vee \ldots \vee B_n]$ is not derivable, we cannot maintain that this disjunction is *true*; all we can say is that it is *not false*. It can be indeterminate. This will be the case if no observation of the particle at one of the slits has been made. [26], pp. 162–163

Were the disjunction true, one could deduce that the particle must have gone through one of the slits, and this deduction would lead to the well-known anomaly that opening a new slit affects the probability that a particle going through one of the old slits will reach a given point on the screen (as discussed by Reichenbach in [26], Section 7). Hence the importance of showing that the disjunction need not be true.

But the analysis quoted above can be paralleled by a semantic analysis as follows. Let $r_1, \ldots, r_n$ be the positions of the slits; let q be the measurable physical magnitude which is the position of the particle at the time of its passage through the diaphragm, corresponding to the operator Q. Since q has $r_1, \ldots, r_n$ as its only possible values, these are the eigenvalues of Q; and we have:

(15) At most one of $U(q, r_1), \ldots, U(q, r_n)$ is true; if all but $U(q, r_i)$ are false, then $U(q, r_i)$ is true,

provided 'false' is construed in accordance with alternative (IB). The assumption of bivalence leads from (15) to: one of $U(q, r_1) \ldots, U(q, r_n)$ must be true – and hence to anomaly. But since 'false' in (15) is construed in accordance with (IB), this assumption of bivalence does not hold. This recognition of the failure of bivalence as a necessary condition for the correctness of (15) blocks the anomalous inference in just the same way as Reichenbach's analysis.

Moreover, since '$U(q, r_i)$ is true' can here be translated into '$Qx = r_i x$', we see that this semantic analysis can in turn be paralleled by a purely quantum-theoretic analysis of the experimental situation. Thus the suppression of the anomalies is also independent of the choice between alternatives (IA) and (IB).

5.2. *Lambert's Use of Supervaluations*

Karel Lambert has recently sought to implement Reichenbach's approach (choice of (IB) and (IIB) in [19] by using not matrices but *supervaluations*, a device introduced by the present author in [35] and [33]. The advantages of doing so are that the connectives are automatically the classical ones, and classical logic remains sound for them. The semantic relation of incompatibility has no object-language counterpart, but as we have argued in the preceding section, this cannot be considered a weakness.

Just as did Reichenbach, Lambert discusses the semantics in terms of measurement. We shall again reformulate this work as pertaining to a completion of the language L in accordance with alternatives (IB) and (IIB). Since supervaluations are a much less well-known device than matrices, we shall do so in some more detail. To the syntax of L we add:

(16a) If A and B are sentences of L then $\sim A$ and $(A \vee B)$ are complex sentences of L. Every sentence of L is either an elementary statement of L or is a complex sentence constructed out of elementary statements of L in this manner.

And to the semantics we add:

(16b) (i) A *classical valuation* over a model $K = \langle X, f \rangle$ is a function v which assigns to each sentence of L the value T or F, subject to the conditions:
1. if $Mf(X) = rf(X)$ then $v(U(m, r)) = T$
2. if $Mf(X) = r'f(X)$ and $r' \neq r$, then $v(U(m, r)) = F$
3. $v(\sim A) = T$ if and only if $v(A) = F$
4. $v(A \vee B) = T$ if and only if $v(A) = T$ or $v(B) = T$

(ii) The *supervaluation* induced by the model K is the function s such that
1. $s(A) = T$ if and only if $v(A) = T$ for every classical valuation v over K
2. $s(A) = F$ if and only if $v(A) = F$ for every classical valuation v over K
3. $s(A)$ is not defined otherwise.

(iii) The sentence A is *true in the model* K if it is assigned T by the supervaluation induced by K, and is *false in the model* K if it is assigned F by that supervaluation, and otherwise is neither true nor false in K.

It must be noted that clause (16b) (iii) does not contradict clause (6b) (iv). The search for the correct logical system is made simpler by the following useful result:

(17) A sentence A is valid in L (as completed by (16)) if and only if A is assigned T by all classical valuations over all models.

This means that the semantic relation of incompatibility plays *no* role with respect to the validity of sentences; also that classical propositional logic is sound with respect to L.

The relation of incompatibility now plays a role only with respect to the validity of arguments (semantic entailment). All classically valid arguments are still valid, but there are further arguments which are valid due to the relation of incompatibility. We shall here discuss one example of this, which leads to a semantic characterization of incompatibility for *L*.

Consider again the operators *P* and *Q* governed by the Heisenberg exchange relation

(18) $QPx - PQx = i\hbar x$

It is clear that if $Qf(X) = rf(X)$ in the model $K = \langle X, f \rangle$, then $f(X)$ is not an eigenvector of *P*, and vice versa. From this it follows that some classical valuation *v* over *K* will assign *F* to the conjunction

$$U(q, r) \ \& \ U(p, r')$$

where & is defined in terms of $\vee$ and $\sim$ as usual. Hence the supervaluation induced by *K* cannot assign *T* to this conjunction. In other words, this conjunction is not true in any model. From this it follows that it semantically entails any other sentence of *L*. Thus the argument

$U(q, r) \ \& \ U(p, r')$
hence $\sim (U(q, r) \ \& \ U(p, r'))$

is valid in *L*, though the corresponding conditional is not a valid sentence.

It is interesting to note that, conversely, any sentence which semantically entails its own negation is not true in any model. But it will not be false in every model, unless the corresponding conditional is valid.[11]
So the following is a reasonable generalization of the notion of incompatibility to all the sentences of *L*:

(19) *A* and *B* are incompatible with each other if and only $(A \ \& \ B)$ semantically entails $\sim(A \ \& \ B)$, but $(A \ \& \ B) \supset \sim(A \ \& \ B)$ is not a valid sentence of *L*

– where $\supset$ is defined as usual in terms of $\vee$ and $\sim$.

5.3. *The Limitations of Reichenbach's Approach*

As we have seen, the choice of alternatives (IB) and (IIB) can be implemented in various ways (of which Lambert's seems the most direct). However what we have considered so far represents a rather shallow analysis of the semantic structure of the set of elementary statements. And these are surely more interesting in themselves than any complex sentences which may be constructed out of them by means of extraneous linguistic devices. Let me hasten to add that the bare presentation of Reichenbach's and Lambert's work, reformulated as pertaining to our language L, cannot hope to do justice to the philosophical discussions which accompanied their formalization. But it is only to be expected that the choice of alternative (IIA) will lead to a closer, deeper, analysis of the semantic relations among elementary statements.

6. Von Neumann's Approach to Quantum Logic

John von Neumann chose alternatives (IA) and (IIA) in the earliest work in quantum logic in the section called 'Projections as Propositions' in [37].[12] This initiative was pursued in different ways by Birkhoff and von Neumann [7] and Strauss [27], as well as some later writers. We shall discuss these in turn, but first we wish to explain the basic idea of 'Projections as Propositions'.

Let M be a Hermitean operator with the eigenvalues $r_1, \ldots, r_k, \ldots$.[13] Then there is a special set of Hermitean operators $P_1, \ldots, P_k, \ldots$ each of which has only the eigenvalues 0 and 1, and $Mx = r_i x$ if and only if $P_i x = x$. These form a disjoint family of *projective operators* in the sense that

(20)(a) $P_i P_i x = P_i x$

(b) $P_i P_j x = 0$ when $i \neq j$

(c) $x = P_1 x + P_2 x + \cdots + P_k x + \cdots$

From this it follows that the elementary statements $U(m, r_i)$ and $U(p_i, 1)$ semantically entail each other (given only the conditions on L given by (6)).[14] When alternative (IA) (bivalence) is adopted, that means that $U(m, r_i)$ and $U(p_i, 1)$ are semantically entirely equivalent. Thus we can *without loss* regard every elementary statement as concerning the value of a quantity corresponding to a projective operator.

One point is in order to avoid confusion in the discussion of negation below. The statements

$$Mx = r_i x \quad \text{and} \quad P_i x = x$$

are equivalent (and hence so are $Mx \neq r_i x$ and $P_i x \neq x$). But the statements

$$Mx = r'x \quad \text{for some} \quad r' \neq r_i$$
$$P_i x = r'x \quad \text{for some} \quad r' \neq 1 \quad (\text{i.e. } P_i x = 0)$$

are not equivalent.

There are several ways to give some intuitive content to this formal possibility. First, we may see the transition from the Hermitean operator M to the projective (i.e. idempotent Hermitean) operators P_i as a transition from

quantity m has the value r in system X

to

system X has the *property* that the quantity m has the value r in it. (this property being a measurable quantity with possible values 1 and 0)

A second point concerns the calculation of probabilities. Since $PPx = Px$, it follows that Px is always an eigenvector of P corresponding to the eigenvalue 1, or the zero vector. Thus the expansion (20c) can be given in the form

(21) $$x = c_1 x_1 + c_2 x_2 + \cdots + c_k x_k \ldots$$

where $P_i x = c_i x_i$, and x_i is a unit eigenvector of M corresponding to the value r_i. A basic postulate of quantum mechanics is that (after normalization) the *probability* that a measurement for m upon a system with state vector x will (would) yield the value r_i, equals $c_i^* c_i$. It is easily shown that this probability equals 1 (resp. 0) if and only if $c_i^* c_i = c_i = 1$ (resp. 0). Remembering that $P_i x = c_i x_i$, we see that $U(p_i, 1)$ – respectively $U(p_i, 0)$ – may be read as: the probability that a measurement for m will (would) yield the value r_i equals 1 – respectively 0.[15] (We may just note that this makes $U(p_i 0)$ equivalent also to 'the probability that a measurement for m will (would) yield a value other than r_i equals 1' – which does *not* entail that $U(m, r')$ is true for some value $r' \neq r_j$.)

In the section 'Projections as Propositions' von Neumann explicitly considered the sentential connectives of negation, conjunction, and disjunction, suggesting two ways to implement the choice of (IA) and (IIA). These two ways were worked out in some more detail by Strauss and by Birkhoff and von Neumann respectively.

6.1. *Strauss' Use of Projection Operators*

Martin Strauss formulated the elementary statements of his language Q in [27] in such a way that they contain an explicit reference to a system. This means that one can refer to several systems in the same language (though no single sentence of Q refers to more than one system). We shall continue to consider only the simple case in which elementary statements concern the same system.

To emphasize that all the sentences of Q are elementary statements (choice (IIA)), Strauss considers not sentential connectives but what he calls *predicational connections*; that is, he uses the form 'Y is (F and G)' rather than 'Y is F and Y is G', for example. This constitutes no essential difference; in view of the practice of the other writers, we present his work as if he had considered sentential connectives.

But finally, Strauss' elementary statements correspond *only* to those which may be given the form $U(p, 1)$ – where p corresponds to a projective operator P. Now it is easily seen that $U(p, 0)$ is the *choice negation* of $U(p, 1)$, since

$$Px=0x=0$$
$$Px=rx \quad \text{for some} \quad r\neq 1$$

are equivalent for the projective operator P. Strauss points out (following von Neumann) that the set of elementary statement of form $U(p, 1)$ is still closed under choice negation. For the operator P defined by

(22) $\bar{P}x=x-Px \quad (\text{or}: \bar{P}=I-P)$

is again a projective operator, and is such that

(23) $\bar{P}x=x$ if and only if $Px=0$
$\bar{P}x=0$ if and only if $Px=x$

Hence the statement $U(\bar{p}, 1)$ is equivalent to $U(p, 0)$ and is the choice negation of $U(p, 1)$. Henceforth writing '$U(p)$' for '$U(p, 1)$', we see that every elementary statement can be given the form '$U(p)$', and:

(24) Every elementary statement $U(p)$ has a *choice negation* $\neg U(p)$ namely the elementary statement $U(\bar{p})$.

Next we may consider conjunction. The conjunction of $U(p)$ and $U(q)$ must be true if and only if both of these are true. So we must find projective operator R such that $Rx=x$ if and only if $Px=Qx=x$. Here the situation is that the calculus of operators defines such an operator R for the operators P and Q, namely their product PQ, but only if they commute. So the set of elementary statements is not closed under conjunction on this approach, though conjunction is defined for an important subset:

(25) The pair of elementary statements $U(p)$ and $U(q)$ has a *conjunction* $(U(p) \& U(q))$, namely $U(r)$ where r is the quantity corresponding to the operator PQ, if and only if P and Q commute.

Disjunction can be introduced definitionally as $\neg(\neg U(p) \& \neg U(q))$ and is defined only if P and Q commute: similarly, of course, any other classical connective can be introduced by definition. It must be noted that choice negation is the only significant familiar connective under which the set of elementary statements is closed for Strauss' approach.

6.2. *Birkhoff and Von Neumann's Use of Subspaces*

The well-known joint paper of Birkhoff and von Neumann [7] provides a way to adopt alternatives (IA) and (IIA), and yet have the language closed under the usual statement connectives. This is not done by broadening the set of elementary statements. Rather it is based on the observation that the projective operator stand in a one-to-one correspondence to the subspaces (closed linear manifolds) of the Hilbert space, but that the calculus of projections is not isomorphic to the calculus of subspaces.

To make this clear, we have to say more about the geometry of Hilbert space. For any vectors x, y of the Hilbert space, the following are defined:

the vector sum $x+y$, a vector
the inner product (x, y), a scalar

and the vectors x and y are *orthogonal* to each other if and only if $(x, y)=0$. A *linear manifold* is a set of vectors which is closed under addition of vectors and multiplication by scalars (if x and y belong, so does $rx+r'y$). If a linear manifold is topologically closed, then it is a *subspace*; a subspace of a Hilbert space is again a Hilbert space. Each subspace contains the *null-vector* 0; the smallest subspace is the *null-space* 0 which contains only the null-vector. The subspaces of a Hilbert space play a role analogous to the lines and planes through the origin in ordinary Euclidean 3-space. This analogy provides a good intuitive guide to many of the points made below.

If we perform the usual set-theoretic operations of intersection, union, and complementation on subspaces, we find that only the first of these produces new subspaces. But similar to union is *linear union*: the linear union $S \oplus T$ of S and T is the least subspace containing both. Similar to complement is *orthogonal complement*: the orthogonal complement $S^{\perp}$ of S is the set of vectors which are orthogonal to all the elements of S (again a subspace). The subspaces of a Hilbert space ordered by set inclusion, and with the operations of intersection, linear union, and orthogonal complementation form an *orthocomplemented lattice*.

A basic theorem concerning subspaces is that if S is a subspace, then any vector z is equal to a sum $x+y$, where x belongs to S and y to $S^{\perp}$. The *projection on the subspace* S is then defined by

$$P(x+y)=x \quad \text{where} \quad x \in S, \quad y \in S^{\perp}$$

Above we defined a projective operator as an idempotent Hermitean operator. It can now be shown that every projection on a subspace is a projective operator and vice versa. Specifically:

(26) P is the projection on S if and only if P is the projective operator such that $S=\{x:Px=x\}$.

The operator $\bar{P}$ defined by $\bar{P}x=x-Px$ is clearly the projection on $S^{\perp}$ in this case; equivalently, $S^{\perp}=\{x: Px=0\}$.

In the preceding section we remarked that (given alternative (IA)) every elementary statement $U(m, r)$ may be given the form $U(p)=U(p, 1)$ with p corresponding to a certain projective operator P. Thus the Birkhoff-von Neumann completion of L can be given by adding to the syntax and semantics:

(27)(a) Every sentence is an elementary statement.

(b) $U(p)$ is true in $K=\langle X, f\rangle$ if and only if $f(X)$ belongs to $h(U(p))=\{x: \mathrm{P}x=x\}$, and false otherwise.

Note that (27b) does not contradict (6b) (iv) but only adds the principle of bivalence. We now have the result that every sentence U of L corresponds to a subspace $h(U)$, the set of vectors which satisfy U, in the sense of Section 2. The connectives will be defined in terms of this mapping h.[16]

The complement $\bar{P}$ of a projective operator P corresponds to the orthogonal complement $S^{\perp}$ of a subspace S, as we remarked above, so we have choice negation as before:

(28) Every elementary statement U has a *choice negation* $\neg U$, namely an elementary statement V such that $h(V)=h(U)^{\perp}$.

Considering now conjunction, recall that if P and Q commute, then PQ is again a projective operator with the property:

$$PQx=x \quad \text{if and only if} \quad Px=Qx=x.$$

But that means that PQ is the projection on the *intersection* of the subspaces on which P and Q project. Accordingly, Birkhoff and von Neumann adopt the following generalization:

(29) Every pair of elementary statements U and V has a *conjunction* $(U \& V)$, namely the elementary statement W such that $h(W)=h(U)\cap h(V)$.

The question is now what the conjunction means if P and Q do not commute. We still have: $(U \& V)$ is true if and only if U and V are both true, for all cases. Hence it is entirely correct to call this connective *conjunction.* For a pair of incompatible propositions, recall our discussion of the operators corresponding to the X-coordinates of momentum and position respectively. These have *no* eigenvectors in common. Hence for two projective operators which correspond to specific values of position and momentum, the intersection of the two subspaces is as small as it possibly can be: it is the null-space. Thus the conjunction

electron X has position r and momentum r'

which was meaningless for Bohr and Heisenberg, *always indeterminate-or-false* for Reichenbach, and *not well-formed* for Strauss, is *always false* for Birkhoff and von Neumann. This assimilates it to such other necessarily false propositions as

the table is red and green all over;
The bar is warmer than itself;
The cupola is both round and not round.

This is the obvious move if one accepts bivalence and also wishes the language to be closed under conjunction. For there is no eigenstate of both position and momentum – hence the conjunction in question cannot be true. Bivalence entails then that it is false.

If we introduce disjunction $\vee^*$ again definitionally through de Morgan's law, the principle

$$(S^{\perp} \cap T^{\perp})^{\perp} = S \oplus T$$

for subspaces S and T shows that

$$h(U \vee^* V) = h(U) \oplus h(V).$$

This is like ordinary disjunction in that each disjunct semantically entails the disjunction.

Nevertheless, this is by no means ordinary disjunction, as is shown by the fact that $U \vee^* V$ can be true even if neither U nor V is true ($S \cup T$ is in general a proper subset of $S \oplus T$). Reminiscent of Aristotle's discussion of future contingents is the fact that the law of excluded middle ($U \vee^* \neg U$) holds, although neither U nor $\neg U$ may be true in a given model.

The character of the logic adequate with respect to the language L thus construed is of course determined by the character of the lattice of subspaces. This is in all cases a complete orthocomplemented lattice. If the dimension of the Hilbert space H is 0 or 1, the lattice satisfies the *distributive law:*

$$(30) \qquad A \cap (B \oplus C) = (A \cap B) \oplus (A \cap C)$$

and hence is a Boolean algebra. If H is finite-dimensional but of dimension greater than 1, the lattice is not distributive but still satisfies the weaker *modular law:*

$$(31) \qquad A \oplus (B \cap C) = (A \oplus B) \cap C \quad \text{provided} \quad A \subseteq C.$$

However, in quantum mechanics infinite-dimensional Hilbert spaces are used, and here even the modular law breaks down ([16], p. 22) Birkhoff and von Neumann were quite content to have distributivity fail, but they considered changes in quantum theory to preserve modularity (see especially [6]).

6.3. *Extensions of Birkhoff and Von Neumann's Approach*

The work of Birkhoff and von Neumann certainly influenced every subsequent writer on the subject. Besides the ones we have already discussed there has been important further work by Février [11] and [12], Fuchs [14], Emch and Jauch [10], Varadarayan [36], Mackey [21], Kochen and Specker [18], and Suppes [28] and [29]. Of these, we shall briefly discuss the contributions of Février and Fuchs.

Paulette Février first attempted to develop a quantum logic by means of three-valued matrices; however, the values were *true, contingently false*, and *necessarily false* (or: false by virtue of incompatibility). She soon linked the truth-values with the subspaces of Hilbert space, and Birkhoff regarded her work as a continuation of von Neumann's and his own ([6], p. 157). Most interesting from our point of view is that she eventually attempted to combine alternatives (IIA) and (IIB) (following, she said, a suggestion by Beth – see [11], p. 382). Thus one would have one kind of connective which leads only from elementary statements to elementary statements, and another kind of connective which leads to sentences in general not even logically equivalent to elementary statements (such as exclusion negations). Other such combinations might be interesting – for example the choice of (IB) and (IIA).

W. Fuchs' short article [14] is notable mainly for his attempt to introduce a kind of implication, which is not merely the definitional analogue of the material conditional. Using our notation, he introduces his arrow by:

(32) $A \to B$ is the proposition which says that $h(A) \subseteq h(B)$.

He then discusses restrictions on the introduction rules for implication to avoid the theorem $A \to (B \to A)$. Such iteration of arrows is well-formed only if the arrow is a sentential connective of L (and is not simply a symbol for the metalinguistic relation of semantic entailment). But then (32) is not satisfactory; since it does not define $h(A \to B)$, it does not tell us how to understand such iteration. A possible definition of $h(A \to B)$ is given by

(33) $$h(A \to B) = \begin{cases} H & \text{if} \quad h(A) \subseteq h(B) \\ O & \text{otherwise} \end{cases}$$

which means that $A \to B$ is again an elementary statement if A and B are. This would yield an analogue to Lewis S_5 strict implication (and $A \to (B \to A)$ is not valid for it).[17]

7. Arguments concerning the significance of quantum logic

The work discussed above has occasioned a number of reviews and discussions; many of these rather critical. Many of the criticisms I consider to have been cogent with respect to the original formulation. But in this paper we have presented a reconstruction of the subject of quantum logic: from our point of view, each attempt at quantum logic has been an attempt to elucidate and exhibit semantic relations among the elementary statements. While I believe the reconstruction to be entirely faithful to the *intent* of the original work, it nevertheless avoids the difficulties noted by the above-mentioned criticisms – as I shall attempt to show.

On the other hand, it has been suggested at various times that the subject of quantum logic throws a radical new light on the foundations of logic. If our reconstruction is correct, the work in quantum logic falls entirely within the scope of the semantic analysis of logic in general, so that this suggestion can hardly be true. But I would maintain that the semantic analysis of logic is of major importance to the foundations of logic, *and* that the attempts to formulate a logic of quantum mechanics constitute pioneering work in this field.

7.1. *Quantum Logic and Quantum Theory*

We may begin with a look at McKinsey and Suppes' very critical review of P. Février's *La structure des théories physiques* [22]. They understood her aim to be a formalization of quantum theory *based on* a non-standard logic. Implicit in the work of Carnap and many of his contemporaries in philosophy of science is the following picture of a physical theory: it is ideally constructed by adding axioms with empirical content to a formalized system of logic and mathematics, say *Principia Mathematica*. The latter consists of standard logic plus axioms for sets. In that sense one might say that the current picture of a physical theory implied that its (eventual) formalization must be 'based on' standard logic. McKinsey and Suppes understood Février to be challenging this picture of a physical theory.

Perhaps that is what she had in mind; in that case it is indeed fair to say that she "does not appear to appreciate the difficulties that would be involved in the heroic task of developing quantum mechanics in the framework of a non-classical logic" ([22], p. 52), and that the mathematical part of this project "would be somewhat analogous to the writing of *Principia Mathematica*, though vastly more onerous" ([22], p. 54).

But from our point of view a logic of quantum mechanics is simply an attempt to give a systematic account of the semantic relations among the elementary statements of that theory. And these semantic relations are to be *deduced from* the quantum theory – *that* is the sense in which this logic is a quantum logic. It is not meant to be the basis for a formalization of the theory, or for a new, non-standard *Principia*.

On the other hand, this does not mean that the research in quantum logic has no value for foundational research in quantum theory. An exposition of the Birkhoff and von Neumann system is part of Ludwig's *Die Grundlagen der Quantenmechanik* [20]. Elegant recent work in this area by Mackey [21] and Zierler [42, 43], is clearly indebted to (or inspired by) that of Birkhoff and von Neumann. But none of these attempts involve the proposal to develop the theory in the framework of a non-standard mathematics.

The point that the relation between quantum mechanics and quantum logic is that the former provides the semantics for the latter, is also relevant to Feyerabend's critique of Reichenbach.

First, Feyerabend sees as a central assumption in the classical point of view the principle that if m is a physical magnitude, then $U(m, r)$ must be true for *some* value of r at any given time. He correctly points out the semantic feature of the elementary statements of quantum mechanics which is the violation of this assumption – and that this is sufficient to dissolve the anomalies exhibited by Reichenbach ([13], pp. 51–53). But of course, Reichenbach's three-valued logic is intended to reflect exactly this semantic feature. Thus it is hardly *à propos* to say that "Reichenbach is one of those thinkers who are prepared to give up... even classical logic because they cannot adjust themselves to" the inadequacy of the classical point of view ([13], p. 53).

In Section 4 of [13], Feyerabend argues that Reichenbach's system violates his own criteria of adequacy. Reichenbach wrote:

When we wish to incorporate all quantum mechanical statements into three-valued logic, it will be the leading idea to put into the true-false class those statements which we call quantum mechanical laws. ([26], pp. 159–160)

Now Feyerabend argues, and Reichenbach admits, that the law of conservation of energy gets the middle value. Reichenbach writes:

The principle requiring that the sum of kinetic and potential energy be constant connects simultaneous values of momentum and position.... It follows that the principle of conservation of energy is eliminated... from the domain of true statements, without being transformed into a false statement; it is an indeterminate statement. ([26], p. 166)

Feyerabend adds:

The same results if we use the statement in the form in which it appears in quantum mechanics. In this form the statement asserts that the sum of various operators, not all of them commuting, will disappear....

The last argument admits of generalization; every quantum-mechanical statement containing non-commuting operators can only possess the value 'indeterminate'. This implies that *the commutation rules* which are among the basic laws of quantum mechanics *as well as the equations of motion*... will be indeterminate.... ([13], p. 54)

These would indeed be disastrous consequences. But from the point of view of our present reconstruction, the discussion went astray at its very beginning.

First, we have sharply distinguished between the language L of elementary statements, and the general language of quantum theory. In the latter we encounter such statements as:

Postulate I. A system with *n* degrees of freedom is completely specified by the normed state vector $\psi_t(q_1, \ldots, q_n)$....
Postulate II. To every observable corresponds a Hermitean operator....

which are found in many texts and discussions of quantum theory (here quoted from Mandl [23]). The above statements are in fact examples of what philosophers of science have generally called *correspondence rules* (assuming that 'system' and 'observable' have counterparts in the 'observation language'); for us they are part of the semantics of *L*; and for the physicist they are basic principles of the quantum theory.

There is an obvious and excellent reason for distinguishing between these two languages: for it is easy to construct an artificial language satisfying the conditions we have placed on *L*, but no one has as yet shown how to formalize a language adequate to the complete development of quantum theory. Were the latter formalized, we could join the two languages (or perhaps the former would already be part of the latter). However, this should not affect the semantic structure of either. Since the development of quantum theory is within the framework of classical mathematics, one would find that

$$Mx = rx$$

is always true or false, even if

$$U(m, r)$$

is sometimes neither true nor false. (This is perfectly compatible with the fact that the one statement is true if and only if the other is true.) In Reichenbach's formulation, these two statements *cannot* be identified, for this very reason.

As we have construed it, Reichenbach's proposal generates indeterminacies *only* among elementary statements. Both Reichenbach and Feyerabend begin by considering the law of conservation of energy stated in its classical form and construed, one is led to assume, in terms of elementary statements and numerical quantifiers. This is not *à propos*, and Feyerabend very rightly turns to the quantum theoretic formalism, in which the law has the form of an assertion that a certain sum of operators equals zero. Some of these operators do not commute. But the important point is that this statement is *not* a sentence of L, but of the language of the theory of linear operators on Hilbert space. Therefore this statement must be true or false. The correct identification of the operators makes it true. Mutatis mutandis for the exchange relations and the equations of motion.

To sum up, the laws of quantum mechanics concern the semantics of the language of elementary statements; they do not belong to it. (This shows equally why these laws are not indeterminate in Reichenbach's approach, and not necessarily false in the Birkhoff-von Neumann approach.)

7.2. *Quantum Logic and the Foundations of Logic*

Turning now to the significance of quantum logic for the philosophy of logic, we find that Février, and Emch and Jauch, see in its development concrete support for Gonseth's point of view that logic is a 'physique de l'objet quelconque'. But once we accept that it is a logic only of the set of elementary statements for a certain physical theory, we must fully concur in Jordan's judgment on this matter:

Man kann [es] nach Birkhoff-Neumann so ausdrücken, dass man von einer *Quantenlogik* im Gegensatz zu einer *klassischen Logik* spricht. Natürlich ist es Geschmacksache, ob man diese Bezeichnung anerkennen will; jedoch ist sie jedenfalls dann naturgemäss, wenn man unter 'Logik' die Gesetze der möglichen Verknüpfungen von Aussagen... über den Zustand eines physikalischen Systems verstehen will – in *dieser* Auffassungsweise ist auch die Logik eine empirische Wissenschaft.... ([17], p. 368)

Only from such a general point of view does a multiplicity of logics make sense; and then the discovery of a non-standard logical structure does not constitute a conceptual revolution.

Gonseth's characterization of logic is somewhat reminiscent of the views of Quine, according to whom logical principles, no more than physical laws, are immune to revision in the face of new experimental evidence:

Revision even of the logical law of the excluded middle has been proposed as a means of simplifying quantum mechanics; and what difference is there in principle between such a shift and the shift whereby Kepler superseded Ptolemy, or Einstein Newton, or Darwin Aristotle? ([25], p. 43)

But this is an entirely misleading way of stating the point. No law is contradicted by the proposal of a non-standard quantum logic. Rather, what is shown is that we can construct languages for which the familiar laws do not hold; hence, what is shown is that standard logic has a limited domain of application. This should come as no surprise; as purely formal possibilities, these non-standard languages have been investigated by logicians since the second decade of this century. The (pleasant) surprise is rather to see that these ideas concerning non-standard logical systems have a natural application in the set of elementary statements of a physical theory. This is well argued by Putnam:

perhaps this is what is meant when it is said that a three-valued logic does not constitute a real alternative to the standard variety: it exists as a calculus, and perhaps as a non-standard way of using logical words, but there is no *point* to this use.

... three-valued logic and other non-standard logics had first to be shown to exist as consistent formal structures ... The only remaining question is whether one can describe a physical situation in which this use of logical words would have a point.

Such a physical situation (in the microcosm) has indeed been described by Reichenbach. ([24], pp. 76–77)

That is, the question is whether there is a philosophically or scientifically interesting language of which we can say: this non-standard logical system is the logic of that language.

Finally, I should like to emphasize what I find philosophically the most significant feature of these attempts to formulate a 'logic of quantum mechanics'. These very diverse attempts can all be understood, as pursuing different but related aims, from the point of view of a semantic analysis of physical theories. I see in this point of view a powerful rival for the concept of a physical theory as a syntactic system with formation, transformation, and correspondence rules, and its emendations by means of coordinative definitions, reduction sentences, and meaning postulates. This strongly syntactic approach certainly led to a great deal of insight into the structure of scientific theories; perhaps the semantic approach, so fruitful in foundational work in logic and mathematics, will prove equally successful in further philosophical inquiry in science.

University of Toronto

NOTES

* Presented at the Philosophy of Science Association First Biennial Meeting, Pittsburgh, October 1968. The research for this paper was supported by NSF grant GS-1566.

1 Birkhoff and von Neuman [7] generalized the term phase space' to cover both cases; Weyl [40] used 'system space'. The term 'state-space' is adopted from systems theory; cf. [41]. Similar conceptions of the structure of physical theories can be found in Destouches [8], [9], Strauss [27] and in Weyl's interesting article [39].

2 While $U(m, r, t)$ may be read as 'A measurement of m at t will (would) find the value r', this must not be understood as an operationalist identification of meaning. The exact relation between $U(m, r, t)$ and the outcomes of actual experiments is the subject of an auxiliary theory of measurement, which is only caricatured by the notion of a set of 'correspondence rules'; cf. [28].

3 Here we can let V be a set of elementary statements in which case $h(V) = \cap \{h(A) : A \in \in V\}$.

4 In this simplified presentation we only consider pure cases, and not mixtures; this appears to be sufficient for a discussion of quantum logic.

5 Cf. [15], pp. 35–36 or [23], pp. 100–101.

6 We add the word 'sharp' since for each state (hence also those which are not eigenstates) there is a determinate *probability* that the value r will be found upon a measurement for m.

7 The principles of the theory should govern the choice of the state-space and of the operators, and/or limit the set of models; superselection rules may set further limitations

8 The terms 'exclusion negation' and 'choice negation' are taken from Mannoury's discussions of the foundations of mathematics, which played an important role in the development of Intuitionism (cf. [2], pp. 20–22). Similar distinctions go back at least to the Arab logicians of the Middle Ages.

9 For a semantic discussion of this notion of presupposition, see [33], Sections I and II.

10 Concerning incompatibility; also needed are axioms to the effect that if $r \neq r'$ and $U(m, r)$ then not $U(m, r')$.

[11] In terms of [33], $U(m, r)$ and $\sim U(m, r)$ presuppose that the state-vector is an eigenvector of M (hence $\sim$ is choice negation), and a conjunction of incompatible sentences presupposes its own denial. We may note in addition that if we are content with the present, relatively shallow analysis of semantic relations among the elementary statements, L can be reformulated as a radical presuppositional language [34], for which metalogical theorems are very straightforward.
[12] The question of bivalence is not raised explicitly; in the historical context this can be taken to signify its acceptance.
[13] We consider only a simple case: M is assumed to have a discrete spectrum and to be non-degenerate. For the extension to other cases we refer to von Neumann's work.
[14] It should also be noted that the statement 'the value of m lies in the interval I' is also expressed by an elementary statement of the form $U(p, 1)$, where p corresponds to a projective operator P; see [37], p. 252. A presentation of quantum theory which follows this approach closely is that of Temple [31].
[15] This suggests, of course, that the set of elementary statements ought really to be broadened to include *all* probability statements concerning what a measurement for m will (would) find. J. von Neumann sketched such an extension in [38].
[16] This will show more perspicuously how L is here a semi-interpreted language in the sense of [32]. The presentation of the subject most directly along these lines is that of Ludwig [20], pp. 54–59.
[17] Cf. my discussion of the relation between S_5 and semi-interpreted languages in [32], Section 5.

BIBLIOGRAPHY

[1] Beth, E. W., 'Analyse Sémantique des Théories Physiques', *Synthese* **7** (1948–49) 206–207.
[2] Beth, E. W., *Mathematical Thought*, D. Reidel Publ. Co., Dordrecht-Holland, 1966.
[3] Beth, E. W., *Natuurphilosophie*, Noorduijn, Gorinchem, 1948.
[4] Beth, E. W., 'Semantics of Physical Theories', *Synthese* **12** (1960), 172–175.
[5] Beth, E. W., 'Towards an Up-to-date Philosophy of the Natural Sciences', *Methodos* **1** (1949), 178–185.
[6] Birkhoff, G., 'Lattices in Applied Mathematics', *American Mathematical Society, Proc. of Symposia in Pure Mathematics* **2** (1961), 155–184.
[7] Birkhoff, G. and von Neumann, J., 'The Logic of Quantum Mechanics', *Annals of Mathematics* **37** (1936), 823–843.
[8] Destouches, J-L., 'Les Principes de la Mécanique Générale', *Actualités Scientifiques et Industrielles* **140** (1934).
[9] Destouches, J.-L., 'Le Role des Espaces Abstraits en Physique Nouvelle', *Actualités Scientifiques et Industrielles* **223** (1935).
[10] Emch, G. and Jauch, J. M., 'Structures logiques et mathematiques en physique quantique', *Dialectica* **19** (1965), 259–279.
[11] Février, P., 'Logical Structure of Physical Theories' in L. Henkin *et al.* (eds.), *The Axiomatic Method*, North-Holland Publ. Co., Amsterdam, Holland, 1959, pp. 376–389.
[12] Février, P., 'Les relations d'incertitude de Heisenberg et la logique', *Comptes Rendus de l Académie des Sciences* **204** (1937), 481–483.
[13] Feyerabend, P., 'Reichenbach's Interpretation of Quantum Mechanics', *Philosophical Studies* **9** (1958), 49–59.

[14] Fuchs, W. R., 'Ansatze zu einer Quantenlogik', *Theoria* **30** (1964), 137–140.
[15] Green, N. S., *Matrix Mechanics*, Noordhoff, Groningen, 1965.
[16] Halmos, P. R., *Introduction to Hilbert Space*, Chelsea, New York, 1957.
[17] Jordan, P., 'Quantenlogik und das Kommutative Gesetz', in L. Henkin *et al.* (eds.), *The Axiomatic Method*, North-Holland Publ. Co., Amsterdam, Holland, 1959, pp. 365–375.
[18] Kochen, S. and Specker, E. P., 'Logical Structures Arising in Quantum Theory. in J. W. Addison *et al.* (eds.), *The Theory of Models*, North-Holland Publ. Co., Amsterdam, Holland 1965, pp. 177–189.
[19] Lambert, K., 'Logic and Microphysics', in K. Lambert (ed.), *The Logical Way of Doing Things*, Yale University Press, New Haven, 1969, pp. 93–117.
[20] Ludwig, G., *Die Grundlagen der Quantenmechanik*, Springer-Verlag, Berlin, 1954.
[21] Mackey, G. W., *The Mathematical Foundations of Quantum Mechanics*, Benjamin, New York, 1963.
[22] McKinsey, J. C. C. and Suppes, P., *Review* of P. Destouches-Février, 'La Structure des Theories Physiques', *Journal of Symbolic Logic* **19** (1954), 52–55.
[23] Mandl, F., *Quantum Mechanics*, Academic Press, New York, 1957.
[24] Putman, H., 'Three-Valued Logic', *Philosophical Studies* **8** (1957), 73–80.
[25] Quine, W. V. O., *From a Logical Point of View*, Harper and Row, New York, 1963.
[26] Reichenbach, H., *Philosophic Foundations of Quantum Mechanics*, University of California Press, Berkeley, 1944.
[27] Strauss, M., 'Mathematics as Logical Syntax – a Method to Formalize the Language of a Physical Theory', *Erkenntnis* **7** (1937–38), 147–153.
[28] Suppes, P., 'Logics Appropriate to Empirical Theories', in J. W. Addison *et al.* (eds.), *The Theory of Models*, North-Holland Publ. Co., Amsterdam, Holland, 1965, pp. 364–375.
[28] Suppes, P., 'The Probabilistic Argument for a Non-Classical Logic in Quantum Mechanics', *Philosophy of Science* **33** (1966), 14–21.
[30] Suppes, M., 'Probability Concepts in Quantum Mechanics', *Philosophy of Science* **28** (1961), 378–389.
[31] Temple, C., *The General Principles of Quantum Theory*, Methuen, London, 1961.
[32] van Fraassen, B. C., 'Meaning Relations Among Predicates', *Nous* **1** (1967), 161–179.
[33] van Fraassen, B. C., 'Presupposition, Implication, and Self-Reference', *Journal of Philosophy* **65** (1968), 136–152.
[34] van Fraassen B. C., 'Presuppositions, Supervaluations, and Free Logic', in K. Lambert (ed.), *The Logical Way of Doing Things*, Yale University Press, New Haven 1969, pp. 67–91.
[35] van Fraassen, B. C., 'Singular Terms, Truth-Value Gaps, and Free Logic', *Journal of Philosophy* **63** (1966), 481–495.
[36] Varadarajan, V. S., 'Probability in Physics and a Theorem on Simultaneous Observability', *Comm. Pure Appl. Math.* **15** (1962), 189–217.
[37] von Neumann, J., *Mathematical Foundations of Quantum Mechanics* (tr. by R. T. Beyer), Princeton University Press, Princeton, 1955.
[38] von Neumann, J., 'Quantum Logics', unpublished; reviewed by A. H. Taub in *John von Neumann, Collected Works* (vol. **4**, pp. 195–197), Macmillan, New York, 1962.
[39] Weyl, H., 'The Ghost of Modality', in M. Farber (ed.), *Philosophical Essays in Memory of Edmund Husserl*, Harvard University Press, Cambridge, Mass., 1940, pp. 278–303.

[40] Weyl, H., *The Theory of Groups and Quantum Mechanics*, Dover, New York, 1950.
[41] Zadeh, L. A. and Desoer, C. A., *Linear System Theory: The State Space Approach*, McGraw-Hill, New York, 1963.
[42] Zierler, N., 'Axioms for Non-Relativistic Quantum Mechanics', *Pacific Jounal of Mathematics* **11** (1961), 1151–1169.
[43] Zierler, N. and Schlesinger, M., 'Boolean Embeddings of Orthomodular Sets and Quantum Logic', *Duke Math. Journal* **32** (1965), 251–262.

EDWARD MACKINNON

ONTIC COMMITMENTS OF QUANTUM MECHANICS

I. SURVEY OF METATHEORIES

In this paper, we are trying to do two things: to present a general theory about the interpretation of scientific theories, and to use quantum mechanics as a test case for judging the validity of this metatheory. By a metatheory of scientific theories, I mean a general theory concerning the structure, functioning, and interpretation of scientific theories. The construction of such metatheories has been one of the primary concerns of philosophers of science in the twentieth century. A brief historical sketch may serve to show some of the limitations in the results so far obtained. Here, a word of caution is in order. Since this historical sketch is simply intended as a background to make a new interpretation intelligible, we will present a schematic view which necessarily oversimplifies the positions considered.

Pierre Duhem, writing at the turn of the century, distinguished two aspects of scientific theories: a representative part, the formal structure, and an explanatory part, the conceptualization of reality implicit in the theory. A scientific theory is explanatory, he felt, inasmuch as it goes beyond the phenomena and gets down to the basic reality underlying the phenomena; i.e., inasmuch as it becomes metaphysical. His historical investigations convinced him that different, even contradictory, metaphysical views could be read into the same theory. From this, he concluded that the success of a physical theory is independent of its explanatory part. This, in turn, implies that the explanatory part of a theory, its conceptualization of reality, is dispensible and, in fact, is really not a part of physics at all. As he summarized his views: "A physical theory is not an explanation. It is a system of mathematical propositions, deduced from a small number of principles, which aim to represent as simply, as completely, and as exactly as possible a set of experimental laws."[1]

Subsequent developments in the twentieth century, particularly the rise of logical atomism and logical positivism, tended to re-enforce this stress on the logical structure of scientific theories while deliberately disregarding the particular conceptualization of reality implicit in the theories treated. The types of interpretations resulting from such a logic-centered approach to scientific systems can be outlined in a rather schematic way.

In a formal approach, a scientific system can be thought of as having three components: a theoretical language, an observational language, and correspondence rules relating the two. The theoretical component may be visualized as running from axioms, at the top of the page through theorems, or general conclusions, in the middle of the page to the particular conclusions at the bottom of the page. Considered as a purely formal theory, it has no physical significance until it is given an interpretation by means of the correspondence rules.

In traditional logical positivism, as well as in the current views of Carnap[2] and Hempel[3], such correspondence rules are essentially a bottom of the page affair. From an *explanans* made up of general laws and specific antecedent conditions, one deduces the *explananandum*, which is a statement of the phenomena. The theory becomes interpreted when such *explananda* are correlated with observation reports. Originally, it was hoped that such observation reports could be expressed in protocol sentences which would be linguistic transcriptions of the given of sensation. When this proved inadequate, the attempt was made to develop an empiricist language which would be essentially theory-free.

In recent years, this type of observation-centered, or bottom of the page, interpretation of scientific systems has been the target of so many attacks that its shortcomings are rather notorious. The idea of a theory-free observation language embodies a myth, the myth of the given, to use Sellars' apt phrase. Such a view effectively denies scientific theories any explanatory role. A theory is essentially an inference mechanism leading from one set of observation statements to others. Also, such an interpretation makes it logically impossible to attach any significant notion of physical reality to theoretical entities; i.e., to entities named in the theoretical language but not in the observation languagc. They can be interpreted only as fertile theoretical constructs.

A step beyond this is a 'middle of the page' interpretation of scientific theories developed in different ways by Campbell[4], Braithwaite[5], Nagel[6], and many operationalists[7]. Here, we shall focus on Nagel's views. In comparing theories and empirical generalizations, he notes that such empirical generalizations as Kepler's laws, Boyle's law, the laws for spectral lines, etc. can be formulated in one statement, can be established in a quasi-inductive way, can be verified or falsified in a rather direct and unambiguous fashion, can precede the theories devised to explain them, and can survive their demise. A theory, such as the kinetic theory of gases or the Bohr theory of the atom, admits of no simple inductive justification, has a more complex logical structure, and is confirmed or refuted by experiments only in a partial and rather indirect way.

What this contrast leads to is a three-level view of scientific explanation. The three levels are: axioms, or non-explained explainers, empirical generalizations, or explained explainers, and observation reports, or explained non-explainers. From the axioms of a theory, one deduces general laws which have the same form as inductively established empirical generalizations. Correspondence rules are essentially a middle of the page affair relating these general principles deduced from the axioms to empirical generalizations from which one deduces observation reports. An example of such a correspondence rule is the relation between the mean kinetic energy of molecules, something defined within kinetic theory, and the temperature of a gas, a parameter that plays an essential role in the gas laws.

Wilfred Sellars, whose position in epistemology has had a strong influence on the present paper, has submitted this 'levels' interpretation of scientific theories to a detailed criticism at a previous meeting of the Boston Colloquium and elsewhere.[8] The basic criticism is that, while this mode of interpretation is correct in discerning functionally different levels, it is fundamentally misleading in the significance it attaches to the top level. Theories do not explain empirical laws by deductive subsumption. Rather, theories postulate unobservable entities and then explain why these entities, behaving in the manner specified by the theory, obey the laws they do.

Sellars' criticisms are indicative of a new trend, a top of the page interpretation of scientific theories developed in different ways by Sellars, Quine[9], Feyerabend[10], Putnam[11], and others. Our outline will simply be a rough composite picture supported by some strategy considerations.

The key idea is that an acceptance of theories as explanatory and irreducible entails an acceptance of the entities postulated by the theories.

This is a return to realism but not to a simple or naive realism. Its development represents a fusion of elements drawn from logical positivism, British analysis, and American pragmatism. Knowledge, or at least the public expression of knowledge may be thought of as an interrelated set of linguistic or conceptual systems, such as ordinary language, different scientific theories, etc. Each system has some sort of logical structure. For ordinary language, this is an informal structure clarified by linguistic analysis. Scientific theories can – or so it is generally believed – be reformulated along the lines of a formal axiomatic system; i.e., in terms of an object language supplemented by suitable metalanguages. Each linguistic system involves ontic commitments; i.e., commitments to the existence of the entities one speaks about within the system. For ordinary language, this involves the type of commitment to things and persons as basic categories that Strawson has clarified, a subject we shall return to a bit later. For formal languages, this means a commitment to the entities whose names serve as values of variables in the existentially quantified axioms of the theory. To be is to be the value of a variable.

Rudolf Carnap, who first formulated some of these ideas,[12] sought to avoid the conclusion that science supported philosophical realism by introducing some distinctions that have since become normative. Thus, the question, 'Are there molecules?', considered as an *internal* question within the language of molecular physics must be answered affirmatively. But, as Carnap saw it, such an internal statement is purely analytic, merely an explicitation of what is implicit in the axioms. As an analytic, rather than a synthetic, statement, it carries no commitment to the real existence of molecules. If the same question is re-interpreted as an *external* question – i.e., as a question about, rather than within, the system – then, in Carnap's opinion, the question is incorrectly phrased. The correct external question is: 'Should this system be accepted rather than its competitors?' Such a question calls for a practical decision rather than a commitment on an ontological issue.

It was here that considerations drawn from American pragmatism intervened, leading to a doctrine of philosophical realism. Accepting a system, such as molecular physics, as explanatory, is a pragmatic decision. But it should also be a reasonable decision. If one has good reasons for accepting the system, then one has equally good reasons for accepting the entities postulated by the system. What one has, accordingly, is a theory-centered rather than an observation-centered interpretation of scientific systems. Correspondence rules serve, not to give the theory a physical interpretation, but to relate observation statements, formulated in ordinary language, to the theory that explains these observations at a more basic level. The ontic commitments accepted as basic are those proper to the theory accepted as explanatory rather than those proper to the observation language. If the theory accepted as the most fundamental physical theory exhibits a logic different from classical logic, then this new logic should be accepted as the basic logic of reality, a point Putnam has been stressing.

As promised, this survey was very sketchy but a more detailed exposition may be found elsewhere.[13] My own evaluation is that these recent theory-centered explanations of scientific systems are generally correct in what they are attempting to do but are seriously misleading in the way they are developed. The principal reason for this inadequacy is, in my opinion, the almost exclusive concentration on formal logic and on systems which are or can be constructed in accord with the rules of logic rather than an examination of functioning scientific systems. The implications of this should be considered in more detail.

Philosophers are rather reluctant to work with functioning scientific systems because of their complexity and their tendency towards obsolescence and rapid replacement. If, however, one believes that any scientific system can, at least in principle, be rationally reconstructed along logical lines in such a way as to leave no explanatory residue, then one can ignore particular theories and concentrate on a structural description of the logical form proper to any scientific theory, much the way the linguist does in developing a structural linguistics which prescinds from the differences between English, French, etc. This leads to a view of scientific systems as formal axiomatic systems developed either in accord with the classical logic, as Quine would argue[14], or in accord with quantum logic, as Putnam would argue.[15]

In a formal axiomatic system, the basic terms are implicitly defined by the axioms interrelating these terms. This gives the terms an internal syntactical meaning, something distinct from a physical interpretation. In the type of theory centered interpretation we are considering, these terms are not given physical significance by means of correspondence rules, for correspondence rules serve to relate observations to the theory that explains them, rather than to interpret theories in terms of an observation language. Any change in an axiom or rule, accordingly, necessarily changes the meaning of every basic term in a theory. This approach, accordingly, allows for no core of content or meaning common to competing or successive theories, a point Feyerabend has stressed.[16] This conclusion seems inescapable, granted the principles, but it also seems to be patently false when applied to scientific theories. A statement such as 'The normal helium atom has two electrons' has essentially the same meaning in the Bohr, Heisenberg, Schrödinger, or Dirac theory of the atom. These theories clearly have a common substantive core epitomized in the familiar periodic table of the elements and the wealth of information it contains.

Another significant shortcoming in this metatheory is that it does not afford a basis for explaining the role of conceptual revolutions in scientific development. The conceptualization of reality implicit in a scientific theory is systematically ignored except for the features, such as ontic commitments, which serve as presuppositions in the interpretation of a logical structure. Recent studies by Hanson,[17] Kuhn,[18] Jaki,[19] Toulmin,[20] and other historians of science have made it clear that the key factor in scientific advancement, through theory replacement, is a change in the conceptualization of reality proper to a theory. Some philosophers, particularly Wartofsky and Agassi,[21] have attempted to relate the role that such conceptualizations play to formal logical structures. Yet, any explanation of scientific theories, developed at the level of a structural description of the logical forms common to all theories, necessarily prescinds from the conceptualization or reality proper to any particular theory.

II. TOWARDS A THEORY OF SCIENTIFIC THEORIES

The preceding survey was intended to specify the problems which a metatheory of scientific theories must meet. As I see it, such a metatheory must preserve the advances made by recent theory-centered interpretations of science in clarifying: the truly explanatory rather than the inference-mechanism role of logical structures; the functional realism implicit in the use of scientific theories, particularly the commitment to the real existence of the theoretical entities the theory discusses; and the clarification of the epistemological problems involved in interpreting scientific theories. But an acceptable metatheory should also explain the role that a conceptualization of reality plays in the development and interpretation of a theory and how differing theories can contain a common conceptual core. In what follows, I'll try to develop a rough first approximation to such a metatheory and then use quantum mechanics as a test case in judging its validity.

We can begin with the idea of a scientific system containing two components linked by correspondence rules. I believe that this idea is essentially correct but that each component, as well as the rules interrelating them, has been misinterpreted. Accordingly, rather than speaking of these components as an 'observation language' and a 'theoretical language', I will speak of them as a 'physical language' and a 'mathematical language', a terminology adapted from Dirac. By a 'physical language', I mean essentially a transformed extension of ordinary language, a transformation that requires some explanation.

The basic ideas on the role of language, which I will simply summarize rather than justify, stem from the later work of Wittgenstein as well as from the work of Strawson[22] and Sellars. The meaningfulness of language is essentially public and only derivatively private. As a public symbolic vehicle, language can be used to refer, describe, narrate, and explain only if it implicitly contains a conceptualization of reality. This language-dependent conceptualization is something essentially different from a visualizable model though many philosophers have tended to blur Kant's clear distinction between conceptual and perceptual representings.

The view of reality or descriptive metaphysics, implicit in ordinary language – at least, in ordinary Indo-European languages – may be summarized as follows. The world is an interrelated collection of bodies which are extended in space and perdure through time. These objects are endowed with both primary (e.g., extension, shape) and secondary (e.g., color, taste) qualities which are ontologically real properties of bodies. The categorization of such properties has a hierarchical ordering reflected in Aristotle's ordered table of categories: substance, quantity, quality, etc. Thus, 'whatever is colored is extended' is an *a priori* proposition of conceptual necessity while its converse is not because secondary qualities (or proper sensibles) are conceived or as adhering in substances through the mediation of primary qualities or common sensibles. Things, or substances, are of different ontological types with persons as a unique class of which both mental corporeal attributes may be predicated. Among persons, I – to speak like a phenomenologist – have a unique role. My bodily location anchors the space-time framework for me, making the world 'my world'.

This, of course, is the familiar common sense view of reality. Labelling it 'descriptive metaphysics' emphasizes the point that it is *not* intended as a theory of reality as it exists independently of our knowing but as an explicitation of what our language implicitly says reality is. Any changing conceptualization of reality must be embedded in and transmitted through ordinary language or its specialized extensions. A consideration of such extensions involves two interrelated questions: how does a language dependent conceptualization of reality relate to a mathematical formalism, and how can this conjunction serve as the vehicle for the development and interpretation of new theories? In attempting to answer these questions, I am adapting some ideas that have been developed in more detail elsewhere.[23]

To explain the relation between a conceptualization of reality and a mathematical formalism, it is necessary to make explicit and, to some degree, to regiment some formal aspects in the descriptive metaphysics of ordinary language. We will call any predicates designating properties of physical objects 'first order predicates'. Thus, 'white', in 'This paper is white', or even a two-place predicate (or a relation) such as 'taller than', would count as a first order predicate. A second order predicate is one designating a property of a property. Thus, 'invariant' in 'The quantity of a liquid is invariant' is in respect to a change of shape. Usually, it is more convenient to use a material language (in Carnap's sense) which speaks of properties and properties of properties rather than a formal language which speaks of predicates and predicates of predicates. But we consider the two ways of speaking equivalent. What we are doing here is analysis and not metaphysics. That is, we are not attempting to discuss what bodies are in themselves but to make explicit what our language implicitly says bodies are.

The first order properties of bodies can by *symmetric* (if John is the *brother of* James, then James is the brother of John), or *anti-symmetric* (if John is the *father of* James, then James is not the father of John); *transitive* (if *A* is *taller than B* and *B* is *taller than C*, then *A* is *taller than C*), or intransitive, etc. By introducing a slight regimentation which makes formal definitions of terms normative for their use, the requisite second order properties may be defined more explicitly. In these definitions, we presuppose a domain of objects, A, with properties and relations specified by first order predicates. Then, the higher order predicates which are our immediate concern can be defined as follows (where 'iff' means 'if and only if'):

Assymetry: A relation, R, is assymetric iff

$$(x)\,(y)\,(x \in A\ \&\ y \in A\ \&\ xRy\ \&\ yRx)$$

Transitivity: A relation, R, is transitive iff

$$(x)\,(y)\,(z)\,((x \in A\ \&\ y \in A\ \&\ z \in A\ \&\ xRy\ \&\ yRz) \supset Rxz)$$

Connected: A relation, R, is connected iff

$$(x)\,(y)\,((y \in A\ \&\ y \in A\ \&\ x \neq y) \supset (xRy \vee yRx))$$

A relation, R, which fulfills these three conditions automatically groups the objects, $x, y, z, \ldots$, of the domain, A, into equivalence classes which are isomorphic to the ordinal numbers. Equivalently, one might say that the objects so related represent an interpretation or a model of the ordinal numbers.

Ordinality represents a minimum correspondence between a conceptualization and a mathematical system. To introduce cardinality, two further requirements must be fulfilled. There must be a defined unit for the extensive quantity being measured and there must be a method of applying this unit in a measuring process. Thus, in measuring length, the unit, the meter bar or a replica based on it, is a physical length which is chosen as a matter of convenience. Expressing an arbitrary length in terms of this unit depends on a physical operation; for example, juxtaposing a ruler and the object to be measured and observing coincidences. If it is possible to define a physical unit and a physical process of addition which leaves the extensive magnitude invariant, then one has an isomorphism between the real numbers (plus dimensions) and specifiable properties of physical entities.

Granted such a basic isomorphism between real numbers and an extensive magnitude, one can extend the applicability of mathematics by establishing a more complex correspondence between physical properties and operations and mathematical entities and operations. An example may be more helpful than a formal definition. Consider what is involved in mensuration, in applying arithmetic to lengths, areas, and volumes. The property 'long' may be expressed in terms of an applicable unit measure while the relation 'longer than' is assymetric, transitive, and connected. For lengths, there are two basic joining operations to which all others may be reduced. One may join lengths in a straight line to form another length or may join them perpendicularly to form an area. Using '$+$' as a symbol of linear joining, '$\times$' as a symbol of rectilinear joining, and '$\circ$' to cover either operation, the joining operations have the properties of:

(1) *Associativity:* $(a \circ b) \circ c = a \circ (b \circ c)$
(2) *Commutativity:* $(a \circ b) = (b \circ a)$
(3) *Distributivity*: $(a \times (b + c)) = (a \times b) + (a \times c)$

The simplified notation used in (1)–(3) is intended as a short-hand formulation for statements of the form:

(1a) Given a domain, A, of objects having the properties, a, b, $c, \ldots$, which belong to the property class, L, a physical operation, $+$, which relates these objects with respect to these properties is associative iff
$(a)\,(b)\,(c)\,(a \in L\ \&\ b \in L\ \&\ c \in L\ \&\ ((a+b)+c) = (a+(b+c)))$,
where (a) stands for universal quantification over objects possessing the property a, etc.

Formulas (1)–(3) also admit of a different interpretation. If 'L' stands for the domain of real numbers, 'a', 'b', 'c' ..., for individual numbers, '$+$' for addition, 'x' for multiplication, and '$\circ$' for either operation then formulas (1)–(3) specify the behavior of real numbers under the combined operations of addition and multiplication. Because of this isomorphism between properties of mathematical entities, in this case numbers with dimensions, and properties of physical properties, mathematical entities can stand for physical properties.

To clarify the significance of this correspondence, let us return to the schematism of a physical language (PL) and a mathematical language (ML) related by different types of connecting rules. PL is basically ordinary language but with some regimentation of meaning through the insistence that key terms – e.g., the specification of extensive magnitudes – be used only in accord with explicit definitions. The descriptive metaphysics of this language is not a theory of what things are in themselves but an explicitation of the conceptualization of objects, properties, and relations which is used in categorizing, referring, describing, and explaining. This conceptualization involves formal structures which are, or can be adjusted to be, isomorphic to the formal structure of mathematical systems. If we use Roman letters to stand for terms in PL and Greek letters to stand for terms in ML the operative correspondence can be schematized as follows:

CHART A

PL		ML
Properties (s', t', u'...) of physical properties defined with respect to	⇔	Properties 'σ', τ', υ'...) of mathematical entities defined with respect to
Physical Operations (L, M,...) on	⇔	Mathematical Operations (Λ, M,...) on
Properties (a, b, c,...) of Physical bodies	↔	Mathematical entities (α, β, γ,...)

In PL, the order of conceptualization is from the bottom up. That is, speaking of properties presupposes bodies which have these properties. This is the only aspect of being a body that enters into the correspondence. No metaphysics of substances need be presupposed or entailed. Though 'a', 'b', etc. stand for properties such as length, area, etc., they could be unpacked as a series of propositions attributing a property to a body. In referring to mathematical entities in ML, we are not presupposing platonism but simply using symbols α, β, etc., to stand for the type of things mathematicians speak about; e.g., real numbers, complex numbers, vectors.

If mathematical entities are to stand for properties of physical objects, two types of conditions must be met. First, there are the *theory-dependent conditions of correspondence*. Here, the basic condition is that the properties which physical properties have with respect to specifiable physical operations must be isomorphic to the properties which mathematical entities have with respect to corresponding mathematical operations. The double arrows on Chart A signify this isomorphism between second order physical properties and first order mathematical properties and between physical operations and mathematical operations. In the extension of science to new domains, this type of higher order isomorphism plays a key role, a point that has been developed in different ways by Sellars[24] and by Wigner[25].

There are also *context-dependent conditions of correspondence*. The most basic one is the requirement that the aspects of physical reality not included in the formalism do not play a determinative role in the physical situation. Thus, numbers are invariant and can correspond to lengths only if the physical operations involved leave lengths invariant. If the bodies in question were rubber bands subject to varying tensions, the correspondence would not hold.

If these conditions are fulfilled, then theory-dependent physical implication can be represented by an adaption of Körner's formula,

$$\text{(K)} \qquad [(a_i \,\&\, c_j) \,\&\, (a_i \leftrightarrow \alpha_i) \,\&\, (\alpha_i \underset{\text{ML}}{\rightarrow} \beta_k) \,\&\, (\beta_k \leftrightarrow b_k)] \underset{\text{PL}}{\rightarrow} b_k .$$

In (K), we have used single-headed arrows with language indicating subscripts for entailment and double headed arrows for correspondence. To unpack (K), it states that from a conjunction of physical premises, a_i, which specify the properties of bodies, boundary conditions, c_j, a correspondence between the PL premises a_i and a set of ML premises, α_i, one may infer by the rules proper to ML a set of mathematical conclusions, β_k. If there is a set of PL propositions, b_k, corresponding to the ML propositions, β_k, then b_k is a set of theory-dependent physical implications of the initial premises and boundary conditions. Thus, in the simple example given earlier, one could establish a correspondence between the length, height, and width of a box and a set of three real numbers, perform the mathematical operation of multiplying the three numbers together and conclude that the product stands for the volume of the box. As will be seen, (K) is routinely exemplified in the solution of any problem in quantum mechanics.

III. SOME HISTORICAL CONSIDERATIONS

Before adapting this schematism to quantum mechanics, we should consider, at least briefly, whether it fits normal scientific reasoning and can come to grips with the difficulties mentioned at the conclusion of Section I. An initial difficulty stems from the central role assigned to second order propositions in PL or to propositions expressing the behavior of properties under transformations. Can this be considered an explicitation of what is implicit in ordinary language about the properties of bodies and the quantitative reasoning based on this language?

Fortunately, Piaget's psychology, one of the least disputed aspects of it, clarifies what is involved in conceptualizing the properties of bodies. Most of you are probably familiar with his simple experiment of pouring water from a tall, thin glass into a normal glass and then into a short, squat glass, each of which holds exactly the same amount of water. When children of four or five are asked which glass holds more water, they tend to pick out one of the extremes and answer that the tall glass holds more water because it is higher. After an intermediate period of intuitive compensation, the growing child finally comes to the realization that each glass holds the same amount because the volume of a fluid is not dependent on the shape of the container. That is, he has conceptually assimilated the property of quantity only when he has grasped the properties of this property, invariance under transformations in shape. Piaget, Inhelder, and others[26] have extended this in showing that first order properties of bodies are not really understood until one has grasped the properties of these properties under approate transformations.

We have two basic components involved in interpreting scientific theories: physical language embodying a conceptualization of reality, and a mathematical language. Any extension of ordinary language to new domains, even extensions involving conceptual revolutions, must necessarily preserve the framework features and organizing principles that make a consistent conceptualization possible. The principles we will be concerned with are those governing predicates used to characterize properties of physical objects. A historical example may be more helpful than an abstract analysis in clarifying the role of such principles.

Both Descartes and Newton, building on the work of Galileo, accepted inertia as a basic concept in mechanics and both, more or less, understood 'inertia' in the sense now familiar from Newton's first law. The way this concept could be related to a mathematical system, however, depended on the way bodies were conceptualized. For Descartes, extension was *the* basic property of physical objects, with all other properties explained by reduction to extension. Inertia, accordingly, was represented mathematically as the product of the volume of a body and its speed. Descartes' theory of collisions, developed on this basis, proved to be quite inaccurate.[27]

Newton conceived of mass, rather than extension, as the basic property of material bodies, defined 'inertia' or 'momentum' in terms of the product of mass and velocity and, on this basis, worked out a mechanics which could give an adequate treatment of elastic collisions. The new conceptualization of material reality that emerged from the new mechanics had a basic continuity with the descriptive metaphysics of ordinary language in that physical reality was conceived of as bodies with properties rather than in terms of an ontology of events, or fields, or monads, or a Spinozistic all-pervading substance. The conceptual revolution, which presupposed this background continuity, came in with the way bodies were conceptualized as well as in the new concepts of absolute space and time. The properties of bodies considered basic were mass, inertia, and gravitational attraction while secondary qualities, accepted as real in ordinary language, were either disregarded or effectively relocated by reduction to primary qualities. This could be done while preserving a basic conceptual continuity because secondary qualities depend on primary qualities, while primary qualities do not depend on secondary.

A further aspect of this interrelation between a physical conceptualization and a mathematical formalism can be seen in the development of thermodynamics. At the end of the eighteenth century, Farenheit and Celsius developed temperature scales, a basis for applying numbers to degrees of heat. The attempts to develop a science of heat on this simple basis failed because temperature was not a simple additive property under the operation of mixing bodies of differing temperatures. Joseph Black supplied the key concepts needed: a distinction between degrees of temperature and quantity of heat, a distinction between overt and latent heat, and the notion of a specific heat characterizing differing substances. At this time, heat was thought of as a fluid with both chemical and thermal properties. The conceptualization became clearer after Lavoisier rejected the phlogiston theory of combustion and, in his monumental *Traité Élémentaire de Chimie* (1789), listed caloric as a basic element.

Caloric was thought of as an indestructible, uncreatable, weightless, elastic fluid whose particles repel one another (which explains the diffusion of heat) and are attracted by particles of ordinary matter, the magnitude of the attraction being different for different substances and different states of the same substance. This model served as a basis for a mathematical theory of heat which was quite successful. Since the quantity of heat is a measure of the caloric present and caloric is uncreatable and indestructible, heat is a conserved quantity and can be characterized by numbers obeying additive laws.

Though the mathematics was correct in its treatment of heat exchange and related phenomena, the conceptualization of heat as caloric was incorrect, something Count Rumford established. The problem, then, was to develop a new conceptualization of heat which would preserve the valid mathematical laws established on the basis of the caloric theory, while modifying the theory. Mayer, Joule, Kelvin, Maxwell, and others did this by establishing the ideas that heat is a form of energy, that energy is conserved, and that it can exist in latent (or potential) as well as overt forms. The first order predicates those which embody a conceptualization of what heat is, changed. But they changed in such a way as to preserve the second order predicates that ground the correspondence with a mathematical formalism. The applicability of numbers to characterize the quantity of heat in a system requires a conservation law for heat and this, in turn, requires the notions of specific heats and latent heats.

This developmental process combining conceptual revolution with an underlying continuity in mathematical theories could be traced in the evolution of other theories such as those of electricity and gravitation. In both cases, the conceptualization of the phenomena in question, characterized by a set of first order predicates changed from electricity as a fluid to an electro-magnetic field, from gravitation as interaction at a distance to gravitation as a warping of space. In both cases, the mathematical formalisms had a basic continuity in the sense that the new laws reduced to the old laws in the domain in which the old laws were valid. This mathematical continuity underlying a conceptual revolution is grounded in the fact that both the new conceptualization and the old share a set of second order predicates, predicates whose primary function is to express laws of conservation and invariance under certain transformations or, as Wigner characterized them, laws about the laws of nature.[28]

IV. LOGICAL ANALYSIS OF DIRAC'S QUANTUM MECHANICS

The problem of interpreting quantum mechanics can be approached from various directions, depending on the questions one is asking. Here, we are concerned with the nature of scientific explanation and, in particular, with the role of theoretical entities in such explanations. That is, we are concerned with the type of questions raised by the philosophers considered in the introductory survey rather than with the type of questions philosophically oriented physicists often ask. However, we will not follow the general trend these philosophers set in their method of attempting to answer such questions by beginning with logical structures and rules of interpretation thought to be proper to a rational reconstruction of scientific theories. Instead, we will begin by trying to make explicit the logical structure and rules of interpretation already operative in quantum mechanics as a flourishing scientific discipline. Later, this can be compared, at least briefly, with projected reconstructions.

Dirac's *The Principles of Quantum Mechanics*[29] was a synthesis and unification of the developments that preceded it and became the accepted basis for further developments. It may be, as Dirac now believes,[30] that the general formalism he introduced is inadequate for quantum electrodynamics and quantum field theory. But there is no doubt about its adequacy for non-relativistic quantum mechanics, the subject we are considering. In developing this synthesis, especially in the successive revisions it has undergone, he was very much concerned with logical rigor and precision. The logical structures, however, were those proper to theoretical physics rather than formal logic. To make this logical structure explicit, it was necessary to impose a few labels and classifications either not found or not emphasized in the text. But it was not necessary to change anything. Dirac's book represents one of the most rigorously logical developments in the history of scientific thought. An outline of this is contained in the appendix. Here, I'll try to clarify and interpret the principal features. Then, it will be compared with other formulations of quantum mechanics.

As the opening quote in the appendix indicates, Dirac developed quantum mechanics in terms of a physical system, a mathematical system, and connecting rules relating the two. Rather than follow the order of the outline, which is the order of the book, it might be clearer if we consider each system separately. The physical system is based on established facts and reasonable assumptions. The pertinent facts are of two kinds. First, Dirac accepts the existence of definite physical systems; e.g., atoms and particles as something already established. Secondly, he presupposes some knowledge concerning the states of these systems and their characteristic properties. In particular, he assumes that the state of an atomic system can not be completely specified in classical terms. The state of a system must be considered as a superposition of two or more different states in a way that can not be conceived in accord with the ideas of classical physics.

An example may clarify the crucial significance of this superposition principle. If plane polarized photons are passed through a tourmaline crystal photons with a polarization parallel to the crystal axis will pass through while those whose polarization is perpendicular to the crystal will not. If the photons are obliquely polarized, a certain percentage of them will get through where the percentage is precisely specified by the angle, α, involved. The peculiar non-classical significance of this phenomenon becomes clear when one analyzes what happens to a single obliquely polarized photon. It either gets through or it does not. If it does get through, its polarization is parallel to the tourmaline axis. Accordingly, its initial state of oblique polarization may be considered as a superposition of two states, one parallel to the tourmaline axis and one perpendicular, with the numerical factors, $\sin^2\alpha$ and $\cos^2\alpha$. The process of measurement selects one of the superimposed states, while a series of measurements will effectively reproduce the numerical value of α.

A macroscopic analogue illustrates the incompatibility of this principle with the conceptualization of reality proper to classical mechanics. Imagine a man weighing 175 pounds being weighed on scales that only record weights of either 150 or 200 pounds. If the superposition principle applied, the 175 pound man could be considered a superposition of two men, one weighing 150 and one weighing 200 pounds. A given weighing would select one of the two superimposed men. A sufficient number of such weighings would average 175. The difficulty is not just that weighing, or the measurement of a classical property, *does* not produce such results but that it *cannot* without involving a contradiction in the conceptualization involved. Yet, quantum mechanics requires such a principle as a basis for conceptualizing states of systems.

Dirac also makes further factual presuppositions when appropriate for particular problems; e.g., that an atom consists of a nuclear core plus electrons bound to it by an electromagnetic field. Further physical assumptions are of two types. Adapting Dirac, these have been labeled 'physical principles', when ideas or principles developed in classical mechanics are accepted subject to restrictions which will be considered later. The leading principles here are: first, classical physics is valid in the domain in which it has been experimentally verified; and second, the dynamical variables, or measurable properties, which are used to specify a classical mechanical system can also be used to specify a quantum mechanical system.

The second group are assumptions concerning the behavior of physical systems. Dirac calls them 'laws of nature' only when they have been incorporated in the theory.[31] Examples of such physical assumptions which become laws of nature via the formalism are: an undisturbed system develops causally; the states of a system are interdependent; and certain state variables can be measured.

Before considering the relationship between this set of physical assumptions and the mathematical formalism, we should relate these ideas to the earlier considerations of a conceptualization of reality embodied in and transmitted through a language. Ordinary language has its own descriptive metaphysics which is used, though not necessarily affirmed, by normal speakers. In considering the modification of this conceptualization proper to classical, or Newtonian, physics, we focused on the hierarchical categorization of concepts which serves as a basis both for continuity and transformation.

In ordinary language, physical objects are spoken of as substances with primary qualities, or 'quantity' in the Aristotelian sense, and secondary qualities which inhere in substance through the mediation of quantity. The Aristotelian-scholastic philosophy of nature has, in effect, made explicit some aspects of this ordinary language, or common sense, view of reality. The physical language proper to classical mechanics referred to bodies as substances with definite primary qualities, such as mass, extension, and inertia, but without the familiar secondary qualities.

In developing the Copenhagen interpretation of quantum mechanics Bohr and Heisenberg repeatedly insisted that classical mechanics can be considered an extension of ordinary language in a way that quantum mechanics can not be.[32] For a variety of reasons analytic philosophers have generally found this claim unacceptable and, indeed, unintelligible. The differences between ordinary language and mathematical physics are strikingly obvious and apparently irreducible. What is not so obvious is that, until recently, analytic philosophers have not supplied the tools requisite to clarify what Bohr said. And it must be admitted that some of his expressions badly need clarification. The real point at issue is not the difference in syntactic structure or the varying uses to which ordinary language and mathematical physics are put, but the descriptive metaphysics, or conceptualization of reality, implicit in the language used. If a physicist is to relate his system to observable reality, he must be able to use basic category terms in making references. Any new terms so employed must ultimately be embedded in an ordinary language framework, the framework through which we relate ourselves to observable reality. The descriptive metaphysics this entails is retained, in a modified form, by classical mechanics. Thus, the terms 'particle' and 'wave' as used in classical physics presuppose the ordinary language notion of 'substance'. It is this shared notion that quantum mechanics modified.

The physical language proper to quantum mechanics represents a further extension and a more drastic modification of this shared descriptive metaphysics. It is still the language of things and properties, and the properties considered basic in classical physics do carry over. Or, in more technical terms, the dynamical variables used to specify the state of a system in classical mechanics can also be used to specify the state of a quantum mechanical system. But the definite attribution of properties to bodies no longer obtains. This is the significance of the superposition principle.

Recent writings on the quantum theory of measurement have given a somewhat different interpretation of this. The indeterminacy principle, which in Dirac's formulation is a consequence of the superposition principle, is interpreted as a relationship between the scatter in a distribution of coordinate measurements and the scatter in a distribution of momentum measurements on a statistical ensemble of systems all in the same state.[33] Margenau insists that there is no meaningful way to relate the indeterminacy principle to a single measurement.[34] Dirac's formulation did not consider these conceptual complications in the quantum theory of measurement. He simply postulated that the superposition principle must apply to one isolated particle.[35] This postulation brings with it all the difficulties associated with the reduction of the wave packet. But it also entails that classical properties can not be attributed to quantum mechanical systems in a definitive way. The attribution of a state, or maximal set of simultaneously specifiable properties, to a particle need not forbid attributing a different state to the same particle at the same time. This aspect of the descriptive metaphysics implicit in the physical language of quantum mechanics is clearly at variance with the conceptual core shared by ordinary language and classical physics.

One might summarize the progression by saying that the process of language extension and transformation involved a steady decrease in the content and definiteness of the descriptive metaphysics used to characterize physical objects. In quantum mechanics, this conceptualization is effectively the functional minimum necessary to set up and interpret the formalism. But this functional minimum carries the ontic commitments proper to quantum mechanics; i.e., a commitment to the type of entities the theory is about and the properties attributed to them. It is the bearer, in fact, of the core of ontic commitments proper to differing theories of quantum mechanics. This explains why a statement such as 'the normal helium atom has two electrons' can be accepted as true by physicists with differing formal theories. To accept such a statement as true entails accepting 'helium atom' as a term which names a class of entities and which can be used to refer to members of this class. This, in turn, entails what we have been calling a 'descriptive metaphysics'; i.e., a conceptualization of reality implicit in the language used.

In discussing classical mechanics, we indicated the role that the ordering of concepts plays in relating a physical conceptualization to a mathematical formalism. One focuses on the properties of a physical system, such as length, and then considers the properties of these properties with respect to some physical operation, such as juxtaposing lengths. One also has a set of mathematical entities, such as numbers, which have definite properties with respect to mathematical operations; e.g., addition. Then, a rather complex isomorphism is established between physical properties and mathematical entities and physical operations and mathematical operations. The governing factor here is an isomorphism between properties of the physical properties and properties of the mathematical entities.

Or, in formal rather than material language, the set of second order physical predicates are used to make statements which are formally identical with statements made with the corresponding set of first order mathematical predicates.

Quantum mechanics follows this general pattern but makes it a bit more complex. Here, a diagram may be helpful (see p. 276).

In place of the terms 'objects' and 'properties', the more general terms 'system' and 'state' are used, where a state is the set of properties of a system that can be simultaneously specified, and a system may be a collection of objects as well as an individual object. Mathematical entities representing these states must have properties isomorphic to the properties of states. The distinguishing property of quantum mechanical states is the fact that they obey a superposition principle. Accordingly, it is necessary to use mathematical entities that also obey a superposition principle. A vector can be considered as the superposition of two or more different vectors. Therefore, states of a system are represented by directions of a vector; in this case, a vector in complex, infinite dimensional Hilbert space.

Besides this correspondence between states of a system and directions of a vector, there must also be a correspondence between physical and mathematical operations. The state of a system changes when any of the dynamical variables characterizing the state change. These dynamical variables are chiefly those that quantum mechanics takes over from classical mechanics, such as position, momentum, and energy. If the state of a system is represented by the direction of a vector, dynamical variables which induce changes in the state of the system should be represented by some sort of mathematical operator which can operate on a vector to change its direction. Operators which transform a vector into another vector, or equivalently rotate a vector into a new orientation, are called linear operators. Mathematically, these linear operators, represented by small Greek letters, form a group. This gives the second basic connecting rule. Linear operators correspond to dynamical variables.

Further developments lead to technical refinements of this connecting rule. Real dynamical variables correspond to self-adjoint linear operators. Such an operator, operating on a state vector, may produce the same vector multiplied by a number, called the eigenvalue. Correspondingly, a physical operation, the measurement, yields a number which measures

CHART B: Schematic Outline of Dirac's Quantum Mechanics

Physical System		Mathematical System	
Assumptions	Connecting Rules		Axioms
Facts 1. Existence of atoms, particles, etc.	State of a system ↔ $\|A\rangle$ ↔ $\langle A\|$	$\|A\rangle$ – Ket $\langle A\|$ – Bra $\langle I\rangle$ Number $\|IA\rangle\| = \sqrt{\langle A \| A\rangle}$ (length)	1. Superposition of state with self gives same state $c_1 \|A\rangle + c_2 \|A\rangle = (c_1 + c_2) \|A\rangle$
2. States not completely specifiable classically 3. Superposition Principle 4. A measurement correlates a number and a dynamic variable (D.V.)	Dynamical Variable ↔ Linear operator, α Compl. conj. of D.V. ↔ Adjoint op. $\bar{\alpha}$ Real variable ↔ Self-adjoint op. Measurement of D.V. ↔ Eigenvalue Average Value ↔ $\langle A \| \xi \|A\rangle$ Momentum ↔ $-i\hbar\, \partial/\partial x$ Energy ↔ $H = i\hbar\, \partial/\partial t$	$\|F\rangle = \alpha \|A\rangle$ $\langle F\| = \langle A\| \bar{\alpha}$ $\alpha = \bar{\alpha}$ $\xi \| \xi'\rangle = \xi' \| \xi'\rangle$	2. B-K correspond. $\langle A\| \leftrightarrow \|A\rangle$ $(\langle A\| + \langle A'\|) \leftrightarrow (\|A\rangle + \|A'\rangle)$ $\langle A\| \bar{c} \leftrightarrow c \|A\rangle$ 3. $\langle B \| A\rangle = \overline{\langle A \| B\rangle}$ 4. $\langle A \| A\rangle > 0$, $\|A\rangle \neq 0$ $\int \langle \xi' x \| \xi'' x\rangle \, d\xi'' \rangle\, 0$, if $\|\xi' x\rangle \neq 0$
Principles 1. Classical physics is valid in the domain in which it is verified 2. Classical D.V. may specify a Q.M. system. *Laws* 1. Causality 2. Interdependence of states in quantum jumps. 3. Every observable is measurable.	*Classical Physics* — *Quantum Physics* Poisson Bracket $[u, v]$ ↔ Poisson Bracket [,] + commutator ( ) Domain of valid application ↔ $\hbar \to 0$ Hamiltonian in Cartesian coordinates ↔ Hamiltonian (yields operator form.)	Commuting observables $\langle \xi\eta - \eta\xi) IA\rangle = 0$, $\|A\rangle$ any vector	5. Quantum condition $(uv - vu) = i\hbar [u, v]$ 6. Dynamical var. for different degrees of freedom commute. 7. $\|\|Pt\rangle\| = \|\|Pt_0\rangle\|$

the value of the corresponding dynamical variable characterizing the system in a particular state. Here is the effective bottom of the page correspondence. The eigenvalues yielded by the theory must be the same as the measured or measurable values supplied by experiments for the theory to be acceptable.

I've called these 'connecting rules' rather than 'correspondence rules' to avoid confusion with the correspondence principle, a technical term in quantum theory. Such connecting rules specify the type of mathematical formalism necessary to represent the content embodied in the physical language. From here on, the real problem is the development of the formalism. Dirac's development represents a general theory of which the Schrödinger and Heisenberg formalutions are special cases. To explain it is to teach a course in quantum mechanics. Here, I'll simply classify it in terms of terms, axioms, rules, definitions, and theorems and comment briefly on the role each plays.

The basic terms, or mathematical entities, are vectors in a complex, separable, infinite-dimensional Hilbert space, called 'bra' and 'ket' vectors, and the linear operators mentioned earlier.[36] The validity of the mathematics required to handle these is simply assumed. The problem facing the physicist is to get axioms, in addition to the logical and mathematical principles presupposed, which will establish a formalism characterizing the physical system. Therefore, the axioms introduced are given a quasi-justification in terms of plausibility arguments but are not justified within the system in the sense of being deduced. These axioms, labeled 'MA' plus a number in the appendix, are concerned with the interrelation of bra and ket vectors and the numbers formed from the product of a bra and a ket.

Earlier, when discussing contempory theory-centered interpretations of science, we mentioned the emphasis these philosophers give to the ontic commitments implicit in the axioms used in a formal theory. The ontic commitments of the axioms in Dirac's mathematical formalism are to vectors in a Hilbert space rather than to such theoretical entities as atoms and electrons. Such entities are ontic commitments of the physical language, rather than the mathematical language, and are related to the mathematical formalism by the quantum connecting rules. The theory-centered interpreters stressed ontic commitments to justify a realistic interpretation of science but, I believe, did not develop the conceptual tools requisite to justify the type of realism actually operative in scientific practice.

In addition to the quantum connecting rules, there are also classical connecting rules; i.e., rules relating classical and quantum mechanics. These are of three general types. First, there are heuristic devices for using classical formulations of a problem as a means of obtaining the appropriate quantum mechanical formulations. This general principle is eventually specified in terms of a correspondence between Poisson brackets in classical mechanics and commutation relations of linear operators in quantum mechanics. Secondly, there is the general principle that quantum mechanics should merge with classical mechanics in the limit of large quantum numbers (or when any factor multiplied by $\hbar$ is negligible). This is a form of the correspondence principle which played such a large role in the original establishment of quatum mechanics and still serves as a means of interpreting the results of quantum mechanics. Finally, there are some general principles, such as conservation laws, valid in both classical and quantum mechanics. Through such connecting rules, classical mechanics forms part of the interpretative base of quantum mechanics.

The other terms used in classifying the mathematical formalism will simply be defined.

Rules of Manipulation: The bulk of Dirac's book is concerned with mathematical manipulations. Here, I've only selected those rules which play a distinctive logical role. Often, the significance of a rule is captured in a definition that results from it. In such cases, I've generally given the definition and omitted the rule.

Mathematical Principles: Dirac is attempting not to justify mathematics but to specify the type of mathematics needed to express a definite physical theory. Accordingly, I've only singled out those principles that specify or limit the mathematical formalism employed.

Definitions and Theorems: Here again, we have limited the outline to those that play a significant role in the problem of interpretation and omitted theorems or definitions that only play a technical role within the formalism.

Any further technical details will be consigned to the appendix. The basic point is that quantum mechanics fits the metatheory developed earlier. It has two essential components: a physical system which is basically a transformed extension of ordinary language embodying a conceptualization of a domain of reality, a mathematical system and connecting rules relating the two. These connecting rules depend on an isomorphism between second order phsysical properties defined with respect to certain physical operations and first order mathematical properties defined with respect to certain mathematical operations. This higher order correspondence specifies the type of mathematical entities that will be used to represent properties of physical systems. With these connecting rules established, theory-dependent physical implication follows the schema, (K) given in Section II. One specifies the properties, a_i and boundary conditions, c_i proper to the system being considered. By using the connecting rules relating properties, or dynamical variables, to linear operators, α_i, one may set up an equation (or set of equations) proper to the system in question. The solution of such equations yields a set of numbers, the eigenvalues, β_k. By a further connecting rule, these numbers are interpreted as the measurable values characterizing the properties in question. When these conditions are fulfilled, the set of numbers, b_k are the consequences of the original premises in accord with theory-dependent physical implication.[37]

V. REDEVELOPMENTS OF QUANTUM MECHANICS

Though the Dirac formalism considered as a general theory has been accepted as the basic framework for the development of quantum mechanics, it is not unique. Excluding textbook treatments, whose axioms are more pedagogical devices than axioms in a strict sense, there are two general types of alternative formulations of equal generality. The first type is a redevelopment by physicists, whose primary interest is in clarifying and simplifying the physical significance of quantum mechanics and extending it to new problems. The second type is a rational reconstruction of quantum mechanics along more formal axiomatic lines. Here, we shall only indicate the degree to which such reformulations modify the type of conclusions we have drawn.

For the first type of reformulation, we shall consider briefly the development by Schwinger, which retains Dirac's mathematical system while modifying the manner in which it is given a physical interpretation, and the development by Feynman, which introduces a different formalism that is equivalent to the Dirac formalism in its observable consequences.

Schwinger develops quantum mechanics in terms of what he calls "the algebra of measurements."[38] A measurement may be considered a process by which an assemblage of systems is sorted into sub-assemblages characterized by the same set of numbers representing the property being measured. Thus, if A is the property with values, $a', a'' \ldots$ a simple measurement may be symbolized $M(a')$. Because measurements divide the assemblage into sub-assemblages the result of successive mesurements is. $M(a')\, M(a'') = \delta(a', a'')\, M(a')$ where $\delta(a', a'')$ is 1 if $a' = a''$ & 0 otherwise.

This idea of measurement can be extended from properties to states if one measures the maximum number of simultaneously determininble properties. A system so selected is in a state which may be characterized by a'.

A more general measuring process selects systems in one state a' and leaves them in a different state, a''. This may be symbolized $M(a', a'')$ and allows for the possibility that an interaction with the measuring apparatus changes the values of some properties characterizing the state. The type of measurement of interest to quantum mechanics involves two such successive measurements, one called the preparation of state and the other the measurement proper. In general, the result of two such complex measurements would be represented by the formula:

$$M(a'b')\, M(c'd') = \begin{cases} \delta(b'c')\, M(a'd') & \text{in class. mech.} \\ \langle b' \mid c' \rangle\, M(a'd') & \text{in quant. mech.} \end{cases}$$

The product must be proportional to $M(a'd')$ since the combined measurements select systems only in state a' and emit systems only in state d'. Since b' and c' represent complete specifications of states, their product in classical measurement must be 0 unless $b' = c'$; that is, unless the type of states emitted by the first stage of the combined measurement is the same as those admitted by the second stage.

In quantum mechanics, the states b' and c' obey the superposition principle. That is, the state b' is equivalent to a superposition of states and the complex number $\langle b' \mid c' \rangle$, called 'the transformation function' characterizes the probability of state c' being a component of this superposition. In summary, successive complex measurements based on *classical* assumptions are equivalent to a single complex measurement multiplied by a Kronecker delta function $\delta(b'c')$, which has values of 0 or 1. The algebra of such measurements is Boolean. Successive complex quantum mechanical measurements are equivalent to a single complex measurement multiplied by complex numbers which do not form a Boolean algebra. These complex numbers, $\langle b' \mid c' \rangle$, can be considered the dot product of two vectors, $\langle b'|$ and $|c'\rangle$, characterizing the system in the states b' and c' – Dirac's bra and ket vectors. This development, which thus yields the Dirac formalism, also yields the equations of motion, the energy-momentum tensor, and the conservation laws when one considers variations of the transformation function.

This approach also clarifies the physical significance of interpreting quantum mechanical measurements in terms of a set of 'Yes' or 'No' questions. Any physical measurement can be reinterpreted as a set of questions, corresponding to definite physical operations in such a way that they yield only 'yes' or 'no' answers. Thus, a measurement of the length of the desk in centimeters could be interpreted as the set of questions: Is the desk one centimeter long?... two centimeters long... *n* centimeters long? In classical physics, one keeps asking such questions until a 'Yes' answer is received. Schwinger has generalized this in terms of measuring apparatuses which either admit or do not admit systems in a certain state. What this interpretation presupposes are physical systems with measurable properties where the set of all simultaneously measurable properties specifies the state of the system. The distinction between classical and quantum measurements presupposes that the definiteness (only 'Yes' or 'No' answers) characteristic of classical properties and states is modified in quantum mechanics.

Where Schwinger keeps a physical conceptualization to a functional minimal and relies on elaborate mathematical methods, Feynman stresses the role of physical reasoning and deliberately introduces a different conceptualization of the objects treated in quantum mechanics.[39] The significance of this difference may most easily be seen by considering a simple example.

When a beam of electrons is incident upon a diaphragm containing one slit, the pattern of collisions of the electrons with a fluorescent screen behind the diaphragm can be explained by considering the electron as a particle. If the diaphragm has two slits, then the pattern of collisions with the fluorescent screen is not the simple sum of the patterns given by each slit separately. There are interference effects. Accordingly, it is not possible to get a consistent interpretation which retains the two propositions: the electron always behaves as a particle and, therefore, must pass through one of the two slits; and the probability of its hitting a particular point on the screen is the sum of the probabilities from each slit considered separately.

The standard explanation, stemming from Bohr and Heisenberg, rejects the first of these premises and retains the second. That is unobservable alternatives (which of the two slits the electron went through) are not considered meaningful, but probabilities are additive in the classical fashion. In Feynman's interpretation, one retains the first proposition while rejecting the second. The electron is always treated as a particle. As such, it must go through one of the two slits. But the probability of its reaching the same spot on the screen by different paths are not additive. Feynman and Hibbs, accordingly, define 'interfering alternatives' as those that cannot be resoved by an experiment. Each such alternative trajectory yields an amplitude and the absolute square of the amplitudes yields the probability. Here, the addition of probabilities depends on the addition of complex amplitudes.

The conceptualization underlying Feynman's formalism treats electrons photons, and other elementary systems as particles whose trajectories are governed by non-classical laws. This is not simply a different formulation of the Dirac theory; it is a different theory. The demonstrations that the Feynman theory is equivalent to the standard theory do not establish a formal equivalence, but a pragmatic equivalence in terms of observable physical consequences.[40] Yet, this theory shares a central core of commitments with the standard theory. Feynman's particles have standard properties and his states have the standard significance of representing all the simultaneously measurable properties of a system. Though the correlation between properties and the mathematical formalism differs somewhat from the standard formulation, the correlation between states of a system and state functions is essentially the standard correlation.[41]

Rational reconstructions of quantum mechanics along more formal axiomatic lines began with the pioneering work of Von Neumann. More recent axiomatic formulations have been given by Mackey[42], Jauch[43], Marlow[44], Gudder[45], and others. Heelan's paper, included in this volume[46], treats the distinctive logical features of these formulations and the philosophical problems generated by the significance read into such interpretations. Accordingly, I'll confine myself to a few remarks on the role that such reformulations play in interpreting quantum mechanics and the bearing this has on the problem of the nature of scientific explanations. I believe that such rational reconstructions are helpful, even indispensible, in answering certain types of questions, but that an exclusive reliance on such reformulations can obscure the way in which the mathematical formalism is given a physical interpretation. This is the only aspect of the problem I will consider here.

Mackey's formulation utilizes the idea of a series of questions that admit of only 'Yes' or 'No' answers. Jauch's is basically similar but focuses on the answers, the 'Yes' and 'No' propositions given by a measuring apparatus. We will concentrate on Mackey's questions. These are basically of the form: "Would the measurement of an observable for a system in a certain state yield a value within a certain numerical range, or within a certain subset, *E*, or the set of all Borel subsets on the real line?"

Any measurement within either classical or quantum mechanics is equivalent to a series of such questions. In classical mechanics, all such questions concerning the state of a system are, in principle, simultaneously answerable. The answers, a series of 'Yes' and 'No' responses constitutes a Boolean or distributive lattice. In quantum mechanics, there are sets of such questions which are not simultaneously answerable. This outlaws a Boolean lattice but allows a non-Boolean lattice. This lattice, or partially ordered set of all such questions, is isomorphic to the partially ordered set of all closed sub spaces of a separable, infinite dimensional complex Hilbert space.

This axiomatic approach in terms of questions, accordingly, yields either a phase space formulation of classical mechanics or a Hilbert space formulation of quantum mechanics. What emerges as the distinguishing feature of quantum mechanics is the non-Boolean nature of the lattice of questions or propositions proper to quantum mechanics, a lattice based on the fact that, in quantum mechanics, all questions are not simultaneously answerable. If one is interested in an answer to such a basic question (in the ordinary sense of 'question') as: "What is the physical significance of asking and answering the questions on which the quantum formulation is based"?, he will find that Mackey's very formal approach effectively prescinds form this question. Answering a question means, for him, establishing a mapping between values of an observable and measure functions defined on the real line. He simply accepts as an established fact the conclusion that, in quantum mechanics, all questions are not simultaneously answerable.[47] Jauch discusses this in more detail. But his contention that experimental measurements *of themselves* determine that the propositions describing the intrinsic structure of a physical system be an orthocomplemented non-Boolean lattice rests on rather dubious reasoning.[48] Even more misleading, in my opinion, is Punam's contention that quantum logic must be considered the logic of the world in a basic sense.[49] This contention ultimately rests on a combination of a one component interpretation of scientific systems, rather than the two component (PL and ML) interpretation developed here, and a doctrine of strict determinism.[50]

On a functional level, this question has been effectively answered by Schwinger's formulation with its explicit commitment to the conceptual minimum needed for contemporary quantum mechanics; that is, a commitment to systems with measurable properties, where the set of all simultaneously measurable properties specifies the state of the system, and the ascription of properties to quantum mechanical systems has the type of indeterminateness previously discussed.

This is what we have been calling the 'descriptive metaphysics' implicit in the conceptualization of reality embedded in the physical language used in quantum mechanics. In concluding, we can make a few general remarks on the role such conceptualizations play. First, *some* language-embodied conceptualization of the reality being treated is absolutely necessary in setting up, developing, and interpreting any scientific theory. Secondly, particular conceptualizations are revisable. Second order, or metalinguistic, predicates play the key role in establishing an isomorphism with a mathematical formalism. The successful features of this formalism may be preserved and extended while the first order predicates, those characterizing physical reality, are changed.

Yet, it is the set of first order physical predicates and the propositions formed with them that embodies the conceptualization of reality. Our primary task here has been one of clarifying what this conceptualization is and how it functions. The conclusions may be summarized by schematizing this conceptualization in terms of three layers, a basic structure, a fine structure, and a hyperfine structure.

The basic structure is the underlying descriptive metaphysics shared by Indo-European languages, classical mechanics, and quantum mechanics.

This is a conceptualization of reality as an interrelated collection or physical objects with properties moving in a space-time framework. The fine structure is the core of physical commitments shared by otherwise competing theories. Here, it is basically the set of particles accepted as real, together with the properties, relations, and activities attributed to them. The left-hand column of Chart B effectively summarizes the basic ideas proper to this level. In calling these the 'ontic commitments of quantum mechanics', I am admittedly using the term 'ontic commitment' in a away that differs from Quine's original definition. The epistemological justification for this is given elsewhere.[51] Basically, it is a question of gearing the usage of the term to the practice of the scientific community in accepting as true statements of the form 'There are *N*'s'. Ordinarily, this shared acceptance is had only when the statement in question can be detached from particular theories. Finally, there is the hyperfine structure, the commitments and conceptualization of reality proper to one particular formulation; e.g., Feynman's. It was necessary to consider such particular formulations as a means of clarifying the common conceptual core shared by Dirac, Feynman, Schwinger, and others. This common conceptual core generally goes by the name 'the Copenhagen interpretation'. In spite of its difficulties and inadequacies, the Copenhagen interpretation function as a part of contemporary physics, while no competing interpretation has yet been accepted as a functioning part of physics.

This type of conceptualization of physical reality is functional rather than metaphysical. Its function is to serve as the vehicle used by physicists in forming propositions that are reasonably asserted and accepted as true. This does not imply that the conceptualization can cover all problems in a way that is adequate and consistent. Quantum mechanics, it would seem, is neither. Within the framework of the interpretation of quantum mechanics presented here, it does not seem to be possible to give an explanation of the process of measurement that is conceptually consistent without introducing such unacceptable doctrines as psychophysical parallelism. Nor has the extension of quantum mechanics to relativistic phenomena been shown to be either consistent or completely adequate. In the light of these shortcomings, one can not but wonder whether further revisions of quantum mechanics might lead to the abandonment of the linguistic residue of a substance – property ontology shared by ordinary language, classical mechanics, and the present accounts of quantum mechanics.

I believe this might well happen but can only specualte about what might occur. Three types of ontology seem to be contenders competing with a substance-property ontology in speculations about the physics of the far future: an ontology of events, favored by some who think that the theory of relativity is more basic than quantum mechanics; an ontology of bare particulars, favored by those with a nostalgic yearning for the promised land of the *Tractatus*; and an ontology of fields, the most likely candidate. If an adequate and successful quantum field theory of fundamental particles could be developed, then fields would be considered the basic entities with particles as stable or metastable interactions of fields. If this were to occur, then the descriptive metaphysics proper to the quantum mechanics we have outlined would have to be considered phenomenological relative to this more basic ontology.[52] A final possibility is that the techniques of the present quantum mechanics will be retained while the requirement, met by the Copenhagen interpretation, that the physical conceptualization have at least a minimal functional consistency will be abandoned. This is the so-called 'Bootstrap' approach to fundamental particles, a label that effectively indicates the low regard had for consistency in a physical conceptualization.

Finally, what is the relation between such types of descriptive metaphysics and what one might like to think of as a science of metaphysics, or the relationship between what a particular conceptualization of reality implicitly says is real and the really real? Here, one has the critical question in a contemporary form and no amount of descriptive metaphysics, based on conceptual analysis, serves to answer it, though such analysis is a necessary propaedeutic to any reasonable attempt to answer this question. Descriptive metaphysics is, in Husserl's terminology, ontology within a natural framework and is essentially precritical. How one should go beyond this to answer the critical question and develop a science of metaphysics, I am not altogether sure. But I am sure that nothing I've said this evening constitutes an answer to this more basic question.

California State University, Hayward

NOTES

[1] Pierre Duhem, *The Aim and Structure of Physical Theory* (transl. by Philip P. Wiener), Princeton University Press, Princeton, 1956, p. 19.

[2] Rudolf Carnap, 'The Methodological Character of Theoretical Concepts', in H.

Feigl and M. Scriven (eds.), *Minnesota Studies in the Philosophy of Science*, Vol. I, University of Minnesota press, Minneapolis, 1956, pp. 38–76.

[3] Carl G. Hempel and Paul Oppenheim, 'The Logic of Explanation', in H. Feigl and M. Brodbeck (eds.), *Readings in the Philosophy of Science*, Appleton Century Crofts, New York, 1953, pp. 319–52. For Hempel's later views, see his *Aspects of Scientific Explanation*, Free Press, Glencoe, Ill., 1965, pp. 331–489.

[4] Norman R. Campbell, *Physics: The Elements*, England, Cambridge, 1920.

[5] R. B. Braithwaite, *Scientific Explanation* (pb. ed. Harper, 1960), esp. Chap. XI.

[6] Ernest Nagel, *The Structure of Science: Problems in the Logic of Scientific Explanation*, Harcourt, Brace and World, New York, 1961, Chaps. V and VI.

[7] See, for example, Stephen Toulmin, *The Philosophy of Science: An Introduction* (Harper pb. ed., 1960), esp. the discussion in Chap. iv on theories as maps or inference mechanisms.

[8] W. Sellars, 'Scientific Realism or Irenic Instrumentalism', in R. Cohen and M. Wartofsky (eds.), *Boston Studies in the Philosophy of Science*, II, Humanities Press, New York, 1965, pp. 171–204. See also his essay, 'The Language of Theories', in his *Science Perception and Reality*, Humanities Press, New York, 1963, pp. 106–26. The first article cited introduced the idea of the role of second order predicates in the extension of conceptual frameworks, an idea which has been exploited in the present paper.

[9] W. V. O. Quine, *Word and Object*, The MIT Press, Cambridge, Mass., 1960. See also his articles 'On What There Is', in his *From a Logical Point of View* (Harper pb. ed., 1963), pp. 1–19; and 'Ontological Relativity', *The Journal of Philosophy* **65** (1968), 185–212.

[10] His views have been developed in a series of articles: 'Problems of Empiricism', in R. Colodny (ed.), *Beyond the Edge of Certainty: Essays in Contemporary Science and Philosophy*, Prentice-Hall, Englewood Cliffs, N.J., 1965, pp. 145–260; 'Problems of Microphysics', in R. Colodny (ed.), *Frontiers of Science and Philosophy*, University of Pittsburgh Press, Pittsburgh, 1962, pp. 189–283; 'Explanation, Reductionism, and Empiricism', in *Minnesota Studies in the Philosophy of Science*, III, University of Minnesota Press, Minneapolis, 1963, pp. 28–97.

[11] H. Putnam, 'Is Logic Empirical?', in R. Cohen and M. Wartofsky (eds.), *Boston Studies in the Philosophy of Science*, Vol. V, 1969, p. 216.

[12] R. Carnap, 'Empiricism, Semantics, and Ontology', published as an appendix in his *Meaning and Necessity* (rev. ed.; University of Chicago Press, Chicago, 1956), pp. 205–21.

[13] Edward MacKinnon, *Philosophical Problems of Scientific Realism*, Appleton-Century Crofts, 1969, 1971.

[14] Quine, *Word and Object*, Chaps. v–vii.

[15] H. Putnam, *op. cit.*

[16] A more detailed criticism of Feyerabend's views on this point may be found in my 'The New Materialism', *The Heythrop Journal* **8** (1967), 5–26.

[17] N. R. Hanson, *Patterns of Discovery*, Cambridge University Press, Cambridge, Eng., 1958.

[18] Thomas Kuhn, *The Structure of Scientific Revolutions*, University of Chicago Press, Chicago, 1965 (pb. ed.). I believe that the criticisms which D. Shapere has made against the ambiguity in Kuhn's idea of paradigms (contained in his review in *The Philosophical Review* **73** (1964), 383–94) would not have the same force against the formulation presented here which links conceptualizations explicitly to language systems.

[19] Stanley Jaki, *The Relevance of Physics*, University of Chicago Press, Chicago, 1966, esp. Chaps. i–iii.
[20] Stephen Toulmin, 'Conceptual Revolutions in Science', in R. Cohen and M. Wartofsky (eds.), *Boston Studies in the Philosophy of Science*, III, Humanities Press, New York, 1968, pp. 331–47.
[21] Marx Wartofsky, 'Metaphysics as Heuristic for Science', in *Boston Studies* (Note 8), pp. 123–72. See also Joseph Agassi, 'The Nature of Scientific Problems and Their Roots in Metaphysics', in M. Bunge (ed.), *The Critical Approach to Science and Philosophy: In Honor of Karl R. Popper*, Free Press, Glencoe, Ill., 1964. Both men use 'metaphysics' in the context of the type of conceptual analysis practiced here in discussing descriptive metaphysics rather than in the doctrinaire sense of a system to be defended as an elucidation of the really real.
[22] P. F. Strawson, *Individuals: An Essay in Descriptive Metaphysics* (Doubleday pb. ed., 1963).
[23] The relation between a conceptualization of reality embedded in a language and a mathematical formalism is treated by Marx Wartofsky in his *Conceptual Foundations of Scientific Thought*, Macmillan, New York, 1968, Chaps. vi and vii; and in Henry Kyburg, Jr., *Philosophy of Science: A Formal Approach*, Macmillan, New York, 1968, Chap. iii. The idea of theory-dependent and context-dependent conditions of correspondence and the formula for theory-dependent physical inference were developed by Stephan Körner in his *Experience and Theory*, Humanities Press, New York, 1966. The epistemological problems involved in the interpretation given here are discussed in more detail in my forthcoming *Truth: The Hecker Lectures, 1968*, Paulist-Newman Press, New York, Chaps. ii and iii.
[24] W. Sellars, 'Scientific Realism of Irenic Instrumentalism', *op. cit.*
[25] E. G. Wigner, *Symmetries and Reflections*, Indiana University Press, Bloomington, 1967, esp. Chaps. i–iii.
[26] B. Inhelder and J. Piaget, *The Growth of Logical Thinking From Childhood to Adolescence* (transl. by A. Parsons and S. Milgram), Basic Books, New York, 1958. See also Jerome Bruner *et al.*, *Studies in Cognitive Growth*, New York, 1966, esp. Chaps. xi–xiv.
[27] This is adapted from an account given in Richard Blackwell, *Discovery in the Physical Sciences*, University of Notre Dame Press, Notre Dame, 1969, Chap. i.
[28] E. Wigner, *op. cit.*
[29] P. A. M. Dirac, *The Principles of Quantum Mechanics*, (4th ed.), Clarendon Press, Oxford, 1958.
[30] In the Dirac formulation, the relationship between the Heisenberg picture and the Schrodinger picture is formulated:

$$\xi_s = e^{iHt/\hbar}\,\xi_H e^{iHt/\hbar}$$
$$|A_s\rangle = e^{iHt/\hbar}\,|A_H\rangle$$

Here, ξ_H and $|A_H\rangle$ are operators and state vectors in the Heisenberg picture and ξ_t and $|A_s\rangle$ are the corresponding operators and vectors in the Schrödinger picture. These transformations are meaningful only if the exponential can be represented by an expansion which is convergent. This is possible in relativistic quantum mechanics, as in non-relativistic quantum mechanics, until one attempts to include in the expansion of H terms of order α^5, α^6, etc. where α is the fine structure constant. This introduces the weak infinities of quantum electrodynamics, and Dirac considers renormalization an inadequate solution to the theoretical problems they present. If such transformations do not exist, then Hilbert space is inadequate. Dirac has been attempting to formulate quantum mechanics in a more general space but has not yet achieved satisfactory results. (This summary is based on a lecture Dirac gave under the sponsorship of Yeshiva University on November 16, 1964. He has discussed the problem frequently since then but seems to have made no significant change in these ideas.)

[31] "We have made a number of assumptions about the way in which states and dynamical variables are to be represented mathematically in the theory. These assumptions are not, by themselves, laws of nature but become laws of nature when we make some further assumptions that provide a physical interpretation of the theory. Such further assumptions must take the form of establishing connections between the results of observations, on one hand, and the equations of the mathematical formalism on the other." (Dirac, p. 34).

[32] Neils Bohr wrote and spoke on this from 1927 until his death in 1962. One of his clearest presentations is 'Discussion with Einstein on Epistemological Problems in Atomic Physics', in *Albert Einstein Philosopher-Scientist*, Harper reprint, New York, 1959, pp. 201–41. An excellent summary lf Bohr's views may be found in Aage Peterson, 'The Philosophy of Niels Bohr', *Bulletin of the Atomic Scientist* **8** (Sept., 1963), 8–14. For Heisenberg's views, see his *Physics and Philosophy: The Revolution in Modern Science*, Harper, New York, 1958.

[33] See Jeffrey Bub, 'Hidden Variables and the Copenhagen Interpretation – A Reconciliation', *British Journal for the Philosophy of Science* **19** (1968), 185–210; J. Jauch, E. Wigner, and M. Yanase, 'Some Comments Concerning Measurement in Quantum Mechanics', *Il Nuovo Cimento* **48** (1967), 144–51; and the summary in R. Schlegel, *Completeness in Science*, Appleton-Century-Crofts, New York, 1967, Chap. x.

[34] Henry Margenau, 'Measurement and Quantum States', *Philosophy of Science* **30** (1963), 1–16, 138–57.

[35] See Dirac, p. 9.

[36] In developing his quantum mechanics, Dirac deliberately refrained from using the term 'Hilbert space' and simply postulated a space with the properties his development required. The basic difficulty was that the special functions Dirac introduced (MD 14 in the Appendix) could not be considered functions in a Hilbert space. After L. Schwartz reinterpreted these δ functions as distributions rather than as functions in the mathematical sense, Dirac's vector space could be called a complex Hilbert space. See A. R. Marlow, 'Unified Dirac-Von Neumann Formulation of Quantum Mechanics', *Journal of Mathematical Physics* **6** (1965), 919–27.

[37] We have been using the terms 'implication' and 'entailment' rather loosely. The complicating factor here is that the reasoning process leading from premises to conclusion in solving problems in quantum mechanics not only involves formal deduction; it also involves physical assumptions; e.g., neglecting certain terms as not physically significant, the use of models, and approximations. This is something more than what the logician means by 'entailment'.

[38] These ideas have been developed in seminars, some of which have been published as lecture notes and privately circulated: Julian Schwinger, *Quantum Mechanics: Lecture Notes from a Course given at the Université de Grenoble, L'École D'Été de Physique Theorique, Juillet, 1955*; and (an earlier version of the same ideas) Julian Schwinger, *Quantum Dynamics*, National Bureau of Standards Report, Washington, D.C., 2188, 1952. A general formulation of quantum mechanics on the basis of the transformation principle and the extension of this basis to field theory may be found in his article, 'The Theory of Quantized Fields, II', *Physical Review* **91** (1953), 713–28.

[39] Feynman's original development was in a series of articles in *Reviews of Modern Physics* **20** (1948), 367ff. for non-relativistic quantum mechanics; and *The Physical Review* **76** (1949), 749–56, 769–74 for relativistic quantum mechanics. A systematic development of quantum mechanics along these lines is contained in R. P. Feynman and A. R. Hibbs, *Quantum Mechanics and Path Integrals*, McGraw-Hill, New York, 1965.

[40] Feynman-Hibbs demonstrate this equivalence by showing that the Feynman formalism yields the Schrodinger equation. In this demonstration, plausible physical reasoning plays a large role; e.g., in deciding which terms in a mathematical expansion may be ignored. Earlier, F. J. Dyson had shown that the Schwinger and Feynman formulations of quantum electro-dynamics are practically equivalent in the sense that one can begin with the Tomonaga-Schwinger formalism and derive Feynman's rules for calculating matrix elements. See his 'The Radiation Theories of Tomonaga, Schwinger, and Feynman', *Physical Review* **75** (1948), 486–502, esp. Sects. vi and vii.

[41] If a property, G, is correlated with a complex wave function, $g(x)$, then $g(x)$ is interpreted as the amplitude that *if* the system has G it is at the location x, while $g^*(x)$, the complex conjugate of $g(x)$, is interpreted as the amplitude that *if* the system is at x, then it has G. If $g(x)$ is a state function – equivalent to a Dirac ket vector with $g^*(x)$ the equivalent of a bra vector – then, it includes the location among the simultaneously measurable properties. See Feynman-Hibbs, Chap. v.

[42] G. Mackey, *The Mathematical Foundations of Quantum Mechanics*, Benjamin, Inc., New York, 1963.

[43] J. M. Jauch, *Foundations of Quantum Mechanics*, Addison Wesley, Reading, Mass., 1968.

[44] A. R. Marlow, 'Physical Axiomatics' and 'Physical Systems and the Quantum Mechanical Model', *Journal of Mathematical Physics* (to be published).

[45] Stanley P. Gudder, 'Hidden Variables in Quantum Mechanics Reconsidered', *Reviews of Modern Physics* **40** (1968), 229–31, redevelops the Jauch-Piron axiom system to take account of objections brought by Bohm and Bub.

[46] Patrick Heelan, S.J., [this volume - Ed.].

[47] See the discussion on Axiom 7 in Mackey, p. 71ff.

[48] As this contention is developed in Jauch, Chap. v, it is the conclusion from a set of premises. First, a physical law is a condition-effect relation, in quantum mechanics a probability relation, which has the general form: "If a system S is subject to conditions $A, B, \ldots$ then the effects $X, Y, \ldots$ can be observed" (p. 71). Secondly, operational definitions are an adequate basis for explaining theoretical terms: "If we wish to determine the physical characteristics of a proton, we must perform a series of experiments which then, in their ensemble, will be a full operational equivalent for the construct *proton*" (p. 72). Third, a set of idealized measuring apparatuses would reproduce not only the values characterizing states of a system, but also the logical properties of combinations (union and intersection) of propositions. Finally, the structure properties (complete, orthocomplemented modular lattice) of this set of propositions reveal the *intrinsic structure* of a physical system; i.e., the set of propositions about a physical system which are independent of the state of the system.

Each of these contentions is open to serious objections. On the difference between conditional statements and physical laws, see Stephan Körner, *Experience and Theory: An Essay in the Philosophy of Science*, *op. cit.* On the impossibility of explaining theoretical constructs exclusively by operational definitions, see Arthur Pap, *An Introduction to the Philosophy of Science*, Free Press, Glencoe, Ill., 1962, Chap. iii.

The difficulties with the third and fourth premises are discussed by Patrick Heelan, S.J. in 'Quantum Logic Does not Have to Be Non-Classical', in *Boston Studies in the Philosophy of Science* (to be published), and by D. Bohm and J. Bub in 'On Hidden Variables – A Reply to Comments by Jauch and Piron and by Gudder', *Rev. Mod. Phys.* **40** (1968), 235–6.

[49] H. Putnam, *op. cit.*

[50] For his doctrine of determinism, see Hilary Putnam, 'Time and Physical Geometry', *The Journal of Philosophy* **64** (April 27, 1967), 240–47. A refutation of Putnam's view is given by Howard Stein in 'On Einstein-Minkowski Space-Time', *The Journal of Philosophy* **65** (January 11, 1968), 5–23.

[51] See my *Truth: The Hecker Lectures*, *op. cit.*, Chap. iii, Sect. iii.

[52] These ideas are based on Julian Schwinger, 'Field Theory of Particles', in *Lectures on Particles and Fields: Brandeis Summer Institute, 1964*, Prentice-Hall, Englewood Cliffs, N.J., pp. 145–288. See especially the introductory remarks to Chap. v, pp. 267–8.

APPENDIX: THE AXIOMATIC STRUCTURE OF DIRAC'S QUANTUM MECHANICS

The new scheme becomes a precise physical theory when all the axioms and rules of manipulation governing the mathematical quantities are specified and when, in addition, certain laws are laid down connecting physical facts with the mathematical formalism, so that, from any given physical conditions, equations between the mathematical quantities may be inferred and vice versa. (P. A. M. Dirac, *Quantum Mechanics*, 4th ed., p. 15. Future references will simply cite page numbers.)

Dirac followed this logical ideal very carefully but, since he was concentrating on physics, did not make the logical structure obvious. To make the logical structure actually employed more explicit, it is necessary to label and categorize. Following his division into a physical system, a mathematical system, and connecting rules, we will use the following notation:

PHYSICAL SYSTEM

PF *Physical facts* which are either accepted as already established or assumed as plausible.

PP *Physical principles* (e.g., principles of class. phys.) These need not be specified in detail since they are all included in the general principle that classical physics is valid in the domain in which it is experimentally verified.

PL *Physical laws* (or laws of nature). These are assumptions made concerning the behavior of physical systems or the interaction of such systems with a measuring apparatus.

PC *Physical conclusions*; i.e., conclusions concerning the physical system derived by means of the math. system.

MATHEMATICAL SYSTEM

MP *Primitive* mathematical terms.

MA Theoretical axioms.

MR *Rules* of manipulation. The bulk of Dirac's book is concerned with mathematical manipulations. Here, we shall select only those rules which play a distinctive logical role. If the significance of a rule is captured in a definition which results from it, we shall give the definition and omit the rule.

MM *Mathematical* principles. We shall single out only those principles which specify or limit the math. formalism employed.

MD *Theoretical definitions.* To avoid excessive complexity, we shall try to include only those definitions that have a significant role in the general interpretation.

MT *Theorems.* Here, too, we shall select only those theorems which play a significant role in the general interpretation.

CONNECTING RULES

CQ *Quantum connecting rules*; i.e., those rules used to establish a correspondence between the physical systems presupposed by quantum theory and the math. formalism.

CC *Classical connecting rules*, rules connecting quantum and classical physics.

This set of categories is neither exhaustive in scope nor completely precise in its application. However, it should be adequate for our limited purposes. In general, we shall follow the order of Dirac's development, occasionally gathering together items which are scattered in different sections. As a general rule, we have opted for completeness rather than selectivity whenever there seemed to be a conflict between the two norms.

PF 1. *The existence of particles and atoms.*

In Sections 1 and 2 the existence of atoms, molecules, and particles is simply accepted as established. Later (p. 152), he assumes "An atom consists of a massively positively charged nucleus together with a number of electrons moving around under the influence of the attractive force of the nucleus and their own mutual repulsions."

PL 1. *Causality*

"Causality applies only to a system which is left undisturbed. If a system is small, we cannot observe it without producing a serious disturbance and, hence, cannot expect to find any causal connection between the results of our observations" (pp. 4 and 108). Here, 'causality' is used in the same sense as in classical mechanics; i.e., for predictive determinism.

PF2. *The state of an atomic system can not be completely specified in classical terms.* (Section 4)

The meaning of this statement in the context depend on the precise meaning accorded to the terms. "A state of a system may be defined as an undisturbed motion that is restricted by as many conditions or data as are theoretically possible without mutual interference or contradiction." The specification of a state is exclusively a question of the measurements, actual or possible, that might be performed: "Only questions about the results of experiments have a real significance and it is only such questions that theoretical physics has to consider" (p. 5). PF 2 is a particular instance of the inadequacy of classical concepts in supplying a description of atomic events (p. 3).

PF 3. *The superposition principle*

The basic idea expressed here is that any given state of a quantum mechanical system may be regarded as a superposition of two or more new states in a way that cannot be conceived on classical ideas (Section 4).

PP 1. Classical physics is valid in the domain in which it has been experimentally verified (p. 14).

This is not so much a formal principle as it is a general boundary condition for the application of other principles.

MP 1. *Ket vector* $|\rangle$ (p. 16).

Vectors are mathematical quantities which satisfy a superposition principle. In introducing these terms D. does *not* presuppose the formal properties of Hilbert space but simply introduces a mathematical term which is capable of representing systems which satisfy the superposition principle. The axioms which govern these vectors are introduced later. Here, he simply lists the properties which these vectors must have if they are to represent a q.m. system.

(a) A ket is a vector in a space with a finite or an infinite number of dimensions.
(b) A ket may be multiplied by a complex number.
(c) Ket vectors may be added.
(d) Kets may be integrated: $\int |\mathbf{K}\rangle \, \mathrm{d}x = |Q\rangle$.

MD 1. *Dependent and independent ket vectors.*

A ket which is expressible lineary in terms of other ket vectors is a dependent ket vector. Otherwise, it is independent.

CQ 1a. *Correspondence between states of a system and ket vectors* (p. 16).

Each state of a dynamical system at a particular time corresponds to a ket vector, the correspondence being such that, if a state results from the superposition of certain other states, its corresponding ket vector is expressible linearly in terms of the corresponding ket vectors of the other states and conversely. If A, B, and R represent kets and c_1 and c_2 numbers, then

$$C_1 \, |A\rangle + C_2 \, |B\rangle = |R\rangle.$$

(The converse of this, CQ 1b cannot be given without presupposing terms yet to be developed.)

MT 1. *The order of superposition is unimportant* (p. 16).

MA 1. *The superposition of a state with itself does not give a new state,* (p. 17)
$C_1 \, |A\rangle + C_2 \, |A\rangle = (C_1 + C_2) \, |A\rangle$

MT 2. *If the ket vector corresponding to a state is multiplied by any complex number, not zero, the resulting ket vector will correspond to the same state.*

MA 1 and its corrolary MT2 imply that a physical state is specified only by the *direction* of a ket vector, not by its length.

MM 1. *Correspondence between ket vectors and numbers* (p. 18).

The ordinary mathematical properties of vectors imply that it is possible to establish a correspondence between ket vectors and numbers which has the following properties:

(a) The correspondence is one to one (or an isomorphism).
(b) The correspondence is linear, i.e.,

if $|A\rangle \leftrightarrow \varphi$ and $|A'\rangle \leftrightarrow \varphi'$
Then $(|A\rangle + |A'\rangle) \leftrightarrow (\varphi + \varphi')$ (φ, φ_a'-numbers)

MP 2. *Bra vectors*, $\langle|$ (p. 19)

Bra vectors are not simply postulated as primitive terms. Rather, D. argues from MM 1 that there musts be a vector dual to $|A\rangle$ such that the scalar product of this dual and $|A\rangle$ is φ. We have labeled them primitives because bra vectors and ket vectors are on a par. Either type can be used to represent the state of a system and the postulation of either one, plus the appropriate mathematics (MM 1), implies the other.

The significance of the notation introduced can be summarized as follows (p. 19):

(a) A complete bracket expression, $\langle\ \rangle$, denotes a number.
(b) An incomplete bracket expression denotes a vector:

$|\rangle$ denotes a ket vector.
$\langle|$ denotes a bra vector.

MA 2. *Correspondence between bras and kets* (p. 20).

There is a one to one correspondence between the bras and kets, such that the bra corresponding to $|A\rangle + |A'\rangle$ is the sum of the bras corresponding to $|A\rangle$ and $|A'\rangle$ and the bra corresponding to $c\,|A\rangle$ is $\bar{c}$ times the bra corresponding to $|A\rangle$, $\bar{c}$ being the conjugate complex number to c.

Because of this correspondence, the same labels may be used. Thus, the bra corresponding to $|A\rangle$ will be labelled $\langle A|$ and one will be called the 'conjugate imaginary' of the other.

CQ 1b. *Correspondence between states of a system and bra vectors.*

On account of the one-one correspondence between bra vectors and ket vectors, any state of our dynamical system at a particular time may be specified by the direction of a bra vector just as well as by the direction of a ket vector (p. 21).

MA 3. $\langle B \mid A\rangle = \overline{\langle A \mid B\rangle}$, (p. 21).

MT 3. $\langle A \mid A\rangle$ Must be real.

In MA 3, substitute $B = A$. Any number equal to its own complex conjugate is real.

MA 4a. $\langle A \mid A\rangle > 0$ except when $|A\rangle = 0$.

MD 2. *Orthogonality of vectors* (p. 21).

A bra vector, $\langle B|$ and a ket vector $|A\rangle$ are orthogonal when $\langle B \mid A\rangle = 0$, $\langle B| \neq 0$, $|A\rangle \neq 0$.
Two bras, $\langle B|$, $\langle C|$ are orthogonal when

$$\langle B \mid C\rangle = 0 \quad \text{or} \quad \langle C \mid B\rangle = 0, \qquad \langle B| \neq 0, \qquad \langle C| \neq 0,$$

Similarly for two kets.

MD 3. *Length of a vector*: $|A\rangle =_{\mathrm{df}} \sqrt{\langle A \mid A}$.

Since the direction rather than the length is the important factor, the lengths will be normalized; i.e., set equal to one.

MP 3. *Linear operators* (pp. 23, 24).

A linear operator (usually denoted by a small Greek letter) is specified by the following conditions:

(a) A linear operator transforms a ket vector into another ket vector, $\alpha\,|A\rangle = |F\rangle$.

(b) Linear operators satisfy the additive and multiplicative conditions of linearity.

$$\alpha\{|A\rangle + |A'\rangle\} = \alpha\,|A\rangle + \alpha\,|A'\rangle$$
$$\alpha\{C\,|A\rangle\} = C\alpha\,|A\rangle,\ C - \text{a complex number.}$$
$$\{\alpha + \beta\}\,|A\rangle = \alpha\,|A\rangle + \beta\,|A\rangle$$
$$\{\alpha\beta\}\,|A\rangle = \alpha\{\beta\,|A\rangle\}$$

However, it is *not* assumed that linear operators are commutative under multiplication; i.e., it is not true in general that $\alpha\beta\,|A\rangle$ equals $\beta\alpha\,|A\rangle$. In mathematical terms, the linear operators form a group.

MT 4. *Linear operators may also operate on bra vectors* (p. 25).

MR 1. *A linear operator operates on a ket vector from the left,* $\alpha\,|A\rangle$ *and on a bra vector from the right,* $\langle A|\,\alpha$.

PP 2. *Specification of a system by dynamical variables* (p. 26).

The dynamical variables which are used to specify a system in classical mechanics; e.g., position coordinates, components of velocity, momentum, and angular momentum and functions of these quantities can also be used to specify a q.m. system.

CQ 2a. *Linear operators correspond to dynamical variables.*

The state of a system at a given time corresponds to the direction of a vector. Changes in the system are due to changes in the dynamical variables. A linear operator can change the direction of a vector so that it corresponds to a new state. CQ 2a states the general principle but does not indicate which operators correspond to which dynamical variables.

MD 4. *Adjoint of a linear operator* (p. 26).

This is derived from the correspondence between bras and kets (MA 2). If $\alpha\,|A\rangle = F$, then there must be an $\bar{\alpha}$, called an 'adjoint operator', such that $\langle F = \langle A|\,\bar{\alpha}$.

MT 5. *For any linear operator* $\bar{\bar{\alpha}} = \alpha$. (p. 27).

CQ 2b. *The adjoint of a linear operator corresponds to the conjugate complex of a dynamical variable.*

MD 5. *A self-adjoint operator is an operator which equals its own adjoint; i.e.,* $\alpha = \bar{\alpha}$ (p. 27).

CQ 2c. Real dynamical variables correspond to self-adjoint operators (p. 27).

Because of this correspondence, self-adjoint linear operators will often be referred to as 'real' linear operators.

MR 2. *Formation of conjugate complex and conjugate imaginary.*

The conjugate complex or conjugate imaginary of any product of bra vectors, ket vectors, and linear operator is obtained by taking the conjugate complex or conjugate imaginary of each factor and reversing the order of all factors (p. 28).

MD 6a and 6b. *Eigenvalues and eigenvectors* (p. 29).

A linear operator ξ operating on a vector may reproduce the same vector multipled by a number; e.g., $\xi|A\rangle = a|A\rangle$, $\langle B|\xi = b\langle B|$. The vectors satisfying such equations will be called eigenvectors (eigenkets or eigenbras) and the numbers that result will be called eigenvalues. To make the correspondence between a particular linear operator; e.g., ξ, and the eigenvalues associated with it explicit the eigenvalues will be labelled ξ', ξ'', etc., where a primed small Greek letter denotes a number.
MT 6a, b, c. *Properties of eigenvalues* (p. 31).
From CQ 2c and MD 6a, b, it follows that only eigenvalue equations involving *real* linear operators are of any physical interest. For such eigenvalue equations, the following properties may be established.

(a) The eigenvalues are all real numbers.

(b) The eigenvalues associated with eigenkets are the same as the eigenvalues associated with eigenbras.

(c) The conjugate imaginary of any eigenket is an eigenbra belonging to the same eigenvalue and conversely.

MT 7. Two eigenvectors of a real dynamical variable belonging to different eigenvalues are orthogonal (p. 32).

PF 4. A physical measurement, or an observation, correlates a real number with a dynamical variable.

CQ 3a. If the dynamical system is in an eigenstate of a real dynamical variable ξ, belonging to the eigenvalue ξ' then a measurement of ξ will certainly give as result the number ξ' (p. 35).

CQ 3b. Conversely, if the system is in a state such that a measurement of a real dynamical variable ξ is certain to give one particular result (instead of giving one or other of several possible results according to a probability law, as is in general the case), then the state is an eigenstate of ξ and the result of a measurement is the eigenvalue of ξ to which this state belongs (p. 35).

CQ 4a. Any result of a measurement of a real dynamical variable is one of its eigenvalues (p. 36).

CQ 4b. Conversely, every eigenvalue is a possible result of a measurement of the dynamical variable for some state of the system (p. 36).

According to CQ 3, *if* a measurement yields an eigenvalue then an operation on the state vector with the corresponding linear operator yields the same eigenvalue. CQ 4 adds that the *only* results which a measurement can yield are eigenvalues.

PL 2. If a certain real dynamical variable is measured with the system in a particular state, the states into which the system may jump on account of the measurement are such that the original state is dependent on them (p. 36).

MD 7. A complete set of states is a set such that any state is dependent on them (p. 36).

MD 8. An observable is a real dynamical variable whose eigenstates from a complete set (p. 36).

An observable is a part of the physical system, not the mathematical system. Because of the correspondence between the two systems, however, an observable is defined in terms of the mathematical properties of the self-adjoint linear operator corresponding to the real dynamical which is observed.

PL 3. In principle, every observable can be measured (p. 37).

MD 9. *Expansion of a vector* (p. 37).

If ξ is an observable with continuous eigenvalues, ξ_c' and discrete eigenvalues, ξ_d', then the corresponding eigenkets (or eigenbras) may be labelled $|\xi_c'\rangle$ and $|\xi_d'\rangle$. Since these form a complete set (MD 8) *any* ket $|P\rangle$ can be expressed as an integral plus a sum of this complete set:

$$|P\rangle = \int |\xi_c'\rangle \, d\xi_c' + \sum_r |\xi_d^r\rangle .$$

MA 4b. For eigenvectors associated with continuous eigenvalues ξ', ξ'', proper to an observable ξ, MA 4 takes the form (p. 40).

$$\int \langle \xi' x \mid \xi'' x \rangle \, d\xi'' > 0 \quad \text{if} \quad |\xi' x\rangle \neq 0$$

MT 10. The expansion of a ket (or bra) in terms of a complete set is unique (p. 40).

MD 10. Function of an observable (p. 41).

Given an observable ξ with eigenvalues ξ', we may define a function of an observable, '$\int(\xi)$', as that linear operator which satisfies the equation $f(\xi)\,|\xi'\rangle = f(\xi')\,|\xi\rangle$.

CQ 5. Using MD 10, it is possible to give a meaning to any function of an observable, f, provided only that the domain of the function of the real variable $f(x)$ includes all the eigenvalues of the observable (p. 43).

This connecting rule is semi-derived from the preceding ones by a plausibility argument.

CQ 6. Representation of average values (p. 46).

If the measurement of the observable ξ for the system in the state corresponding to $|x\rangle$ is made, a large number of times the average of all the results obtained will be $\langle x|\ \xi\ |x\rangle$ prvided $|x\rangle$ is normalized.

The preceding developments make this assumption plausible but do not yield it deductively.

MD 11. Commuting observables (p. 49).

Two observables ξ and η commute if

$$(\xi\eta - \eta\xi)\,|A\rangle = 0, \quad \text{where} \quad |A\rangle \text{ is any vector.}$$

MT 11. Commuting observables and complete sets (pp. 49–50):

(a) If two observables commute, there exists so many simultaneous eigenstates that they form a complete set.

(b) If two observables are such that their simultaneous eigenvalues from a complete set, then these observables commute.

CQ 7. Any two or more commuting observables may be counted as a single observable; i.e., the measurement of one does not interfere with the measurement of the other (p. 52).

This is a derivative connection rule in that it gives the meaning of simultaneous measurements.

SPECIAL REPRESENTATIONS

The general framework of the Dirac theory is now complete. Further developments are of three kinds. First, he wants to show that the Heisenberg and Schrödinger representations emerge as particular representations of his general system and also show how they are interrelated. Secondly, using these representations, he wishes to explain the relation between classical mechanics and quantum mechanics to solve some basic problems. Thirdly, he wishes to develop a relativistic theory.

Our basic concern here is the way in which the formal theory is interpreted. This requires a clarification of the logical structure of representations as particular expressions of the general theory. Particular problems, including the development of the uncertainty relationship are solved by the use of special representations. We shall not treat any individual problems as such, but concentrate on the strategy used in solving them and in interpreting the solution. The relativistic treatment presents novel problems which will not be considered.

MD 12. A *representation* is any way in which the abstract quantities employed by the theory; i.e., bra vectors, ket vectors, and linear operators can be replaced by numbers.

MD 13. Basic bras of a representation (p. 53).

The basic bras of a representation are a complete set of bra vectors; i.e., a set such that any bra can be expressed linearly in terms of them. If all these bras are independent, the representation is called *orthogonal.* Such a set is adequate to determine a representation completely.

MT 12. Commuting Observables and Complete Sets (pp. 53–54).

Any set of commuting observables can be made into a complete set by adding certain observables. Given such a set, it is possible to set up an orthogonal representation in which the basic bras are simultaneous eigenbras of the complete set.

MD 14. Dirac delta function (pp. 58–61).

This is introduced as a means of treating infinities. It is defined by the formulas:

$$\int_{-\infty}^{+\infty} \delta(x)\,\mathrm{d}x = 1 \quad \& \quad \delta(x) = 0 \quad \text{for} \quad x \neq 0.$$

The general definition of representations has been given and the method of obtaining representations of vectors has been indicated (an expansion through the use of basic bras). Two techniques for obtaining representations of linear operators are introduced, one of these specifying the Heisenberg representation and the other the Schrödinger.

IV.1. Heisenberg Representation

MD 15. Matrix Representation of a Linear Operator (p. 67).

(a) Given a single linear operator ξ which forms a complete set by itself and has *discrete* eigenvalues ξ', then the representative of any operator, α, is the set of discrete numbers, $\langle\xi'|\,\alpha\,|\xi''\rangle$, which may be written as the matrix.

$$\begin{matrix} \langle\xi^1|\,\alpha\,|\xi^1\rangle & \langle\xi^1|\,\alpha\,|\xi^2\rangle & \langle\xi^1\,\alpha\,|\xi^3\rangle\ldots\ldots\ldots\ldots \\ \langle\xi^2|\,\alpha\,|\xi^1\rangle & \langle\xi^2|\,\alpha\,\xi^2\rangle & \langle\xi^2|\,\alpha\,|\xi^3\rangle\ldots\ldots\ldots\ldots \\ \langle\xi^3|\,\alpha\,|\xi^1\rangle & \langle\xi^3|\,\alpha\,|\xi^2\rangle & \langle\xi^3|\,\alpha\,|\xi^3\rangle\ldots\ldots\ldots\ldots \\ \ldots\ldots\ldots & \ldots\ldots\ldots & \ldots\ldots\ldots \end{matrix}$$

(b) In the case of *continuous* eigenvalues where it is impossible to set up a matrix representation using discrete numbers one uses a generalized matrix where the elements are defined by the definitions:

Representative of ξ, $\langle\xi'|\,\xi\,|\xi''\rangle = \xi'\,\delta(\xi' - \xi'')$.
Representative of $f(\xi)$, $\langle\xi\,|\,f(\xi)\,|\xi''\rangle = f(\xi')\,\delta(\xi' - \xi'')$.

MD 16. Hermitian Matrix (p. 68).

A Hermitian matrix, A, is a matrix whose elements satisfy the rule, $a_{i1} = \bar{a}_{1i}$ (where $^{-}$ indicates complex conjugate).

MT 13. Real dynamical variables are represented by Hermitian matrices.

MD 17. Probability Amplitude (p. 73).

Given a complete set of commuting observables, ξ, which has both a discrete and a continuous spectrum, the probability for each $\xi_r (r=1, 2, \ldots, v)$ having a value ξ_r' (discrete values) and for each ξ_s $(s=v+1, \ldots, u)$ lying in a specified small range ξ_s' to $\xi_s'+\mathrm{d}\xi_s'$, for the state represented by the normalized ket vector, $|x\rangle$ is

$$P_{\xi'_1 \ldots \xi'_v} \mathrm{d}\xi'_{v+1} \ldots \mathrm{d}\xi'_u = |\langle \xi'_1 \ldots \xi'_a \mid x \rangle|^2 \mathrm{d}\xi'_{v+1} \ldots \mathrm{d}\xi'_u$$

This is a precise generalization of the probability interpretation of quantum mechanics first introduced by Max Born.

The general theory of transformation functions which is developed next will be omitted here. Before treating the relation between classical and quatum mechanics D. introduces one notion proper to the Schrödinger representation, the wave function.

MD. 18. Wave Function.

Given a complete set of commuting observables $\xi_1 \ldots \xi_u$ (symbolized by ξ), any ket, $|P\rangle$, will have a representative, $\langle \xi' \mid P \rangle$. Since this is a definite function of the observables ξ, we may write $|P\rangle = \Psi(\xi)$, where $\Psi(\xi)$ is called the *wave function* and is understood as the product of a linear operator $\Psi(\xi)$ and a standard ket, $\rangle$. Whenever explicit recognition of the latter feature is significant, the wave function will be written as $\Psi(\xi)\rangle$.

Similarly, we may write $\varphi(\xi) = \langle Q|$ where $\varphi(\xi)$ is understood to be multiplied from the left by a standard bra, $\langle|$ (and will be represented as $\langle \varphi | \xi)$ whenever the bra nature of this wave function is significant).

The chief function of the standard bras and kets in the new notation is to specify rules for the manner of operating with these functions; e.g., whether one should operate on them from the right or the left.

IV.2. Classical Mechanics and Quantum Mechanics

The general scheme has not yet been applied to any problems. For such an application, it is necessary to specify the dynamical variables that characterize the system. From a formal point of view, the feature which distinguishes the present linear operators; e.g., ξ and η, from the corresponding dynamical variables in classical mechanics; e, g, u, and v,is that the latter always commute while the former do only for special cases. Accordingly, what is needed for a new mechanics is a general means of determining the value of $(\xi\eta - \eta\xi)$. The equations that determine this are called the *quantum conditions.*

To determine the quantum conditions, Dirac introduces what he calls the method of classical analogy. Since classical mechanics is valid in the limiting case of systems large enough so that they are not significantly disturbed by observation, some of the formal properties of classical mechanics should also be valid in classical mechanics. The formal properties Dirac is particularly interested in are those that will allow him to determine the quantum conditions.

MD 19. *Poisson Bracket* (p. 85).

A poisson bracket is a standard term in classical mechanics. It enters as a definition in the present system under the assumption that the same definition is also valid for quantum systems under conditions to be given.

The classical definition applies to a system which can be specified by means of a set of canonical coordinates, q, and moments, p. Then, for any dynamical variables, u, v, characterizing the system, the P.B. (poisson bracket) may be defined.

$$[u, v] = \sum_r \left\{ \frac{\partial u}{\partial q_r} \frac{\partial v}{\partial p_r} - \frac{\partial u}{\partial p_r} \frac{\partial v}{\partial q_r} \right\}.$$

MT 14. (u, v) is independent of the choice of canonical coordinates and moments (p. 85).

CC 1. *Classical and Quantum Poisson brackets* (pp. 86–87).

For quantum systems with a classical analogue (e.g., a collection of interacting particles), it is assumed that the quantum P.B. has the same from as the classical P.B. However, a quantum P.B. is assumed to exist even for systems that have no classical analogue (i.e., a system for which canonical coordinates and momentum can not be specified). In this sense, the quantum P.B. is more general than the classical.

MR 3. Any expansion of a quantum P.B. must preserve the order of the terms u and v (p. 86).

MA 5. *Determination of the quantum conditions* (p. 87).

A plausibility argument is given to show the relation between P.B. and the quantum conditions. However, this merely shows that the form assumed is reasonable. Accordingly, the formula must be considered an axion rather than a theorem.

If u and v are dynamical variables characterizing a quantum system, (u, v) is the quantum P.B. and $\hbar$ is Planck's constant divided by $2\pi\hbar$ $= 1.05 \times 10^{-23}$ erg s), then

$$uv - vu = i\hbar\,[u, v].$$

MT 15. *Fundamental Quantum Conditions* (p. 87).

If $q_r, q_s, \ldots$ represent canonical coordinates and $P_r, P_s, \ldots$ represent canonical momenta, then MA 5 yields:

$$q_r q_s - q_s q_r = 0\,, \qquad P_r P_s - P_s P_r = 0$$

$$q_r P_s - P_s q_r = i\hbar\delta_{rs}\,, \quad \text{where} \quad \delta_{rs} = \begin{cases} 1\,, & r = s \\ 0\,, & r \neq s\,. \end{cases}$$

CC 2. Classical mechanics may be regarded as the limiting case of quantum mechanics when $\hbar$ tends to zero. (p. 88)

The expression '$\hbar$ tends to zero' is to be interpreted in a purely relative sense. $\hbar$ has a fixed numerical value. If, in a given problem, any factor multiplied by $\hbar$ is negligibly small compared to other factors, then the system can be treated by classical mechanics. When $\hbar$ tends to zero the quantum conditions, given by MA 5, reduce to the classical case where all variables commute.

CC2 is essentially the 'correspondence principle' first introduced by Neils Bohr and given a sharper meaning in the light of Heisenberg's thesis that the non-commutation of operators is the distinguishing feature of quantum systems.

MA 6. Dynamical variables referring to different degrees of freedom commute (p. 88).

This is a generalization of the first two equations in MT 15 for all quantum systems including those that have no classical analogue.

IV.3. SCHRÖDINGER REPRESENTATION

We assume a system which can be described in terms of canonical coordinates, q_s, and momenta, p_r, and also assume that the q's form a complete commuting set. For such cases, Dirac can include Schrödinger's formal theory as one special representation of Dirac's own general system. It is the representation which is most useful in solving problems.

MT 16. The Differential Operator $\partial/\partial q_r$ (p. 90).

The differential operator $\partial/\partial q_r$ considered as a linear operator operating on wave functions is a pure imaginary. This implies that:

$$\frac{\partial}{\partial q_r}\psi > = \frac{\partial\psi}{\partial q_r} >, \qquad < \varphi \frac{\partial}{\partial q_r} = - < \frac{\partial\varphi}{\partial q_r}.$$

On the left hand side of each equation $\partial/\partial q_r$ is regarded as a linear operator which can operate to the right (on a ket) or to the left (on a bra). The right hand side of each equation is a ket or bra vector formed by differentiating the mathematical function $\Psi(q)$ and multiplying the result by the standard ket or bra.

MT 17. $P_r = -i\hbar(\partial/\partial q_r)$.

This characterizes the Schrödinger representation and is a conclusion from Dirac's general formalism except for an arbitrary (and insignificant) phase factor. The only supplementary assumption required to derive MT 17 is the condition that the wave function vanish at infinity.

MT 18. *The uncertainty principle* (p. 98).

If $\Delta q'$ is the width of the wave packet in which q has an appreciable value, and $\Delta p'$ is the width of the momentum space wave packet in which p has an appreciable value, then:

$$\Delta q' \Delta p' = h.$$

MD 20. $|Pt\rangle =_{\text{df}} |P\rangle$ at time t (p. 108).

MA 7. $|Pt\rangle$ has the same length as $|Pt_0\rangle$ (p. 109).

From the superposition principle, Dirac is able to argue that $|Pt\rangle$ must be a linear function of $|Pt_0\rangle$ but is not able to determine the length of the vector. The assumption that the length of the state vector representing a system remains constant is really an assumption about the behavior of the physical system represented. It is essentially a further specification of the meaning of PL 1 (an undisturbed system develops causally). Though only the direction of a vector represents a system directly, its length is significant in any superposition of vectors.

MD 21. $i\hbar(\mathrm{d}|Pt\rangle/\mathrm{d}t)=H(t)|Pt\rangle$. (P. 110)

This is a definition of $H(t)$ considered as a linear operator. The purpose of the definition is to get a linear operator which plays a role analogous to that of the Hamiltonian in classical mechanics.

CQ 7. $H(t)$, as defined in MT 21 represents the total energy of the system (p. 110).

This is made plausible by the classical analogue and by relativistic considerations.

PL 5. The total energy of the system is always an observable (p. 110).

MT 19. Schrödinger's Wave Equation,
$i\hbar\, \partial/\partial t \Psi(\xi t)\rangle = H\Psi(\xi t)\rangle$

REPRESENTATIONS AND PICTURES

A representation was defined in MD 12 as any way in which the abstract quantities employed by the theory can be replaced by numbers. MD 13 introduced a general method for representing state vectors. Two methods of representing operators were presented. The Heisenberg representation represents linear operator by matrices while the Schrödinger representation uses differential operators.

Dirac now introduces 'picture' as a semi-technical term. A 'picture' is any way of looking at the fundamental laws which makes their self-consistency obvious (p. 10).

MD 22. *Schrodinger picture* (p. 111).

The characteristic feature of the Schrödinger picture is that the operators are fixed (i.e., do not change in the course of time) while the state vector (or wave function) does change. Accordingly, the development of a physical system in the course of time is reflected in the mathematical system by rotations of the state vector which represents the system.

MD 23. *Heisenberg picture* (pp. 111–112).

In this picture, the state vectors are fixed while the operators change in the course of time.

MT 20. Heisenberg Form of the Equations of Motion. (p. 112)
$i\hbar \, \mathrm{d}V_t/\mathrm{d}r = v_t H_t - H_t v_t$

Here, v_t and H_t, the dynamical variables in the Heisenberg picture, are related to the Schrödinger dynamical variables, v, H by the formulas:

$$v_t = T^{-1} v T, \quad H_t = T^{-1} H T$$

where T is a linear operator corresponding to time displacement and T^{-1} is its inverse.

Since the corresponding classical equation has the same form, this parallel supplies the classical analogue presupposed in CQ. 7.

MD 24. The Hamiltonian (p. 114).

A quantum mechanical Hamiltonian is H (as defined in MD 21) expressed as a function of dynamical variables whose commutation relations are known. In principle, every quantum mechanical system has a Hamiltonian

CC 3. If a quantum mechanical system has a classical analogue, one can usually assume that the Hamiltonian is the same function of p and q in quantum mechanics as in classical mechanics when p and q are expressed in Cartesian coordinates (p. 114).

This is a heuristic assumption used as a guide in solving particular problems.

MD 25. Constants of the Motion (p. 115).

A dynamical variable is called a constant of the motion if it satisfies the equation $[v_t, H] = 0$, which implies that it does not change in time.

JOHN STACHEL

COMMENTS ON 'ONTIC COMMITMENTS OF QUANTUM MECHANICS'

Dr. MacKinnon's paper is part of the recent trend to try to bring philosophy of science closer to the actual practice of scientists; to get away from the attempt to tell physicists what they are doing (or should be doing) – for example, by constructing elaborate formalisms for their theories; but rather to try to observe and conceptualize what they actually do. He voices his skepticism about excessive formalization, and his desire to bring philosophy of science into a position where it not only can look at a given theory considered as a finished product; but also allow for the discussion of the relationship between competing or successive theories, and conceptual revolutions in science. He is concerned with what ground such theories have in common, as well as their differerences. I can only join him in this desire to view science in its evolutionary perspective; and on this score the only questions I shall raise will be about whether he goes far enough in this direction.

I would argue that it is really only in an evolutionary perspective that one can evaluate any stage in the development of human knowledge.[1] Perhaps we live on the threshold of an epoch when such a perspective may introduce radically new possibilities in the development of our understanding of man in his relation to the cosmos, and thus contribute to our understanding of the relationship of human knowledge to the known. My bias in approaching problems in philosophy of science, and about the nature of human knowledge in general, leads me to discern a historical trend towards the transformation of many such problems into problems which the sciences themselves must treat sooner or later. Of course, this implies recognition of the existence of a philosophical component of scientific theories, rather than a crude empiricist approach to science.

Our whole understanding of the relationship of human knowledge to the universe around us may well take a new turn with the growing ability of science to situate human life in the cosmos as the product of neither sheer accident nor transcendental design; but rather as an evolutionary product of a certain stage of development of the universe. Naturally, since we are only at the earliest beginnings of such an attempt (assuming I am right) I can only shadow forth darkly a few hints of such a trend.

It has been conjectured, for example (by Leon Rosenfeld, among others), that the reason why we function at the classical (or 'common sense' level) is that only systems of the size and complexity of the nervous system ('brain', loosely) are capable of sufficient knowledge storage, stability of operation, self-consciousness, etc. Certainly an atom does not have the requisite complexity to develop quantum theory, for example. Thus, the scale at which we find ourselves in relation to that of the micro-world would be no accident, but a subject for scientific study.

In relation to the cosmic scale also, there has been some speculation (notably by Brandon Carter) that the epoch of the universe we live in, with the current value of the Hubble constant, say, is also no accident; but that this is just that stage in the evolution of an expanding universe when we could expect that a large number of stars with sufficiently cool planets for the evolution of life would arise.

There have also been recent attempts to put the nature of our perceptual apparatus into an evolutionary perspective; and investigate, for example, the survival value of various features of our visual perception patterns. Abner Shimony spoke at earlier Colloquium on the relation of some of the work on the nature of perception to the role of the 'common sense' language in which supposedly all experience must be formulated. He has called attention to the need, in discussing the relation between theory and observation, for further scientific study of the nature of human perception, rather than taking it as a given whole in rather crude form.

When MacKinnon talks about 'physical' language as "a transformed extension of ordinary language," and notes later that "any changing conceptualization of reality must be embedded in and transmitted through ordinary language or its specialized extensions," I would take this as a call for investigation of how far ordinary language may be extended. I am pretty sure that a lot of things we take as 'ordinary language' today, would not have struck thinkers of the past as so ordinary. One can learn to think in terms of the relativistic concepts of space-time, for example, so well that they become quite 'ordinary' for a relativist; but he will then have to remember to translate back the results of his thought into the usual space *and* time terminology to talk to an 'ordinary' physicist in a way which is 'ordinary' for the latter.

I am sure that Dr. MacKinnon would agree with me that, in general, much more scientific investigation is needed – physiological, psychological, sociological, historical and anthropological – before we can say with any assurance how far the limits of ordinary language can be stretched, as a tool for conceptualizing our deepening understanding of the world.

Dr. MacKinnon formulates a model or metatheory for the interpretation of scientific theories, involving: formulation of physical properties of the system in the physical language; formulating mathematical properties of the mathematical entities in mathematical language; and an isomorphism between the second order predicates (properties of physical properties) and the properties of the mathematical entities. The formulation of the physical properties commits us to the existence of corresponding physical entities; the formulation of the mathematical properties commits us to the existence of the mathematical entities; and these are the ontic commitments of the theory. While I am far from prepared to evaluate all the ramifications of his point of view, particularly on such short acquaintance with the paper, I would like to raise several questions about his general scheme, and particularly about its application to quantum mechanics.

To start with a minor point, since Dr. MacKinnon commits himself to the existence of mathematical entities, such as vectors in Hilbert space (p. 125), one would like some discussion of the relationship of mathematical existence to existence in the sense of the physical language. Do these mathematical entities exist outside of space and time, or as relationships between ordinary bodies in space and time, or how? It seems to me that a discussion of the ontic commitments of a theory cannot avoid taking some position on this question.

Secondly, in discussing the transition from one theory to another (one of the main aims of his metatheory is to enable the discussion of such relationships), he states that "... This mathematical continuity underlying a conceptual revolution is grounded in the fact that both the new conceptualization and the old share a set of second order predicates, predicates whose primary function is to express laws of conservation and invariance under certain transformations, or as Wigner characterized them, laws about laws of nature." (p. 117). He gives some examples to bear out his point. Clearly no list of examples could establish a general result like this without some general argument. Yet, if I understand the point correctly, counter-examples exist in abundance. For when we go over from one transformation group to another, invariants of the group often disappear or change drastically. In Newtonian mechanics, for example, the absolute time difference between two events is an invariant under the whole Galilei group. When we go over to the Lorentz group, the nearest equivalent we have is the interval, which is quite a different property of two events than was the absolute time difference between them – for example, it may not even be timelike. Indeed, we even have cases where predicates simply disappear with the invariance group used to characterize them. For example, in Newtonian and special relativistic dynamics energy and momentum are associated with invariance under temporal and spatial translations. But in general relativity, no such invariance generally exists, and the concepts of localised energy and momentum just are not useful because of this lack of invariance. At best, they may exist as global, asymptotically defined quantities in a certain class of solutions. So I would have to see a lot more discussion of this issue before I could accept such a rather dogmatic statement of the form that the relation between new and old theories must take.

My third and most important point arises in connection with the application of his general metatheory to quantum mechanics. In connection with "the physical language proper to quantum mechanics," which "represent a further extension and modification" of 'ordinary language,' we are informed that "This language still has the descriptive metaphysics of substance and properties." ... "One might summarize the progression by saying that the process of language extension and transformation involved a steady decrease in the content and definiteness of the descriptive metaphysics used to characterize physical objects" (p. 122). We must restrict ourselves to a "functional minimum necessary to set up and interpret the formalism" in our application of the classical properties (p. 122) "In place of the terms 'objects' and 'properties' the more general terms 'system' and 'state' are used, where a state is the set of properties of a system that can be simultaneously specified and a system may be a collection of objects as well as an individual object" (p. 123). These states (or pure states, to clarify the point) obey the superposition principle and must be mathematically represented by directions of vectors in Hilbert space (rays).

Now I claim that it is just such an interpretation of the state of a quantum mechanical system, as no more than the indefinite or partial attribution of classical properties to an atomic system, which otherwise represents substance in the sense of ordinary language or classical physics as usually interpreted – it is just such an interpretation that we are precluded from adopting in quantum mechanics. I claim no novelty here, but merely follow the now 'classic' ideas of Bohr on the interpretation oı quantum theory. Paul Feyerabend has particularly emphasized how Bohr was forced to elucidate the *relational* character of the state vector in order to interpret the Einstein-Rosen-Podolsky (EPR) paradox.

But before getting to that, it should be emphasized that one cannot even regard an atomic system as always having a pure state function, just because it was initially characterized by one. If a system originally is characterized by a state vector, and then interacts with another system, after the interaction is over and they are separated the system is no longer in a pure state, characterized by a state vector, but is in what is called a mixed state of the second kind. d'Espagnat, who has discussed this point in his book *Conceptual Foundations of Quantum Mechanics*, indicates there at length just how and why this prevents us from taking any simple realistic point of view about the state vector of a system. If one wanted to take the state vector as a property of the microsystem in the classic sense, since any system keeps interacting with more and more systems as time goes on, and their state vectors keep getting enmeshed by the interaction, one would inevitably be led to introduce the concept of the state vector of the universe – (Everett, Wheeler, DeWitt) a concept which has its own quite serious problems, and certainly leads one far from the classical conception of substance and properties. Rather than pursue this topic here, let us return to the relational aspect of the state vector. The EPR set-up shows that the state vector characterizing a system may change in ways that cannot be attributed to its own internal evolution nor its current interaction with anything else. Bohr, answering Einstein, was led to point out: "even at this stage there is essentially the question of an influence on the very conditions which define the possible types of prediction regarding the future behavior of the system." The nature of the state function is to define a relation between the macroscopic set-ups for the production and detection of microsystems which enables us to determine the statistics of the microprocesses taking place in between the two – i.e., between the preparation defining the emission and the registration (measurement) defining the reception of the micro-system. We use the word system here, but clearly the micro-system cannot be regarded as something in the nature of a *classical* substance with independent properties – at least not in the standard quantum mechanics we are here discussing. Various hidden variable theories have been proposed, largely in the effort to get around this problem by a hypothetical return to a classical ontology at a 'deeper' level. But since Dr. MacKinnon confines himself to standard quantum mechanics, I think there is a problem here he cannot avoid in a discussion of the ontic commitments of quantum theory. By the way, it should be emphasized that no taint of anthropomorphism need or should enter (I don't say it doesn't enter in many presentations of the subject) into this formulation of quantum mechanics as a theory of the observations of microprocesses made by macroscopic measuring devices.

Groenewold has gone a long way towards eliminating any vestiges of the anthropomorphic from the concept of macroscopic observation in quantum mechanics, by formulating the theory in terms of a series of successive quantum measuring acts, which could be performed and recorded by automated measuring devices, say on the spaceship of '2001' after Hal got rid of his human companions. In this way, he shows "... the stochastic relations between the results of a series of successive quantum measuring acts constitute a Markov chain. The significant statements of quantum mechanics can be expressed in terms of conditional probabilities between different elements of this chain. More indirect conceptions of quantum mechanics as e.g. quantum state also have a conditional meaning ... we have got rid of the ambiguous notion of the 'state' in which a system 'is' at a certain moment. A conscientious description in terms of conditional probabilities between successive measuring results or even in terms of conditional probability amplitudes (S-matrices) is commonly less convenient than in terms of quantum states of the system. The latter is harmless as long as one is conscious of the conditional meaning." He shows that the records of these measurements may either be read in the forward temporal sense as usual, or in the reverse direction (reading the records backwards – not time reversal of events is meant here). When we read in the forward sense, the system may be said to be in the appropriate eigenstate of the measuring apparatus *after* the measurements as in the usual discussions of measurement in the text books; but if we read backwards the system must be said to be in that state immediately *before* the measurement. "That is one aspect of the relativity of the quantum state imputed to a system."

Thus, judging MacKinnon by his own test example of quantum mechanics, I think we must say his discussion of ontology is not complete enough. We do indeed "require non-classical concepts of atomic existence and of physical space-time," and it seems to me that a discussion of the ontic commitments of quantum mechanics cannot avoid such issues; the continuity in many aspects of the language of classical and quantum mechanics that he mentions must not be allowed to hide the new problems raised for ontology by quantum mechanics. In particular, I think he would have to address himself to such questions as: does the electron 'exist' and 'have properties' in same sense in (1) the Millikan oil drop experiment; (2) the Stern-Gerlach experiment; (3) the atom; (4) a radioactive nucleus from which it is emitted; (5) a virtual state in quantum field theory?

In the last pages of his talk Dr. MacKinnon raises the question of some other possible models for quantum-mechanical ontology. He mentions ontologies of events, of bare particulars, and of fields. One would like to see a much fuller elaboration of these ideas, particularly since some of the problems of quantum theory do not really show up until one attempts to combine it with relativity in the quantum theory of fields.

Bohr has counseled a form of resignation in the face of the exciting possibilities of other ontological models. Since we can only apprehend reality through the classical categories, any future theory must operate with the classical language, even if it cannot work with the classical ontology. Feyerabend has effectively dealt with the logical fallacies inherent in such a position, if it is presented as a deduction from the scientific aspects of Quantum Mechanics. It is built on a philosophical outlook we are not forced to accept along with quantum theory, unless we are so inclined for reasons outside of physics.

MacKinnon also seems to counsel a form of despair. MacKinnon calls all of the above multitude of possibilities descriptive metaphysics, which get us no closer to a metaphysics of being, or what is really real. One may raise the question: is there anything beyond to get closer to? Why assume we need ever 'touch bottom' in our ever-deeper investigation of the world?

Going back to the evolutionary point of view I started from, it seems only natural (or rather, it is the task of future science to show that it is only natural) that we exist on a certain level of reality (permit me the loose usage of the word 'level' to go along with the loose usage of 'reality') – call it the macroscopic, which is the one at which entities of our complexity of perception and powers of conceptualization and action could come into being. It is natural that humanity's first conceptualizations about reality are based on the properties of this level (although even here we have had a great variety and I don't hesitate to say progress in the development of our concepts: for example mythic, causal, Aristotelian, classical physics). Now, our powers have expanded to the point where we are in a position to begin (just begin – we must remember we are babes in this respect) to undertake the study of other levels of reality. Let us call one the microscopic (electrons, elementary particles) the other the metamacroscopic (cosmos, metagalaxy, universe as a whole). It is only natural that we must first approach these through their interactions with our macroscopic level. Quantum mechanics proves that we are quite able to comprehend and use the properties of the macroscopic level in this way. If we focus on the classical concepts as the starting point, quantum mechanics may look like a restrictive theory (indefiniteness of state descriptions, uncertainty relations, etc.); but who can doubt that it has actually led to tremendous expansion of our power over nature (for good or evil; of course, our problems today are fundamentally social, not due to some presumed limits of knowledge)? Modern developments in cosmology indicate possibilities at other levels. Must we always understand these matters from the perspective or viewpont of our level? I think of this as an open, empirical question; only future research can decide. Don't sell the future short – so long as we have one. As we come to understand the properties of our level better and better, as necessary consequences of its interrelation with other levels, it is just possible we might reach the point where we could use some set of concepts adequate for another level to explain our level, and investigate still further remote levels of reality, if they exist, which may not be directly accessible to us even in the way that atomic events are.

Humanity has come far in a short span of cosmic evolutionary time;

and modern science is only a few hundred years old. The microscopic and metamacroscopic levels have been really opened to scientific exploration for less than a century. Why should we then so prematurely assume there is anything in principle hidden or unknowable to us because of presumed limitations of our ability to conceptualize some aspects of the world?

Boston University

NOTE

[1] Of course, my use of the word 'evolution' is not meant to preclude revolutionary modes of scientific evolution.

PATRICK A. HEELAN

QUANTUM LOGIC AND CLASSICAL LOGIC: THEIR RESPECTIVE ROLES *

Abstract. The paper analyses the claim made by Birkhoff and von Neumann, Finkelstein, Jauch, Putman and others that quantum mechanics implies the use of a non-classical 'quantum logic' (an orthocomplemented non-distributive lattice) on the level of factual sentences. It is shown that this claim is confused by the erroneous assumption that simple theoretical statements about the Hilbert space state vector of a system are logically equivalent to simple empirical statements about the outcome of Yes-no tests. It is shown that simple theoretical statements are equivalent to statements in a meta-context-language which assert the existence, not of events, but of constellations of invariant physical conditions within which events occur. The logic internal to the meta-context-language is a non-classical logic; the logic of quantum event language is or could be classical. The existence of a quantum logic is shown to be related to the general context-dependent character of statements and this claim is illustrated by examples taken from outside the domain of physics.

I. CLASSICAL OR QUANTUM LOGIC?

That quantum mechanics is different from classical mechanics, in what it says about the physical world and how it says it, needs no proof. How precisely to describe and explain these differences, and what significance to attach to them is being continually discussed.

One of the claims that is being made is that the most significant difference between classical mechanics and quantum mechanics is that the latter uses or needs to use a non-classical kind of propositional logic, a logic that has been called a 'quantum logic'. The classical logic is often described as Aristotelian. More accurately, it is the propositional logic of two-valued truth-functional propositions, the logic of classes and the logic of quantification as, for example, these are developed in Russell and Whitehead's *Principia Mathematica* (PM).

G. Birkhoff and J. von Neumann were the first in 1936 to put forward the view that the 'physical quantities' of quantum mechanics constitute an orthocomplemented non-distributive lattice, and not the Boolean algebra of classical PM logic[1]. Other proposals were made about the same time and later by Reichenbach, von Weizsäcker, Heisenberg and others in favor of multi-valued logics in which the classical principle *tertium non datur* is violated. Bas C. van Fraasen has given a brief and systematic survey to the various quantum logics in his paper 'The Labyrinth of Quantum Logics'. I shall not be concerned in this paper with multi-valued quantum logics, but only with the view common to Birkhoff and von Neumann, Segal, Mackey, Finkelstein, Jauch, Putnam[2] and others, that quantum logic is a non-distributive lattice.

This theory says that the basic empirical propositions of quantum mechanics obey, not the axioms and rules of classical PM logic, but those of a logic obtained from classical logic by dropping the distributive laws for 'and' and 'or' and replacing them by what looks like a weaker form of connection. In Jauch's version, the basic empirical propositions of quantum mechanics are those which express the outcome of Yes-No tests

> [Yes-no tests] are observations which permit only *one* of two alternatives as an answer (hence the name Yes-no experiment). Such experiments are part of the daily routine of every experimental physicist: for example, a counter which registers the presence of a particle within a certain region of space .... Every measurement on a physical system can be reduced at least in principle to measurements with a certain number of Yes-no experiments .... Each measurable quantity has a certain range of values which may be indicated as a subset of the real line (or perhaps a Euclidean space). A determination of this quantity is obtained by dividing the real line into smaller intervals and then deciding whether the measured value falls within any one of the intervals. By making the intervals sufficiently small, one can determine the value of the quantity to any desired accuracy.[3]

It is claimed that the kind of logic which these propositions obey is an orthocomplemented non-distributive lattice of weak modularity.[4]

In favor of this proposal are two sets of arguments, one positive the other negative. The negative argument is twofold: it is an attack on a priori intuitionistic arguments used to justify the retention of classical logic and it is an attempt to show that the retention of classical logic involves the rejection of quantum mechanics in its present form. The positive argument is the claim that logic, like physical geometry, is subject to empirical confirmation and disconfirmation procedures, and that quantum mechanics has disconfirmed classical logic in the microdomain. It is consistent with Putnam's position and, I think, also with Finkelstein's, that as a consequence of the alleged breakdown of classical logic in the quantum domain, the distributive law of classical logic is to be held, at least suspect, in all domains (except in the postulational one of mathematics). By a kind of correspondence principle, it can be treated as a rough and ready approximate rule in everyday affairs, much as Euclidean geometry remains the rough and ready approximate geometry of the everyday world even after the general theory of relativity has shown that the physical world is in fact non-Euclidean.

II. THE CASE FOR A QUANTUM LOGIC IN QUANTUM MECHANICS

The role which a quantum logic would play in resolving some of the seemingly paradoxical aspects of quantum mechanics can be illustrated with reference to the two-slit experiment (see Figure 1).

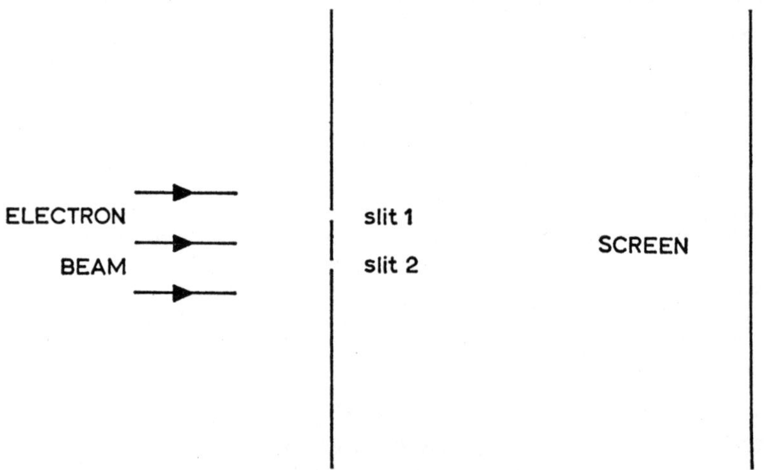

Fig. 1. Diagram of the two-slit experiment.

Let a_1, a_2 and b be the following sentences:

a_1: Electrons of the beam pass through slit 1
a_2: Electrons of the beam pass through slit 2
b : The distribution of electrons at the screen is the arithmetic sum of the distributions obtained when only one slit at a time is open.

It is an experimental fact that b is false.

Consider now the following argument:

Major:[5] $(a_1 \vee a_2) \supset b$
Minor: $\sim b$
Conclusion: $\sim (a_1 \vee a_2)$

The conclusion says that it is not true that electrons of the beam pass through one or both of the slits. Let us suppose this conclusion is unacceptable, since we can suppose that the screen is shielded in such a way that electrons arriving at it could only have reached it via the slits in the diaphragm. We now ask ourselves how one might avoid the conclusion of the argument. One might avoid it by denying the major, or the minor, or the scheme of inference (the *modus tollens*), or by making distinctions in one or both of the premises. Putting distinctions aside, let us consider the three principal ways of voiding the argument.

The minor is the statement of an experimental fact and let us suppose it is untouchable.

One might consider adopting a new logic in which the *modus tollens* was an invalid deductive scheme. None of the writers, however, seems to have considered this possibility. It happens that the *modus tollens* is a valid scheme within the quantum logic of Birkhoff and von Neumann, Finkelstein, Jauch and others. The quantum logic referred to, however, does exclude some important classical schemes, e.g., the scheme,

$$\frac{\begin{array}{c} p \vee q \\ \sim q \end{array}}{p}$$

The third possibility consists in declaring the major premise false. Now the major premise involves three sentences, a_1, a_2 and b, each formulated within the language of classical physics and supposing the truth of the basic conceptual framework of classical physics. These sentences are joined together by two logical functors '$\vee$' ('or') and '$\supset$' ('implies').

Now the most natural solution which suggests itself is that the language framework and concepts of classical physics do not apply to electrons in the two slit experiment. Underlying the belief that $(a_1 \vee a_2) \supset b$ is the set of classical assumptions that each particle is an independent kinematical unit, that each travels along a uniquely determined trajectory from the source to the screen, that each particle consequently passes through one and only one slit. In this model, the set of particles arriving at any area of the screen would be the set theoretic union of two sets of particles, of those that pass through slit 1 and of those that pass through slit 2. In rejecting the validity of classical descriptive assumptions for the 2-slit experiment, one can falsify the major without impugning the validity of classical logic. On the other hand, one might claim that, whether or not classical assumptions are correct, the classical logic of classes which is used in the derivation of *b*, is invalid. This latter position seems to be the residue of the claim that classical logic – or the part of it which is the logic of classes – is not valid in the kinematical description of quantum mechanical systems.

The position just outlined is not a very revolutionary one, even supposing it were unexceptional. Some of the more articulate and public spokesmen on the side of quantum logic, Putnam for example and Finkelstein, have made broader claims for the validity of quantum logic, ones which *are* quite revolutionary. They would argue that the discovery of quantum logic dethrones classical logic in all empirical domains as the discovery of the general theory of relativity dethroned Euclidean geometry in the domain of physics. The argument goes: Just as the general theory of relativity has shown that what one was accustomed to call a 'straight line' is not an Euclidean object, so what one is accustomed to call a 'proposition' is not a PM-classical object, but an object subject to the weaker logic of a non-distributive lattice.

III. ABOUT THE ALLEGED UNIVERSALITY OF QUANTUM LOGIC

It is not my purpose, however, to add my endorsement to the view that quantum logic *in the form proposed* say, by Jauch, Finkelstein and Putnam, is correct but rather to criticize this view and to transform it into what I believe is a better account of the role of various logics in quantum mechanics.

My first move is against the more extreme thesis that the truth of quantum logic in the quantum domain calls into question the truth of classical logic in every empirical domain. The main argument is a reasoning by analogy with the fate of Euclidean geometry in general relativistic physics. The principal author of this argument is Putnam, although it is supported by Heisenberg, von Weizsäcker, Finkelstein and others. It is based upon the analogy between physical geometry and logic. Putnam puts down the proportional equation:

$$\frac{\text{Euclidean geometry}}{\text{General theory of relativity}} = \frac{\text{Classical logic}}{\text{Quantum mechanics}}$$

In his paper 'Is Logic Empirical?' given to this colloquim, Putnam wrote: "I regard the analogy between the epistemological situation in logic and the epistemological situation in geometry as a perfect one.'[6] A propos of classical logic, he writes: "Quantum mechanics explains the approximate validity of classical logic 'in the large' just as non-Euclidean geometry explains the approximate validity of Euclidean geometry 'in the small'."

To explore the analogy between geometry and logic, it is necessary to state Putnam's views about geometry. They are summarized in the following points. (i) Arguments in favor of Euclidean geometry based upon intuitionistic a priori reasons are unacceptable and false. (ii) Without empirical evidence there is no truth. The one physically correct geometry is to be determined by objective, empirical, scientific criteria. (iii) Conventionalism, or the view that a variety of geometries is admissible depending upon the rule of congruence conventionally chosen, is false to the extent that it allows human arbitrariness and subjective bias and prejudice to enter where objective criteria are or should be the sole relevant criteria.

One can have a great deal of sympathy with the thrust of these arguments, without agreeing with Putnam's interpretation of conventionalism and the consequence he draws from his rejection of it, namely, the belief that there is a unique objective physical geometry, one which is altogether removed from free conventions. I prefer to think that his views about conventionalism spring from certain options as to what he believes knowledge and objectivity imply or should imply. Putnam's belief in a unique empirically determined geometry is carried over into his belief that there is a unique kind of empirically determined logic. Since there are already a sufficient number of Putnam's adversaries in the field of geometry, I shall accept a challenge made by Putnam to present an elegant refutation of Putnam by Putnam.[7]

Putnam is committed to the defence of the position that only those schemes of inference are valid which are derivable from quantum logic. Among the invalid schemes of inference is one I have already noted:

$$\begin{array}{c} p \vee q \\ \sim q \\ \hline p \end{array}$$

(To illustrate in what sense this scheme is held by quantum logicians to be invalid in quantum mechanics, one might consider the following example:[8]

$p \vee q$: The electron has spin up or spin horizontally to the left
$\sim q$: The electron has not spin horizontally to the left
p: The electron has spin up

The major is taken to mean that the spin state of the electron is some linear superposition of spin up and spin horizontally to the left. Such a set of spin states spans the entire spin space. Hence, the only legitimate conclusion from the minor is that the spin of the electron is horizontally to the right. This solution satisfies the major and the minor, but it is different from p, the only conclusion authorized by classical logic. The classical scheme breaks down whenever it is a question of complementary propositions p and q.)

Now it is interesting to note that the principal argument of Putnam's paper 'Is Logic Empirical?' is one whose natural expression falls into the scheme I have just referred to. After an exposition of quantum logic,[9] the structure of the argument in support of quantum logic goes this way: Either quantum logic is correct or classical logic is correct; but it is not true that classical logic is correct (since this would involve the use of hidden variables, the observer-observed cut, a perturbation theory of measurement, etc.); hence quantum logic is correct. The scheme of inference for this argument is the one I have just mentioned and it is formally invalid in quantum logic if the two members of the disjunction are complementary. Putnam does not discuss whether or not his disjuncts are complementary: he seems to be unaware he is using a scheme which is no longer universally valid. I believe the propositions are complementary, but the proof depends on accepting the conclusions of this paper – namely, that one can have a quantum logic on the level of a meta-context-language of conditions together with a classical logic on the level of quantum event language.

What I have done in this section is to overturn the argument used by Putnam to support that aspect of his universal thesis which says that quantum logic is the kind of logic to be used on all occasions. I am not taking issue with that other aspect of his thesis which says that empirical evidence is relevant to the kind of propositional logic one endorses. This part of the thesis, I think, is true. My complaint, however, is that

Putnam draws unwarranted conclusions and that he fails to be sufficiently attentive to the quality of the empirical evidence available.

IV. QUANTUM LOGIC IN QUANTUM MECHANICS

My next move is to examine the substance of the more modest claim that in the domain of quantum physics, quantum mechanical propositions obey a special empirical logic, the so-called quantum logic, which is an orthocomplemented non-distributive lattice with weak modularity.

In all the recent discussions of quantum logic, the argument has taken the following form: If the received version of quantum mechanics entails the use of quantum logic, then the rejection of quantum logic entails the rejection of the received version of quantum mechanics, and this consequence is further taken to entail the belief that the true theory of microphysical phenomena will be one of a classical kind not yet formulated, with 'hidden variables'.

When one turns to the recent works on quantum logic, one naturally looks for the sense in which it is novel, exciting, significant and controversial to say that quantum mechanics entails the use of a special quantum logic. I do not find a clear answer in the literature. For in the development of the argument, the antecedent is usually taken trivially to do nothing more than to label a well-known structural property of quantum mechanics related to its Hilbert space formulation. The real thrust of the argument is in the consequent, which is not really about logic, but about quantum mechanics and the alternatives to quantum mechanics.

Now the question of hidden variables or the observer-observed cut or the perturbation theory of measurement are not new problems and merely to express them in new terminology is neither novel, exciting, significant nor controversial. It is not my intention in this paper to speak about the alternatives to the present version of quantum mechanics. I am not challenging quantum mechanics in its present form, but I am concerned with the kinds of logic compatible with or required by the present form of quantum mechanics.

I am claiming that the recent defenders of quantum logic are quite confused about some fundamental things – in what they conceive to be the objects of a quantum logic within quantum physics, and in what the true significance of the new quantum logical operations consists.

Most of the current writers on quantum mechanics give as its domain the domain of those categorical propositions expressing the outcome of Yes-no tests. I say this claim is simply false. I do not deny, however, that a special quantum logic has a place within quantum physics but I put it on the level of a meta-context-language about the conditions under which particular quantum event-languages are applicable, and not, as the writers on quantum logic are accustomed to put it, on the level of quantum event-language itself.

The account writers give of what constitutes the class of quantum mechanical propositions which is subsumed under quantum logic are subject to at least one systematic ambiguity which conceals the real significance and true role of quantum logic. Let me first of all try to make clear the different types of propositions involved in this discussion before taking on the role of critic.

V. VARIETIES OF QUANTUM MECHANICAL PROPOSITIONS

Let me start with a class of propositions each of which states the outcome of a Yes-No experiment.[10] I shall call the categorical statement asserting or denying the outcome of a Yes-No experiment, a *simple empirical proposition*. Let me list the necessary and sufficient conditions which warrant a simple empirical proposition.

A simple empirical proposition presupposes a fully specified kind of measurement situation, and one in which only a definite part of the range of the observables is considered. If the quantum system relative to the observables in question falls into the part of the range being considered, a positive symbol '1', standing for Yes! is produced; if the quantum system does not fall within the range considered, a negative '0' standing for No! is produced. The sentence token standing for the Yes-propositions will have some form like: 'The observables X, Y for the electron now in the apparatus fall within the range $\Delta_{xy} \in B(R_x \otimes R_y)$' where $B(R_x \otimes R_y)$ is the set of Borel sets in the direct product of the ranges R_x and R_y of the observables X and Y. Let us call this proposition a. The No-proposition will have the form: 'The observables X, Y for the electron now in the apparatus do not fall within the range $\Delta_{xy} \in B(R_x \otimes R_y)$'. Since in both cases, the measurement situation is the same, the latter No-proposition is equivalent to the following Yes-proposition: 'The observables X, Y for the electron now in the apparatus fall within the complement of the range Δ_{xy} in $B(R_x \otimes R_y)$'. Let us call this proposition $\sim a$. The kind of negation in question is called by van Fraasen 'choice negation', as distinct from what he calls 'exclusion negation'. In order to warrant a or $\sim a$, it is necessary and sufficient that the physical measurement situation be fully specified in kind.

Let us suppose, what I presume Jauch and other writers on quantum logic supposed, *that the observables X and Y are compatible observables*, represented by commuting self-adjoint operators. We shall return to criticize this assumption later on in the paper.

There is also a class of more abstract propositions of the following kind: 'The projection operator $P(S)$ on the subspace S of the Hilbert space H preserves the state vector ψ; i.e., $P(S)\psi = \psi$.' Let us call this a 'simple theoretical proposition', and denote it by ϕ. Then the space of ϕ's obey a non-classical logic which is isomorphic with the orthocomplemented non-distributive lattice of subspaces of a Hilbert space. I doubt that it is necessary to prove this assertion. It has been proved by Birkhoff and von Neumann and every writer on two-valued quantum logic has used this theorem since.

Now let us suppose that the simple empirical proposition a implies ϕ ($=_{\text{def}}$'$P(S)\,\psi = \psi$') through the semantical rules or uniformities, where S is the subspace of H spanned by the eigenvectors of X and Y whose eigenvalues lie in the range Δ_{xy}. Then $\sim a$ implies ϕ' ($=_{\text{def}}$'$P(S^{\perp})\psi = \psi$') where $S^{\perp}$ is the orthogonal complement of S in H. In other words,

(1) $\quad a \supset \phi$

and

(2) $\quad \sim a \supset \phi'$

Since both a and $\sim a$ presuppose a fully specified kind of measurement situation, and ϕ does not; neither (1) nor (2) are in general biconditionals. That is, it is not true in general that

$$\phi \supset a$$
$$\phi' \supset \sim a$$

The reason there is no biconditional relation between a and ϕ in general is that, as long as X and Y are supposed to be compatible observables, the subspace S might possibly be spanned by other observables, say, U and V, one or both of which are incompatible with X and Y. For example, if the proposition a is 'The spin of the electron is $\frac{1}{2}$ and polarized vertically up or down', the subspace S corresponding to this proposition is the entire spin space. The physical situation, however, also includes an apparatus for measuring vertically polarized spin. Consider, now, the proposition b 'The spin of the electron is $\frac{1}{2}$ and polarized horizontally to right or left'. The subspace corresponding to b is also S. However, in this case the physical situation includes an apparatus for measuring horizontally polarized spin. Both a and b imply ϕ, but ϕ does not imply a or b separately, although it does imply $a \vee b \vee \cdots$ where the disjunction includes all possible simple empirical propositions which imply ϕ.

I do not know to what to attribute the oversight in recent literature of the problem that maybe there is not in general a biconditional relation between a Yes-no test and a simple theoretical proposition which the Yes-no test implies. Birkhoff and von Neumann seemed to be aware of the problem: they spoke of an equivalence class of experimental propositions. But none of the recent literature mentions it. Perhaps, the authors thought it was too trivial a matter to point out. Trivial though it may seem to be, however, big and important consequences follow from it.

In the general case, then, there is at least one other simple empirical proposition b, of type a but incompatible with a, which also implies ϕ. In other words, $a \supset \phi$, $b \supset \phi$, but a is not compatible with b. Either a or b gives full empirical warrant to the proposition ϕ, but the truth of ϕ, does not justify a nor does it justify b, although it does justify

$$a \vee b \vee \cdots$$

where the disjunction includes all possible simple empirical propositions of type a which imply ϕ.

Now the empirical propositions which constitute the space of quantum logic are not propositions of type a, simple empirical propositions (since they are not in biconditional relation to ϕ) but they are propositions of the type

$$(3) \qquad a \vee b \vee \cdots =_{\text{def}} a^* (\phi)$$

I shall call $a^* (\phi)$ the 'empirical interpretation of the simple theoretical proposition ϕ'. A biconditional relation exists between ϕ and the chain of disjunctions $a^*(\phi)$, that is,

(4) $a^*(\phi) = (a \vee b \vee \cdots)$ if and only if ϕ

The $a^*(\phi)$'s then constitute a space of propositions isomorphic with the space of ϕ under the logical functors 'and', 'or', 'implies' and 'not' *as these function within quantum logic.*

We have distinguished then three types of propositions in quantum mechanics: simple empirical propositions (or propositions of type a), simple theoretical propositions (propositions of type ϕ) and the empirical interpretations of simple theoretical propositions (or propositions of type $a^*(\phi)$).

A simple empirical proposition a implies the fulfillment of the following conditions, both necessary and sufficient in their respective orders: (i) in the linguistic order: that there be an event-language L_a with the names of objects (or other referring expressions), descriptive predicates corresponding to the observables of the commuting set X and Y, and the logical grammar of classical logic. Note that L_a does not contain the descriptive predicates for observables U or V which are not compatible with X and Y. (ii) In the physical order: some standard measuring conditions for measuring the ranges Δ_x and Δ_y of the observables X and Y in question. Let me call this the 'physical context of the system'. (iii) In the mathematical order: that the subspace $S \leqslant H$ be coordinatized by the eigenvectors of the commuting self-adjoint operators which represent the physical observables X, Y mentioned in (ii), and which the descriptive predicates mentioned in (i) stand for.

Returning to the language L_a. Let us identify it with the set of those sentences which can be correctly formulated, asserted or denied in the standard measuring conditions described in (ii) above. L_a is a *picturing language.* Out of its resources a variety of linguistic pictures can be put together. One of these is a; another is $\sim a$. But b, which is a proposition incompatible with a, and which implies its own language L_b, is not one of these pictures. That is, $a \in L_a$, but $b \notin L_a$. Similarly, $b \in L_b$, but $a \notin L_b$. The set of linguistic pictures that can be formulated in L_a is the mapping of the manifold of possible physical events that could occur within the limitations of the physical milieu described in (ii) above.

The standard measuring conditions of a physical context are themselves invariant relative to the manifold of possible events that could occur within that physical context. The constancy of the standard measuring conditions is correlated with the domain of applicability of language L_a, L_a being the linguistic invariant of the manifold of possible pictures. There is nothing special about the logic intrinsic to L_a or to L_b separately; it is or could be the classical propositional logic of PM, a Boolean algebra of statements under the operations of 'and', 'or', 'implies' and 'not'. The logic of simple empirical propositions is or could be classical.

The second type of proposition I mentioned is a simple theoretical proposition or a proposition of type ϕ. The space of these propositions constitute an orthocomplemented non-distributive lattice of the kind called a quantum logic. This was proved by Birkhoff and von Neumann in 1936 for quantum mechanics in what we now call its received form.

Are the ϕ's, however, the basic propositions about quantum mechanics to which the quantum logicians refer when they make their claim that quantum mechanical propositions obey a special non-classical logic? It would be easy to quote passages from the writings of Jauch, Finkelstein, Putnam and others to show that whatever else they had in mind this at least was part, and possibly even the whole of what quantum logic was all about. Jauch, for example, writes: "Elementary propositions of quantum mechanics are represented by projection operators, or equivalently by closed linear subspaces."[11] But if this is all the quantum logicians intend to say, then I have no criticism to make. All I would say is that their claim is neither novel, exciting, significant nor controversial.

In fact, however, additional claims are made, that these simple theoretical propositions are also *empirical propositions*, in the sense that they represent the answer to Yes-No experimental questions. Jauch writes: "We shall refer to Yes-No experiments simply as *propositions* of the physical system."[12] But Yes-No experiments are not isomorphic to simple theoretical propositions, as I have shown. A simple theoretical proposition is correlated with a disjunction of simple empirical propositions expressing the outcome of competing and generally mutually incompatible Yes-No tests. In this fact lies the empirical relevance of ϕ. But is ϕ itself a statement of empirical fact? To state an empirical fact is generally to *picture* a state of affairs in a picturing language. A picturing language generally has no more than the following resources: names of objects (or other referring expressions), empirical predicates and a logical vocabulary. The empirical picturing predicates of quantum mechanics are the observables. To say of a quantum system that its state vector lies in a certain subspace of a Hilbert space is then not to picture the quantum system since no observable predicate has been used.

Perhaps, however, it is the implicit but esoteric claim of the quantum logicians that ϕ is a *picturing*.[13] That would indeed be a novel, exciting, significant and controversial claim. But if that is the significant claim they want to make, they seem not to have noticed its unusual character. For if ϕ were a true picturing, then each subspace of H would have to correspond to a unique empirical predicate. There is at present no warrant for this. The traditional interpretation of quantum mechanics permits and indeed implies two different kinds of *picturings*, (i) of the sentences within a particular language L_a which picture the manifold of events in a physical context and (ii) of the set of competing languages $L_a, L_b, \ldots$ which themselves picture the manifold of possible physical contexts. (Note: A physical context constitutes the conditions for a definite manifold of events; it is then invariant relative to the events of the manifold. In other words, two different events of the same manifold share the same physical context; two events from different manifolds do not share the same physical context.)

I said above that the empirical interpretation of ϕ, authorized by ϕ, is:

$$a^* (\phi) = (a \vee b \vee \cdots)$$

We are now in a position to amplify and correct this assertion. Since $a \in L_a$, but $a \notin L_b$ and $b \in L_b$, but $b \notin L_a$, the members of the disjunction $a^* (\phi)$ are sentences belonging to different languages. Let us group together in $a^* (\phi)$ the sentences that belong to each language, repeating a sentence, if necessary, so that each language group is complete. We write,

$$(5) \qquad a^* (\phi) = (a_1 \vee a_2 \vee \cdots) \vee (b_1 \vee b_2 \vee \cdots) \vee (c_1 \vee c_2 \vee \cdots) \vee \cdots$$

where $a_1, a_2 \ldots$ all belong to L_a, $b_1, b_2 \ldots$ all belong to L_b and so on. What is entailed then by ϕ, is not a disjunction of sentences within a common language, but a *disjunction of languages*. Thus, the correct revised form of $a^* (\phi)$ is:

$$(6) \qquad a(\phi) = (\text{`}L_a\text{'} \vee \text{`}L_b\text{'} \vee \text{`}L_c\text{'} \vee \cdots)$$

where 'L_a', 'L_b', etc. represent statements within a higher-order meta-context-language $m(L)$; 'L_a', for instance standing for the statement

'This physical context is an 'L_a'-type context' or some equivalent formulation. $a(\phi)$ is then a statement in $m(L)$; not in any one of the event languages L_a, L_b, etc.

Let us look at the lattice structure of $m(L)$. A partial ordering relation '$\rightarrowtail$' (read 'implies') can be introduced into a set of propositions $a(\phi)$: let

$$a(\phi_1) \rightarrowtail a(\phi_2) \quad \text{if and only if} \quad \phi_1 \rightarrowtail \phi_2$$

then, since $a(\phi)$ is isomorphic with ϕ, the logic of the $a(\phi)$'s. i.e., of $m(L)$, is isomorphic with the logic of the ϕ's; that is, isomorphic with the non-distributive lattice of the subspaces of a Hilbert space. We can define greatest lower bounds (g.l.b.) of two sentences, $a(\phi_1) \otimes a(\phi_2)$ (read '$a(\phi_1)$ *and* $a(\phi_2)$'), least upper bounds (l.u.b.) of two sentences, $a(\phi_1) \oplus a(\phi_2)$ (read '$a(\phi_1)$ *or* $a(\phi_2)$') and complements, $a'(\phi_1)$ (read 'not-ϕ') in the usual way:

$$\begin{array}{lll} a(\phi_3) = a(\phi_1) \otimes a(\phi_2) & \text{if and only if} & \phi_3 = \phi_1 \otimes \phi_2 \\ a(\phi_4) = a(\phi_1) \otimes a(\phi_2) & \text{if and only if} & \phi_4 = \phi_1 \oplus \phi_2 \\ a(\phi_5) = a'(\phi_1) & \text{if and only if} & \phi_5 = \phi_1' \\ a(\phi_6) = \emptyset & \text{if and only if} & \phi_6 = \phi \\ a(\phi_7) = 1 & \text{if and only if} & \phi_7 = H \end{array}$$

where '$\otimes$', '$\oplus$' and complementation on the right hand side refer to the logical operations of conjunction, disjunction and negation in the domain of simple theoretical propositions. The functors '$\otimes$', '$\oplus$' and '$\rightarrowtail$' on the left hand side can then be interpreted as the logical functors 'and', 'or' and 'implies' of a quantum logic in the meta-context-language $m(L)$, with complementation as negation. The logic of the $a(\phi)$'s on the left hand side is then an orthocomplemented non-distributive lattice isomorphic with the lattice of subspaces of a Hilbert space.

Let me point out some of the attributes of the meta-context-language $m(L)$. In the first place, $m(L)$ is not an event-language; its objects are the manifold of possible event-languages and the manifold of conditions for the correct use of each. Its subject matter is the set of physical contexts in which it is relevant to use one linguistic or conceptual framework rather than another. I have called it a meta-context-language, because its subject matter is the semantical applicability of event-languages. As I have already pointed out, $m(L)$ is a *picturing language*; it pictures the manifold of physical contexts; each of the physical contexts constitutes a domain of possibility for some manifold of events.

It is on the level of $m(L)$ that I believe the real significance of quantum logic lies. For while the logic of the meta-context-language of quantum mechanics is a non-distributive lattice, the analogous context-language of classical physics is a Boolean algebra isomorphic with the logic of subsets of the Euclidean phase-space of classical physics.[14]

The situating of the characteristic logics of classical and quantum physics in a meta-context-language throws light on the root cause of the difference in form between these two sciences. The quantum logic part of quantum physics expresses the context-dependent character of simple empirical propositions in quantum physics. The equation:

$$(7) \qquad a(\phi_1) \oplus a(\phi_2) = a(\phi_1 \oplus \phi_2)$$

can be interpreted to mean that if one were to enlarge the physical context of a quantum system satisfying $a(\phi_1)$ so as to include conditions which satisfy $a(\phi_2)$, then one would obtain a new kind of physical context $a(\phi_1 \oplus \phi_2)$ in which the new manifold of possible events is different from the set theoretic union of the separate manifolds of events which corresponded to the former physical contexts in isolation from each other. That it is the context dependent character of simple empirical statements in quantum mechanics which generates a non-classical logic at the level of the meta-context-language $m(L)$ can be shown by examples constructed outside of the realm of physics.

For example, consider the philosophical languages used by two philosophers – let us call them, Joseph and Abner – each possessing a distinct and different philosophical perspective, but each capable of entering into dialogue with the other without yielding anything of which he considers essential to his perspective. Let L_A be the set of philosophical statements which Abner uses in a context that does not involve dialogue with Joseph; L_B be the set of philosophical statements used by Joseph in a context that does not involve dialogue with Abner, and L_{AB} the common language of dialogue. In order to get a non-distributive lattice all we need to suppose is (i) that L_{AB} contains sentences not in L_A or in L_B and, (ii) since within L_{AB} both Joseph and Abner can preserve their different philosophical perspectives, L_A and L_B are subsets of L_{AB}. We can then define complements of L_A and L_B in the following set theoretic way:

$$L'_A = (L_{AB} - L_A) \cup L_B$$
$$L'_B = (L_{AB} - L_B) \cup L_A$$

And we can define implication ('→—') in such a way that L_X→—L_Y, if and only if every sentence of L_X is also a sentence of L_Y. Then, the following objects constitute a lattice (L_0, L_A, L_B, L'_A, L'_B, $L_{AB}=1$), where L_0 is the set theoretic intersection of L_A and L_B. The operations of '⊗', '⊕' and '→—' are schematized in the diagram below. We can interpret the objects in the figure below as sentences in a meta-context-language of conditions $m(L)$: for $L_A \oplus L_B$, read 'This is an 'L_A'-type situation and/or an 'L_B'-type situation'; for $L'_B \otimes L_A$, read 'This is an 'L_A'-type situation *and* it is *not* an 'L_B'-type situation'; and so on. Then we have the result that $m(L)$ is isomorphic to the orthocomplemented lattice represented in Figure 2. The arrows represent the partial ordering '→—' (or 'implies')

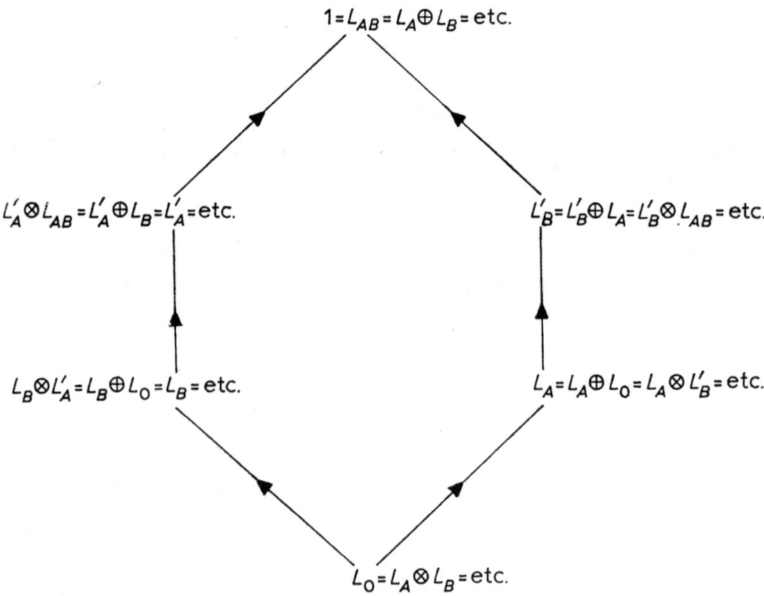

Fig. 2. Lattice of the meta-context language $m(L)$.

among the elements. The l.u.b.'s (representing disjunctions in $m(L)$) and the g.l.b.'s (representing conjunctions in $m(L)$) can be read off from the figure. The lattice is non-distributive as can be seen from the relations.

$$L_A \oplus (L_B \otimes L'_B) = L_A$$
$$(L_A \oplus L_B) \otimes (L_A \oplus L'_B) = L'_B$$

but,

$$L_A \neq L'_B$$

Hence the logic intrinsic to the meta-context-language $m(L)$, which has as its subject matter L_A, L_B, L_{AB} and the conditions for their correct use is an orthocomplemented non-distributive lattice like quantum logic.

The point I am making is that one does not have to look to physics to find quantum logic. One finds it in the meta-context-language of context-dependent statements. You can construct your own examples: the language of work and the language of leisure, the language of esthetic beauty and the language of functional utility. Nor is it necessary to appeal to the actions or emotions of human subjects. In chemistry, there is the language of heterogeneous gases. In quantum physics, there is the language of single slit experiments and the language of double slit experiments.

Since I started this paper with the example of the two slit experiment, I might point out how to read the above diagram as descriptive of the meta-context-language of the two slit experiment. Let L_A be the single slit language relevant to the physical situation of slit 1 open, and let L_B be the single slit language relevant to the physical situation of slit 2 open, and let L_{AB} be the double slit language relevant to the situation of both slits open. Then, the solution to the problem we started with, namely, how to void the argument.

$$\begin{array}{c} (a_1 \vee a_2) \supset b \\ \sim b \\ \hline \sim (a_1 \vee a_2) \end{array}$$

is the following. Either a_1 belongs to L_A and a_2 belongs to L_B, then the major is not well-formed, since we have no rules for disjoining sentences belonging to different languages. Or a_1, a_2 and b are to be interpreted as sentences in L_{AB}. The major of the argument is then false both empirically and theoretically, not, however, logically by virtue of a special logic.

I hope it is clear from these examples that the general character of non-distributivity in the meta-context-language is a direct consequence of the fact that although there are disjoint contexts, say, of Joseph philosophizing to the exclusion of dialogue with Abner and vice versa, the joint interaction context of Joseph and Abner in dialogue has broader possibilities than the sum total of those offered by the disjoint non-interacting contexts. We find just such context dependence in quantum mechanics.

VI. GENERAL QUANTUM MECHANICAL LANGUAGE AND GENERALIZED COMPLEMENTARITY

Remembering that a physical context is those physical conditions which specify, antecedently to the outcome of any experimental observation, the manifold of possible events to which any experimental outcome will belong, and which, consequently, remain invariant throughout the course of any experimental test, one can ask how the separation and union of contexts is represented in quantum mechanics. Consider the following two equations:

$$a(\phi) = (`L_a' \vee `L_b' \vee \cdots) \tag{6}$$

and

$$a(\phi_1) \oplus a(\phi_2) = a(\phi_1) \oplus \phi_2) \tag{7}$$

We can ask; (i) how can one physical context – somehow associated with ϕ – be consistent with a set of mutually exclusive physical contexts like those associated with L_a, L_b, etc.? and (ii) Does it not seem from the form of (7) that ϕ (and not 'L_a' or 'L_b' etc.) *pictures* the physical context of the system?

The answer to the former is that the physical context associated with ϕ is that of the preparation of state, while those associated with 'L_a', 'L_b' etc. are those associated with a subsequent measuring process.

Turning to the latter question: to say that the system is prepared in state ϕ is to say something about the physical world, actually about the preparation of state. *But is it a picturing*? We have already raised this question and the answer given was, No! However, at that time, all we had at our disposal was quantum event-language, i.e., the language of Yes-No tests. The suggestion is now made that possibly 'ϕ' is a picturing in $m(L)$, the meta-context-language that deals with the kind of language appropriate to a physical context rather than with specific quantum events. But if the 'L_a'-type context and the 'L_b'-type context are *absolutely* mutually exclusive, then there is no term in the linguistic resources of $m(L)$ with which to designate a *general physical context*, uniquely correlated with 'ϕ', of which the 'L_a'-type context and the 'L_b'-type context are specific particularizations. Under these circumstances 'ϕ' can only denote a *class* of mutually exclusive contexts – not a general type of physical context.

If, however, the 'L_a'-type physical context and the 'L_b'-type physical context are not absolutely mutually exclusive, a general physical context might exist of which the 'L_a'-type and the 'L_b'-type were specifications (as it were, limiting or special cases of the general case). This would imply the existence of a general language appropriate to this context, say, $L_{abc...}$, of which L_a, L_b, L_c etc. would be subsets corresponding to special or limiting cases. The enlarged possibilities of united contexts then imply that, although $L_a \leqslant L_{abc...}$, $L_b \leqslant L_{abc...}$ etc., it is also the case that

$$L_{abc...} \neq L_a \cup L_b \cup L_c \cup \cdots$$

The meta-context-language, $m(L)$, then, would contain, besides the names 'L_a', 'L_b', 'L_c' etc., the name '$L_{abc...}$' of the general context $L_{abc...}$. It seems plausible to the author that the mutual exclusivity of complementary physical contexts may not be absolute, and that there may exist a general physical context '$L_{abc...}$' uniquely correlated with 'ϕ'. This is suggested by the use which Bohr and Heisenberg make of the Indeterminacy Principle when they apply it to an individual quantum system in one and the same physical context. This cannot be done without the simultaneous attribution of both position and momentum predicates (though evidently not the classical ones) to a quantum system in a general physical context. If $L_{abc...}$ exists, then there is a general physical context pictured by '$L_{abc...}$', and it would indeed become plausible to say that 'ϕ' pictures – not, however, within any of the event-languages L_a, L_b, L_c etc., but within the meta-context-language $m(L)$. The disjunction $a(\phi)$ would then include an extra term,

$$a(\phi) = ('L_a' \vee 'L_b \vee 'L_c \vee \cdots \vee 'L_{abc...}')$$

and since L_a, L_b, L_c etc. are all subsets of $L_{abc...}$, this reduces simply to,

$$a(\phi) = ('L_{abc...}')$$

Under these circumstances, we can drop the assumption made earlier that the observables X and Y referred to in the description of a Yes-No test be commuting observables. Even if they do not commute, they belong to the general language $L_{abc...}$ within which the physical context of the Yes-No test can now be uniquely described.

The situation we have just described suggests that the notion of complementarity be extended to the set of mutually exclusive event-languages L_a, L_b, L_c etc. on condition that there exist a general event-language, $L_{abc...}$ which contains all of them, but is richer in expressive power than L_a, L_b, L_c etc. taken in contextual isolation. This supposes that the contexts A, B, C, etc. which specify the conditions for the valid use of L_a, L_b, L_c etc. are mutually exclusive but not absolutely so. It supposes that there exists a general or synthetic context $ABC...$ to which there corresponds a general language $L_{abc...}$ satisfying the following conditions:

(8) $$L_a \leqslant L_{abc...}$$
$$L_b \leqslant L_{abc...}$$
$$L_c \leqslant L_{abc...} \text{ etc.}$$

and

(9) $$L_{abc} \neq L_a \cup L_b \cup L_c \cup \cdots$$

It further supposes that there is a way of defining complements L'_a, L'_b, L'_c etc. of L_a, L_b, L_c etc. respectively such that $(L_0, L_a, L_b, L_c, ..., L'_a, L'_b, L'_c, ..., L_{abc...})$ constitutes an orthocomplemented non-distributive lattice.

This claim is illustrated and justified by the author in another paper, for the case of two event-languages L_a and L_b.[15] The complements L'_a and L'_b can be defined in the following way:

$$L'_a = (L_{ab} - L_a) \cup L_b$$
$$L'_b = (L_{ab} - L_b) \cup L_a$$

The partial ordering '$\rightarrow$—' ('implies') of the lattice is to be understood in such a way that, $L_x \rightarrow\!\!— L_y$ if and only if every sentence that can be correctly used (either by affirmation or negation) in context X can also be correctly used in context Y. L_a and L_b are said to be *complementary* in L_{ab}. In this situation, the meta-context-language $m(L)$ will be a domain of non-classical logic, even though each of the event-languages L_a, L_b and L_{ab} obey or could obey classical PM-logic.

In the paper referred to, the author worked out some examples taken from quantum mechanics, as, for example, the paradigm case of the relation between quantum mechanical position language (L_a) and quantum mechanical momentum language (L_b) and the general quantum mechanical kinematical language (L_{ab}). The author also uses this logic of contexts (or context-logic) to show the relationship between the Aristotelian and the Augustinian-Platonic languages in the philosophy of Aquinas and suggests the usefulness of the method of context-analysis in the history of philosophy, the history of science and in the metascience of both philosophy and science.[16]

In the case of four event-languages, L_a, L_b, L_c and L_{abc} which satisfy (8) and (9), it may be possible to define appropriate complements so that

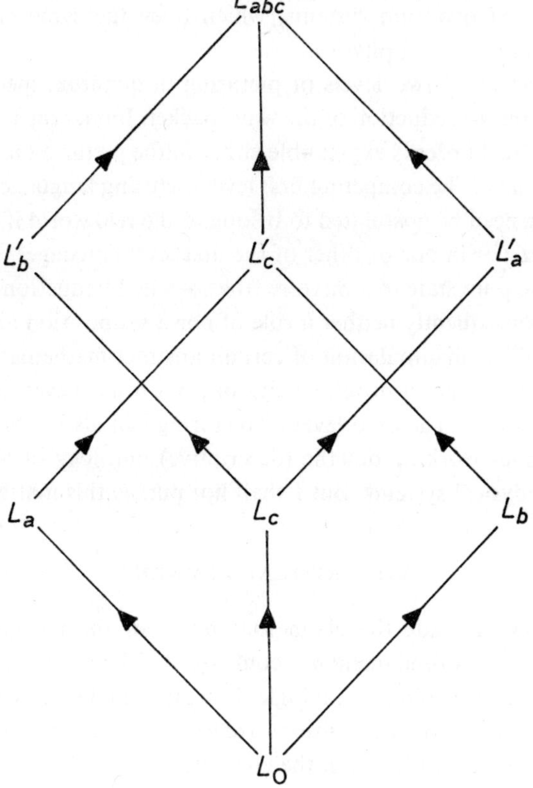

Fig. 3. Diagram of a non-Boolean lattice for three complementary languages L_a, L_b and L_c.

the set (L_0, L_a, L_b, L_c, L'_a, L'_b, L'_c, L_{abc}) constitutes an orthocomplemented non-distributive lattice. In this case, L_a, L_b and L_c are said to be *complementary in* L_{abc}. The structure of such a lattice is represented in Figure 3 below.

In the light of what has been said about the pervasive character of contextual dependence in our experience, classical physics with its context-independence begins to appear in some sense as incomplete, as lacking a contextual dimension one feels every physical statement should have, and one begins to suspect that perhaps this is due to a smoothing, smearing or averaging of many context-dependent results. That this may be so, is suggested by the fact that we can construct certain context-independent variables, called expectation-values, for ensembles of identically prepared quantum systems, which obey the same laws as their counterparts in classical physics.

The separation of two levels of picturing in quantum mechanics also throws light on the reduction of the wave packet. If it is true that whatever concerns the real order is expressible either in the picturing metalanguage $m(L)$ or in one of the competing first level picturing languages, L_a, L_b, ... then nothing need be postulated to belong to the real world if it cannot be stated in $m(L)$ or in one or other of the first level languages. The conversion from the pure state to a mixture (the so-called 'reduction of the wave packet') is consequently neither a rule of nor a proposition about reality, but is part of the manipulation of certain abstract mathematical objects. This, of course, is not a novel conclusion; but it has never been reached before in this way. The dual level of picturing can also suggest how one might go about working out the (descriptive) ontology of the quantum domain of physical systems. But I shall not pursue this matter now.[17]

VIII. CRITICAL REMARKS

In Section IV, I made the claim that much of the recent writing on quantum logic was fundamentally confused and had failed to appreciate the true significance of quantum logic. I ought to make explicit, if it is not already clear, what these confusions are and where their source lies.

The principal confusion is in the characterization of those basic quantum mechanical propositions which, it is claimed, obey quantum logic and not classical logic. Birkhoff and von Neumann describe these propositions as equivalence classes of categorical statements of fact, each statement supposing a definite set of commuting observables. They have the logical form of Equation (3) above, and Birkhoff and von Neumann do not pursue the analysis to the point of deriving Equation (6).

Putnam and Finkelstein, on the other hand, are dealing, not with categorical propositions but with subjunctive conditionals of the form: 'If a certain test were to be made, then the system would pass it'. They are not statements of fact but of the dispositions which a quantum system has. They have the desirable property, however, of being isomorphic with simple theoretical proposition, i.e., if p is one of those propositions, then

$$p \leftrightarrow \phi$$

The logic of these subjunctive conditionals is then a quantum logic isomorphic with the lattice of subspaces of a Hilbert space.

How satisfactory is this solution from the logical point of view? And with what warrant are these subjunctive conditionals called the 'empirical propositions of quantum mechanics'?

From the point of view of either classical or quantum logic a subjunctive or counterfactual conditional is not a clear and definite structure. It suffers from the ambiguities and latent paradoxes which are the subject matter of an extensive and inconclusive literature.[18] Moreover, if the implications of a single subjunctive conditional are not always clear, the logical product of two such sentences is perilous indeed. Consider the sentences: 'If I were drunk, I would dance a jig' and 'If I were sober, I would trim the hedge'. The former implies I am not drunk; the latter implies I am not sober. What then would the conjunction of the two sentences imply? If the basic sentences of quantum mechanics were subjunctive conditionals, the antecedents of the basic sentences would include all the mutually exclusive physical conditions into which the system could be placed. Just as in the example I have chosen and for the same reason, if a conjunction of a set of basic sentences were formed, it would not be clear what such a conjunction would imply.

Again, as statements about the physical constitution of the world, counterfactual conditionals leave open and undetermined, or if you like, to be determined by the subject or by nature (but more often by the subject), the final conditions (or physical context) without which nothing is or can be observed, and no matter of fact established. In Heisenberg's language, they refer to the *potentiae* of nature. One might perhaps be content with this situation if one subscribed to the view espoused by von Weizsäcker and endorsed by Heisenberg, that the subject matter of quantum physics is the subject-object interaction between a human observer and nature. Then, the empirical content of physics might be expected to reflect the initiatives which a scientist-observer takes spontaneously and unpredictably in his relation with the world. But it is not necessary, I believe, and it is certainly contrary to the ideal aims of natural science, that quantum physics renounce its right to make objective and *categorical* statements about the world and not merely about its *potentiae* which are true independently of observers.

Moreover, calling these subjunctive conditionals the 'empirical propositions of science' is unwarranted, for the thrust of work on quantum logic has been and is the attempt to articulate the 'logic of nature' as revealed by quantum mechanics in the sense of the 'interrelatedness of things' in nature. Such a 'logic of nature' would be pictured in empirical propositions of a *categorical* kind, not in counterfactual conditionals. It is true that certain experimental conditions are mentioned which anchor the conditional statement in the real world. These give empirical warrant to the subjunctive conditionals. But wherever there is a warranted subjunctive conditional its warrant is in some set of objective laws linking the antecedent and the consequent.[19] These are the links which express the 'logic of nature', and these are the empirical propositions of quantum mechanics. Now these links are categorical propositions. There are, I claim, two levels of such propositions: the level of quantum event-languages and the level of a meta-context-language $m(L)$ about the conditions for the use of quantum event-languages.

My criticism of Jauch's very elegant treatise *The Foundations of Quantum Mechanics* is that he too seems not to have considered the variety of propositional types employed in quantum mechanics. Sometimes he speaks as if the basic quantum mechanical propositions were simple empirical propositions, the results of Yes-no experiments, sometimes as if they were simple theoretical propositions about subspaces of a Hilbert space, and at other times as if they were counterfactual conditionals. Moreover, while it is understandable that quantum logicians would want to ground their claims in empirical fact, it happens that this empiricist urge sometimes caricatures itself, as we see, for example, in the case of Jauch and Ludwig. They describe the construction of an *infinite experimental filter* for two incompatible propositions $a \otimes b$.[20] One wonders with what right an infinite experimental filter that cannot be made or used is termed 'experimental'.

Department of Philosophy,
State University of New York at Stony Brook

APPENDIX

Some Terms, Definitions, Axioms and Formulas

I. *By classical logic* is meant a *Boolean algebra of sentences.*

A. A Boolean algebra is a set B of at least two distinct elements with two binary operations $\cup$ (cup) and $\cap$ (cap) and one unary operation$'$ (prime) such that B is closed with respect to each of these three operations and for all a, b and c belonging to B the following axioms are satisfied:

A 1 $\quad a \cup b = b \cup a$

A 2 $\quad a \cap b = b \cap a$

A 3 $\quad a \cup (b \cup c) = (a \cup b) \cup c$

A 4 $\quad a \cap (b \cap c) = (a \cap b) \cap c$

A 5 There is an element ϕ belonging to B such that
$a \cup \phi = a$

A 6 There is an element I belonging to B such that
$a \cap I = a$

A 7 $\quad a \cup a' = I$

A 8 $\quad a \cap a' = \phi$

A 9 $\quad a \cup (b \cap c) = (a \cup b) \cap (a \cup c)$

A 10 $\quad a \cap (b \cup c) = (a \cap b) \cup (a \cap c)$

A9 and A10 are the *distributive laws*

Def.: $a \leqslant b$ if and only if $a \cup b = b$

B. A Boolean algebra of sentences is obtained by adding the following interpretation to the formal syntax A:

a, *b*, *c*,	stand for sentences
ϕ	stands for the absurd sentence (always false)
I	stands for the trivial sentence (always true)
$a \cup b$	stands for *a and/or b* (the logical sum)
$a \cap b$	stands for *a and b* (the logical product)
a'	stands for *not-a*
$a \leqslant b$	stands for *a implies b*

II. By a *quantum logic* is meant an *orthocomplemented non-distributive lattice* of sentences.

A. An orthocomplemented non-distributive lattice is obtained (rather than an algebra) if the axioms A 9 and A 10 are dropped in *IA*. Using '$\oplus$', '$\otimes$' and '$\rightarrow$' instead of cup, cap and '$\leqslant$' to distinguish the non-distributive lattice operations from the Boolean operations, we use the following terminology for a lattice:

$a \oplus b$	is called the 'least upper bound' or 'l.u.b. of *a*, *b*'
$a \otimes b$	is called the 'greatest lower bound' or 'g.l.b. of *a*, *b*'
$a \rightarrow\!\!- b$	stands for *a implies b*

B. A quantum logic is obtained by adding to the axioms of II A, the following interpretation:

$a \oplus b$	stands for *a and/or b*
$a \otimes b$	stands for *a and b*
a'	stands for *not-a*
$a \rightarrow\!\!- b$	stands for *a implies b*

where a, b, c, ϕ, I are sentences, ϕ is the absurd sentence and I is the trivial sentence.

III. A diagram representing the relations between the elements of an *orthocomplemented non-distributive lattice* of six elements. The arrows indicate implication.

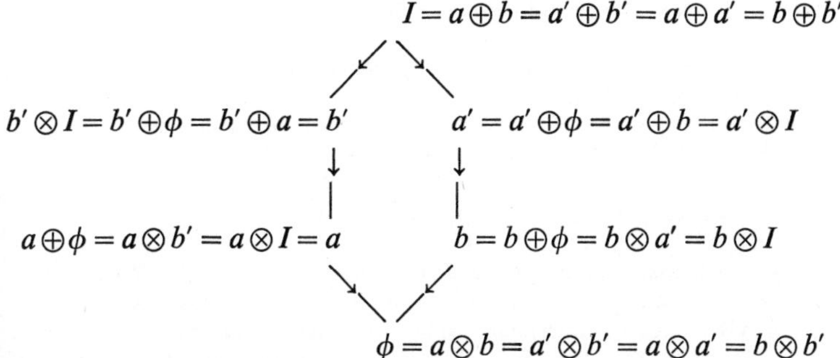

This lattice is non-distributive since:

$$a \oplus (b \otimes b') = a \oplus \phi = a$$
$$(a \oplus b) \otimes (a \oplus b') = I \otimes b' = b'$$
and $a \neq b'$

NOTES

* The original version of this paper was read at a meeting of the Boston Colloquim for the Philosophy of Science, 21 January 1969 under the title 'Quantum Logic Does Not Have to Be Non-Classical'. The author wishes to thank Professor R. S. Cohen and the President of Boston University for the hospitality he enjoyed at Boston University as Visiting Associate Professor of Physics during which time he wrote this Paper. The present version has been much improved due to conversations with Professors R. S. Cohen and A. Shimony of Boston University and especially with my commentator for that occasion, Professor David Finkelstein of Yeshiva University.

1 G. Birkhoff and J. von Neumann, 'The Logic of Quantum Mechanics', *Annals of Math.* **37** (1936) 823–843.

2 See references in the bibliography and also note 4, M. S. Watanabe who shares in his own way the belief that the logic of quantum mechanics is a non-distributive lattice, has based his theory on a different and subtler analysis of the use of ordinary and scientific language than the authors I am expliciting criticizing in this paper. I do not wish to include his views among those I am attacking, although it is possible that some of the positions I take in the latter part of this paper confiict with those of Wanatabe.

3 J. M. Jauch, *The Foundations of Quantum Mechanics*, New York 1968, p. 73.

4 For an exposition of the properties of lattices, see G. Birkhoff, *Lattice Theory*, 2nd ed. 170, Amer. Math. Soc. 1948, and for the application of lattices to quantum mechanics, see J. M. Jauch, *The Foundations of Quantum Mechanics*. In the appendix to this paper will be found a convenient summary of the differences between a Boolean algebra (like PM-logic) and a non-Boolean lattice (like quantum logic).

5 The sign '$\supset$' denotes material implication; '$\vee$' denotes alternation.

6 H. Putnam, 'Is Logic Empirical?', in *Boston Studies in the Philosophy of Science*, vol. V (ed. by R. S. Cohen and M. Wartofsky), New York 1969.

7 At the biennial meeting of the Philosophy of Science Association at Pittsburgh, October 1968, Putnam challenged his audience, among whom the author found himself

on that occasion, to show that he had violated in his argument the principles of quantum logic.

[8] The example given in the first draft of this paper was incorrect as D. Finkelstein pointed out to me. This example is due to him.

[9] Putnam claims that this part of his paper has the value of an inductive argument. The author sees it as being merely of expository value.

[10] See Section I.

[11] J. M. Jauch, *Foundations of Quantum Mechanics*, p. 131.

[12] Jauch, *op. cit.*, p. 73.

[13] W. Sellars in Chapter 3 of *Science and Metaphysics* (London 1968) develops a notion of *picturing* close to what the author would be willing to subscribe to.

[14] The author has treated the interrelationship of language, inquiring behaviour (as a form of life), intentionality, horizon and objectivity in 'Horizon, Objectivity and Reality in the Physical Sciences', in *Internat. Philos. Quarterly* **7** (1967) 375–412. The analysis given there, although expressed in the language of continental philosophy, is nevertheless very applicable here.

[15] P. A. Heelan, 'Complementarity, Context-Dependence and Quantum Logic', *Foundations of Physics* **1** (1970) 95–110.

[16] P. A. Heelan, 'The Role of Subjectivity in Natural Science', *Proc. Amer. Cath. Philos. Assoc.*, Washington, D.C., 1969.

[17] The author has tried to do this in his *Quantum Mechanics and Objectivity* (The Hague 1965) but using a method of intentionality-analysis rather than the analysis of picturing.

[18] For example, N. Goodman, *Fact, Fiction and Forecast*, 2nd ed. (Bobbs-Merrill, NewYork 1965), Chapter I; R. Chisholm, 'The Contrary-to-fact-Conditional', *Mind* **55** (1946) 289–307; S. Hampshire, 'Subjunctive Conditionals', *Analysis* **9** (1948) 9–13; D. Pears, 'Hypotheticals', *ibid.* **10** (1950) 49–62; W. Kneale, 'Natural Law and the Contrary-to-fact Conditional', *ibid.* **10** (1950) 121–125; J. C. D'Alessio, 'On Subjunctive Conditional', *Jour. Philos.* **64** (1967) 306–310.

[19] R. Harré, *Introduction to the Philosophy of Science*, London 1960, pp. 17–24.

[20] J. M. Jauch, *op. cit.*, p. 75.

BIBLIOGRAPHY

Bergmann, G., 'The Logic of Quanta', *Amer. Journ. Phys.* **15** (1947) 497–508.

Beth, E. W., 'Analyse sémantique des théories physiques', *Synthese* **7** (1948–9) 206–7.

Beth, E. W., 'Towards an Up-to-Date Philosophy of Natural Sciences', *Methodos* **1** (1949) 178–185.

Beth, E. W., 'Semantics of Physical Theories', *Synthese* **12** (1960) 172–5.

Beth, E. W., *Mathematical Thought*, D. Reidel Publ. Co., Dordrecht-Holland, 1966.

Birkhoff, G., *Lattice Theory*, 2nd ed. Amer. Math. Soc., Providence, Rhode Island 1948.

Birkhoff, G. and von Neumann, J., 'The Logic of Quantum Mechanics', *Annals. of Math.* **37** (1936) 823–843.

Bunge, M., *Quantum Theory and Reality*, vol. 2., Springer-Verlag, Berlin, 1967.

Destouches, J., 'Les principes de la mécanique génerale', *Actualités scientifiques et industrielles* **140** (1934).

Destouches, J., 'Intervention d'une logique modalité dans une théorie physique', *Synthese* **7** (1948–9) 411–17.

Destouches-Février, P., 'Les relations d'incertitude de Heisenberg et la logique', *Comptes Rendus de l'Académie des Sciences* **204** (1937) 481–83.
Destouches-Février, P., 'Logique et théories physiques', *Synthese* **7** (1948–9) 400–410.
Destouches-Février, P., *La structure des théories physiques*, Paris 1951.
Destouches-Février, P., 'La logique des propositions experimentales', *Applications scientifiques de la logique mathématique*, Paris 1954, pp. 115–118.
Destouches-Février, P. 'Logical Structures of Physical Theories', in *Axiomatic Method* (ed. by L. Henkin, P. Suppes, and A. Tarski), Amsterdam 1959, pp. 376–389.
Dirac, P. A. M., 'The Quantum Algebra', *Proc. Camb. Philos. Sci.* **23** (1926) 412–418.
Emch, G. and Jauch, J. M., 'Structures logiques et mathématiques en physique quantique', *Dialectica* **19** (1965) 259–79.
Falk, G., 'Axiomatics as a Method for the Formation of Physical Theories', *Z. Phys.* **130** (1951), 51.
Feyerabend, P., 'Reichenbach's Interpretation of Quantum Mechanics', *Philos. Studies* **9** (1958) 49–59.
Feyerabend, P., 'Bemerkungen zur Verwendung nicht-klassischer Logiken in Quantentheorie', *Deskription, Analytizität, und Existenz* (ed. by P. Weingartner), Salzburg 1966, pp. 351–9.
Fine, A., 'Logic, Probability and Quantum Theory', *Philos. of Sci.* **35** (1968) 101–111.
Finkelstein, D., 'The Logic of Quantum Physics', *Trans. New York Acad. Sci.*, Ser. 2, **25** (1962–3) 621–37.
Finkelstein, D., 'The Physics of Logic', *Internat. Centre for Theor. Phys.*, Trieste, Report IC/68/35, 621–37.
Finkelstein, D., 'Matter, Space and Logic', *Boston Studies in the Philosophy of Science*, vol. V (ed. by R. S. Cohen and M. Wartofsky).
van Fraasen, Bas, 'Meaning Relation among Predicates', *Nous* **1** (1967) 161–179.
van Fraasen, Bas, 'Presuppositions, Supervaluations and Free Logic', in *The Logical Way of Doing Things* (ed. by K. Lambert), Yale Univ. Press, New Haven, 1969.
van Fraasen, Bas, 'The Labyrinth of Quantum Logics', paper read at the *Biennial Meeting of the Philosophy of Science Association*, Pittsburgh, October 1968.
van Fraasen, Bas, 'Presupposition, Implication and Self-Reference', *Jour. of Philos.* **65** (1968) 136–152.
van Fraasen, Bas, 'Singular Terms, Truth-Value Gaps and Free Logic', *Jour. of Philos.* **63** (1966) 481–495.
Fuchs, W. R., 'Ansätze zu einer Quantenlogik', *Theoria* **30** (1964) 137–40.
Gudder, S., 'Systems of Observables in Axiomatic Quatum Mechanics', *J. Math. Phys.* **8** (1967) 2109.
Guenin, M., 'Axiomatic Foundations of Quantum Theories', *J. Math. Phys.* **7** (1966) 271.
Hack, M., 'Relation Between Measurement Theory and Symbolic Logic', *Nuovo Cimento* **54B** (1968) 147.
Hubner, K., 'Uber den Begriff der Quantenlogik', *Sprache in techn. Zeitalter* (1964), 725–34.
Jauch, J. M., 'Systems of Observables in Quantum Mechanics', *Helv. Phys. Acta* **33** (1960) 711–26.
Jauch, J. M., 'The Problem of Measurement in Quantum Mechanics', *Helv. Phys. Acta* **37** (1964) 293–316.
Jauch, J. M., *Foundations of Quantum Mechanics*, Addison-Wesley, New York, 1968.
Jauch, J. M. and Piron, C., 'Can Hidden Variables Be Excluded in Quantum Mechanics?', *Helv. Phys. Acta.* **36** (1963) 827–37.

Jordan, P., 'On the Axiomatic Foundation for Quantum Mechanics', *Z. Phys.* **133** (1952) 21.
Jordan, P., 'Quantenlogik und das kommutative Gesetz', *The Axiomatic Method* (ed. by L. Henkin, P. Suppes, and A. Tarski), Amsterdam 1959, pp. 365–75.
Kakutani, S. and Mackey, G., 'Ring and Lattice Characterizations of complex Hilbert Space', *Bull. Amer. Math. Soc.* **52** (1946) 727–33.
Kochen, S. and Specker, E. P., 'Logical Structures Arising in Quantum Theory', in *The Theory of Models* (ed. by J. W. Addison *et al.*), pp. 177–189, North-Holland, Amsterdam, 1965.
Kunsemüller, H., 'Zur Axiomatik der Quantenlogik', *Philos. Nat.* **8** (1964) 363–76.
Ludwig, G., *Die Grundlagen der Quantenmechanik*, Springer-Verlag, Berlin, 1954.
Ludwig, G., 'Axiomatic Quantum Statistics of Macroscopic Systems', *Teorie ergodiche* (ed. by P. Caldirola), Academic Press, New York, 1961, pp. 57–132.
Ludwig, G., 'Versuch einer axiomatischen Grundlegung der Quantenmechanik und allgemeinerer physikalischer Theorien', *Zeit. Phys.* **181** (1964) 233–60.
Ludwig, G., 'An Axiomatic Foundation of Quantum Mechanics on a Non-Subjective Basis', *Quantum Theory and Reality* (ed. by M. Bunge), Springer-Verlag, Berlin, 1967, pp. 98–104.
Mackey, G., 'Quantum Mechanics and Hilbert Space', *Amer. Math. Monthly* **64** (1957) 45–57.
Mackey, G., *The Mathematical Foundations of Quantum Mechanics*, Benjamin, Inc., New York, 1963.
McKinsey, J. C. C. and Suppes, P., 'Review of P. Destouches-Février's *La structure des théories physiques*', *Jour. Sym. Logic* **19** (1954) 52–55.
Margenau, H., 'Application of Many-Valued Systems of Logic to Physics', *Phil. Sci.* **1** (1933) 118.
Margenau, H., 'Probability, Many-Valued Logics, and Physics', *Phil. Sci.* **6** (1939) 65.
Miller, M., 'The Logic of Indeterminicy in Quantum Mechanics', *Ph.D. Thesis*, Brown University (1961).
Mittelstaedt, P., 'Untersuchungen zur Quantenlogik', *Sitzungsber. Bayr. Akad. Wiss. Math.-Nat. Kl.* (1959), 321–286.
Mittelstaedt, P., 'Über die Gültigkeit der Logik in der Natur', *Naturw.* **47** (1960) 385–91.
Mittelstaedt, P., 'Quantenlogik', *Fortschr. Phys.* **9** (1961) 106–47.
Mittelstaedt, P., *Philosophische Probleme der modernen Physik*, 2nd ed., Bibliographisches Institut, Mannheim 1966.
Muller, H., 'Mehrwertige Logik und Quantenphysik', *Phys. Bl.* **10** (1954) 151–7.
von Neumann, J., *Mathematical Foundations of Quantum Mechanics*, Princeton Univ. Press 1955.
von Neumann, J., 'On the Algebraic Generalization of the Quantum Mechanical Formalism', Part I, *Collected Works of J. von Neumann* (ed. by A. H. Taub), Oxford 1961, vol. III, pp. 492–561.
von Neumann, J., 'Continuous Geometries with a Transition Probability', *Collected Works of J. v. Neumann* (ed. by A. H. Taub), Oxford 1961, vol. IV, pp. 191–4.
von Neumann, J., 'Quantum Logics (Strict- and Probability-Logics)', *Collected Works of J. v. Neumann* (ed. by A. H. Taub), Oxford 1961, vol. IV, pp. 195–7.
Piron, C., 'Axiomatic Quantum Scheme', *Helv. Phys. Acta* **37** (1964) 439.
Popper, K. P., 'Birkhoff and von Neumann's Interpretation of Quantum Mechanics', *Nature* **219** (1968) 682–5.

Prugovecki, E., 'An Axiomatic Approach to the Formalism of Quantum Mechanics, I and II', *J. Math. Phys.* **7** (1966) 1054–1070.

Putnam, H., 'Three-Valued Logic', *Philos. Studies* **8** (1957) 73–80.

Putnam, H., 'Is Logic Empirical?', *in Boston Studies in the Philosophy of Science*, vol. V (ed. by R. S. Cohen and M. Wartofsky), Humanities Press, N.Y. and D. Reidel Publ. Co., Dordrecht-Holland.

Quine, W. V., *From a Logical Point of View*, Harper and Row, New York, 1963.

Reichenbach, H., *Philosophic Foundations of Quantum Mechanics*, Berkeley, Calif. 1944.

Reichenbach, H., 'Über die erkenntnistheoretische Problemlage und den Gebrauch eine dreiwertigen Logik in der Quantenmechanik', *Z. Naturforsch.* **6a** (1951) 569–575.

Reichenbach, H., 'The Logical Foundations of Quantum Mechanics', *Ann. Inst. Poincaré* **13** (1953) 109.

Richter, E., 'Bemerkungen zur Quantenlogik', *Philos. Nat.* **8** (1964) 225–31.

Scheibe, S., *Die kontingenten Aussagen in der Physik*, Athenäum Verlag, Frankfurt, 1964.

Segal, I. E., *Mathematical Problems of Relativistic Physics*, Amer. Math. Soc., Providence, Rhode Island, 1963.

Sherman, S., 'On Segal's Postulates for General Quantum Mechanics', *Ann. Math.* **64** (1956) 593–601.

Strauss, M., 'Mathematics as a Logical Syntax – a Method to Formalize the Language of a Physical Theory', *Erkenntnis* **7** (1937–8) 147–53.

Suppes, P., 'Probability Concepts in Quantum Mechanics', *Philos. Sci.* **28** (1961) 378–89.

Suppes, P., 'The Role of Probability in Quantum Mechanics', *Philosophy of Science: The Delaware Seminar*, vol. 2 (ed. by B. Baumrin), Wiley, New York, 1962–3, pp. 319–37.

Suppes, P., 'Logics Appropriate to Empirical Theories', *Theory of Models* (ed. by J. W. Addison, L. Henkin, and A. Tarski), North-Holland, Amsterdam, 1965, pp. 364–75.

Suppes, P., 'The Probabilistic Argument for a Non-Classical Logic in Quantum Mechanics', *Philos. Sci.* **33** (1966) 14–21.

Toll, J., 'Causality and the Dispersion Relation; Logical Foundations', *Phys. Rev.* **104** (1956) 1760.

Varadarajan, V. S., 'Probability in Physics and a Theorem on Simultaneous Observability', *Comm. Pure Appl. Math.* **15** (1962) 189–217.

Watanabe, M. S., 'A Model of Mind-Body Relations in Terms of Modular Logic', *Synthese* **13** (1961) 261–302, and *Boston Studies in the Philosophy of Science*, vol. I (ed. by M. Wartofsky), D. Reidel Publ. Co., Dordrecht-Holland.

Watanabe, M. S., 'Conditional Probability in Physics', *Progress of Theor. Phys., Suppl.* (1965) 135–60.

Watanabe, M. S., 'Algebra of Observation', *Progress of Theor. Phys., Suppl.*, Nos. 37–8 (1966) 350–67.

Watanabe, M. S., *Knowing and Guessing*, Wiley, New York, 1969.

von Weizsäcker, C. F., 'Komplementarität und Logik', *Naturwiss.* **42** (1955) 521–9; 545–55.

von Weizsäcker, C. F., Scheibe, E., and Sussmann, G., 'Komplementarität und Logik. III: Mehrfache Quantelung', *Zeit. f. Naturforsch.* **13a** (1958) 705–21.

A. R. MARLOW

IMPLICATIONS OF A NEW AXIOM SET FOR QUANTUM LOGIC

A recent work on the many-body problem in modern physics begins with the following paragraph:

> A reasonable starting point for a discussion of the many-body problem might be the question of how many bodies are required before we have a problem. Prof. G. E. Brown has pointed out that, for those interested in exact solutions, this can be answered by a look at history. In eighteenth-century Newtonian mechanics, the three-body problem was insoluble. With the birth of general relativity around 1910 and quantum electrodynamics in 1930, the two- and one-body problems became insoluble. And within modern quantum field theory, the problem of zero bodies (the vacuum) is insoluble. So, if we are out after exact solutions, no bodies at all is already too many.[1]

This somewhat pessimistic quotation has been chosen to introduce the present, essentially optimistic, paper simply because it leads rather directly to some of the questions to be considered. It is a fact that the quantum mechanical approach to physical reality has not yet yielded a full understanding of any system involving non-electromagnetic interactions between particles, such as, for example, the nucleus or systems containing any of the new high-energy particles and resonances being discovered continually. For this reason, a number of physicists (including among them at least one of the founding fathers of quantum mechanics itself, Paul Dirac) believe that we will never get an understanding of the non-electromagnetic interactions within the mathematical structure of quantum mechanics, that is, within the framework of countably infinite-dimensional vector space theory. These physicists have started on the search for a completely new type of theory, presumably to be formulated in terms of mathematical structures which essentially cannot be fitted into the standard Hilbert space structure of quantum theory. For convenience, we will refer to this point of view as *A*. It is not the majority opinion among mathematical physicists, but it is certainly a reasonable viewpoint in the light of the history of physics, which, if it teaches anything, certainly advises a full willingness to go beyond any theory, no matter how sacred, if it gives signs of inadequacy in the face of new data. Outlook *A* has another decided advantage in its favor – it is extremely progressive in

Boston Studies in the Philosophy of Science, XIII.

its view of the task of present-day theoretical physics, aiming as it does at nothing less than adapting or even creating new realms of mathematics for application in physics. It seems fair to say that this was the attitude of the founders of both relativity theory and quantum mechanics when the inadequacy of classical theories became evident.

The point of view which will be referred to as *B* is the majority point of view at present, and judging from current publication this outlook can count among its exponents Feynman, Gell-Mann, Ne'eman, Wightman, Roman, and most other physicists working toward a unified theory of systems involving other than electromagnetic interactions. These physicists are fully as revolution-minded as those of group *A*, but the revolution they look toward is of a physical and not a methematical nature – a revolution based on new physical insights akin to the basic insight of Copernicus which changed all of astronomical theory from the geocentric to the heliocentric hypothesis and ultimately led to the elegance and simplicity of the Newtonian theory of matter in motion interacting gravitationally, but which did not go outside of the mathematical structure of differential geometry in three-dimensional vector space. Group *B*, then, bases its hopes and expectations on the factual realization that modern physics has by no means even come close to exhausting the extremely capacious resources of mathematical structure available to it within the framework of infinite-dimensional vector space theory and the mathematical entities which can be constructed on the basis of this theory, and until these resources have been exhausted it seems a time and energy consuming shot in the dark to look for a new mathematical language. The difficulties in present physical theory do not lie in the mathematical language employed, but in some physical assumption that is being made, or not being made, within that language.

A third viewpoint can be distinguished – one which will be labelled *C*. This viewpoint (espoused by Bohm, Vigier and their co-workers) can only be described by saying that it starts from the metaphysical and metatheoretical assumption that a physical theory can be satisfactory only if it is of classical type – that is, if it allows only physical questions which can, in principle at least, be answered with a definite yes or no. In other words, indeterminacy principles cannot be an intrinsic part of any satisfactory physical theory. It must be recognized, in all fairness, that one of the great revolutionaries of modern physics, Albert Einstein, subscribed to this opinion. Nevertheless, point of view *C* does reject most of modern mathematical quantum theory as unsatisfactory, and it does so on what seems to be a rather *a priori* basis. Even if mankind were privileged to know in some mysterious way that, as Einstein put it, "God does not play with dice," still this would not really justify the assumption that physics, in its efforts to understand reality, could avoid playing with dice.

Thus, purely on the metatheoretical grounds of the principle of economy – Ockham's Razor – applied in this case in the form 'In constructing a theory, make no more metatheoretical assumptions than are strictly necessary', it would seem that there is a solid reason for tending away from viewpoint *C* and toward either *B* or *C*. The question to be posed now, however, is simply: 'Are there any hard mathematical reasons within the structure of mathematical physics, but not already within the viewpoints outlined above, for choosing one of the approaches over the others?' Until recently it would seem that there were no such reasons, since it has for some time been recognized that von Neumann's proof of the inadequacy of classical type formalisms only goes through on the basis of assumptions which are already clearly within framework *B*, and in the summer of 1966 constructions were published[2], based on a classical formalism, which reproduced enough of quantum mechanical indeterminacy to give renewed hope to the exponents of viewpoint *C*. More recently still, J. M. Jauch and C. Piron[3] have published an extension and strengthening of von Neumann's proof of the exclusion of "hidden variable" models (the models proposed by group *C*) and a renewed dialog between the exponents of viewpoints *B* and *C* has started[4]. The discussion still remains inconclusive, however, since Jauch and Piron base their proof on lattice theoretical assumptions which are rejected as unphysical by the proponents of classical type models. In the opinion of the present author, what has changed the situation now is the existence of a new theorem, proved on the basis of what we will refer to as general physical axiomatics; this theorem, while not excluding any of the three viewpoints outlined above, still gives basic support for the work of group *B*. To get an understanding of the theorem and its implications we will have to consider in a general way what is meant by physical axiomatics; for the full details of the axiomatic structure and a rigorous statement and proof of the results, see references 5 and 6.

What we mean by physical axiomatics dates back at least to Euclid.

Allowing ourselves possibly a large dose of anachronism, we can say in retrospect that he considered a certain class of physical operations, the operations of going from one place to another in the most direct way possible, and set up a consistent logical structure which adequately described the various interrelations among such operations. He did this by selecting three notions as primitive or undefined – 'point' (idealized place), 'line' (idealized direct path between two points), and the relation 'between' – and then putting into his system enough assumptions or axioms concerning these three notions to guarantee that the resulting logical structure actually does mirror the ordinary reality of the spatial milieu in which we come and go. The mathematical structure Euclid discovered is still with us over 2000 years later under the prestigious title *Euclidean geometry.* Two things are often forgotten about geometry, however – that it is essentially physical in its origin and intent, and that it is also somewhat Platonic, or at least idealized (no one has actually produced a point nor traced out a verifiable straight line) – and hence, to say that Euclid's geometry has survived as a mathematical structure is not at all the same thing as to say that it is still the best structure for describing all physical translations from place to place. And of course, in the early part of this century, when physicists became seriously concerned with the actual experience of long range translation, that is, with the operations of sending things or signals over extremely large distances, it was soon realized by Einstein and Minkowski that Euclidean geometry did not fit and that the appropriate mathematical structure needed to describe these operations was the larger structure of four-dimensional pseudo-Euclidean geometry. In other words, special relativity theory of space-time proved to employ a non-Euclidean geometry. This should not really be too surprising: Euclid's geometry reflected the actual experience of only short range translations, even though as a complete mathematical structure it contained points and lines corresponding to arbitrarily large distances – distances for which there was available no actual experience to guide one in choosing suitable axioms.

The point of this excursion into geometry is simply this: in constructing mathematical models of physical situations we must always be aware that the resulting mathematical structures will generally contain elements which are introduced for mathematical or formal reasons – to achieve a complete mathematical structure in which computations can be made.

Some of these mathematical elements may ultimately be able to support an actually measurable physical interpretation (examples of this type would be Euclid's points and lines at large distances), but then we may find that the measurements are inconsistent with the interpretation we have given, so that we have to change either the physical interpretation of the mathematical elements or the mathematical structure itself to accommodate the new facts. Other of these mathematically introduced elements may never bear a measurable physical interpretation and are present simply to give a viable mathematical structure.

We can now describe in general terms what will be meant here by an axiomatic mathematical model or theory of a physical situation. Given a set P consisting of elements which are physically constructible or specifiable in some way or other (that is, things we can talk about in empirical terms and which will form the physical core or nucleus of elements with which the theory has to deal), we will say we have an axiomatic model of P if we have a set U of primitive or undefined elements (we will call U our universe of discourse) and a set $\{A_i\}$ of axioms or assumptions about U which turn U into a consistent mathematical structure M. We understand by M, of course, the class of all structures which can be formed purely mathematically or logically from U on the formal basis of the axioms A_i. In addition, we also need a set $\{R_i\}$ of computational rules on M and a correspondence $E: P \to M$ which assigns to each physical element $p \in P$ a uniquely defined mathematical element $E(p) \in M$ in such a way that the computational rules in $\{R_i\}$, when applied to the elements of $E(P) = \{E(p) \in M : p \in P\}$, yield numerical results which fit the empirical data arrived at by actual laboratory measurements on the physical structures involved in P. Thus, within the mathematical structure M we can compute, and the *minimal* requirement we place on our model is that the computations performed within M should yield numerical results which closely fit the reported empirical data. There are other requirements that we place on a mathematical model, though: it should be, to some extent, simple, elegant and tractable – that is, the mathematical rules of computation should be able to be carried out in practice. These latter requirements are more or less a function of the taste and computational capability of the physical community at any given time; thus what may have been intractable fifty years ago may be quite manageable now with the use of modern computers.

We come now to consider the principal mathematical model of modern physics, quantum mechanics. The study of the axiomatic structure of this model was begun by G. Birkhoff and J. von Neumann [7] in the early stages of quantum mechanics and was recently brought to a stage of relative completeness by G. W. Mackey with the publication of his *Mathematical Foundations of Quantum Mechanics* [8]. At about the same time as Mackey's work was published (1963), an independent axiomatic study of quantum mechanics was being completed in Switzerland by C. Piron [9]. The purpose of these studies was simply to get a mathematically precise statement of the assumptions being made when the quantum mechanical model is employed. The net result of the studies is that the essential distinguishing postulate of quantum mechanics lies in the assumption of a definite mathematical structure for a particularly simple but particularly important class of physical measurements which can be made on any physical system – the class we will call the *physical questions* on a system and symbolize by Q. Before considering the basic assumption of quantum mechanics itself, it will be useful to discuss these physical questions and a general axiomatic structure within which they can be embedded.

By a physical question is meant any actual laboratory measuring process which can have for its possible results only the numbers 1 or 0, interpreted as the answers 'yes' and 'no', respectively. It can be shown[10] that any measuring process on any physical system can be completely characterized in terms only of the physical questioning operations and the set of probabilities for getting the answer 1 (i.e., yes) when any of these questioning operations is performed after some specific preparatory operation on the system. We will symbolize preparatory operations with the Greek letters φ, ψ, ...; these operations are simply arbitrary physical operations which may prepare the system for measurements but are not themselves necessarily measuring operations, that is, unlike the measuring operations preparing operations do not have to result in numbers. Examples would be: going to some specific space-time region, setting up apparatus capable of making a number of distinct measurements, etc. Thus, a mathematical model M for an arbitrary physical system need only give a mathematical realization of the set Q of physical questions on the system and the set S of preparations of the system (or, to use the more standard terminology, states of the system), together with a computational rule for arriving at a numerical function m which will assign to each pair (q, φ) a number

$m(q, \varphi)$ between 0 and 1 expressing the probability that the question q will have the answer yes when it is performed (or asked) on a state φ. The set of numbers $\{m(q, \varphi): q \in Q, \varphi \in S\}$ can be called the *data set* of the system; the computation of the elements of this set is of course the principal aim of any mathematical model. This general statement of the basic results of physical axiomatics is sufficient for our purposes here; for further details of an axiom set that leads rigorously to this basic structure[5]. Without getting too far afield, however, we can remark that these general axioms start off from the undefined notion of a physical operation as such and then build in enough minimal structure so that all the actual operations performed in physics can be dealt with. These axioms are not strong enough to specify any concrete mathematical model of themselves (for example, they do not specify computational rules) but they are general enough so that both quantum and classical models can be built from them by further axioms, and they are minimal enough so that it seems safe to say that any satisfactory models forthcoming from the work of group A will be able to be specified on their basis.

To spell out in detail now the further assumptions necessary for either a classical or a quantum type model, we first have to add an axiom to the effect that the set Q of questions forms a mathematical logic (in technical terms, an atomic, orthocomplemented lattice). Thus, there will be a set of 'atomic', or irreducible, questions or propositions (note that it is immaterial whether we speak of questions or propositions, since every proposition corresponds uniquely to a question and vice versa simply by the proper use of the terms 'It is the case that…' and 'Is it the case that…?), and definite prescriptions for forming more complicated "molecular" propositions from the atomic propositions by using the connectives 'and' and 'or' (in symbols, $\cap$ and $\cup$ respectively) and the operation of negation (symbolized by 'n'). Classical type mathematical models, now, by requiring that all propositions be in principle either definitely true or definitely false – and so limiting the possible probability values $m(q, \varphi)$ in the data set of the system to being 1 or 0 – arrive at ordinary mathematical logic (Boolian or distributive logic) in which the distributive law is valid. This law characterizes ordinary logic, and states that, for arbitrary propositions $p, q, r \in Q, p \cap (q \cup r) = (p \cap q) \cup (p \cap r)$. The logic that quantum theory postulates for the set of physical questions of a system, however, is not of this Boolian type. More positively, we can say precisely what the quantum logic is: it is the logic we get in concrete mathematical form by assigning to each physical question a complete subspace of a countably infinite-dimensional complex Hilbert space. The 'and' of the quantum logic is then taken as intersection of subspaces, the 'or' is subspace union, and the operation of negation corresponds to taking the orthogonal complement subspace. Mackey has shown that the assumption of this 'quantum logic' together with the assumption that there are enough states to guarantee that each physical question can have the answer yes with certainty lead uniquely to the standard quantum mechanical model of physical reality. Thus, along with quantum logic we also get a geometry for physical theory, the geometry of infinite-dimensional Hilbert space.

Now we do not have to go into the full details of the mathematical structure of quantum mechanics[11] to see that these assumptions might put some rather large demands on our physical credibility. First, it is not physically evident that the structure of a logic (either as envisaged by classical or quantum type models) is warranted for the questions on a physical system of arbitrary type[12]. Thus physicists of group A are quite free, as far as known physical requirements go, to investigate models that do not postulate a logic of any sort for the questions. Then, even if the assumption of a logic is accepted, it has not been shown that the postulation of the specific logic required by quantum mechanics is justified on any grounds other than that this postulate has so far successfully explained electro-magnetic interactions. This might be considered quite good evidence, except that it is intrinsically no stronger than the evidence supporting classical theories before the internal structure of the atom was investigated experimentally.

Thus we are led back to the basic question raised earlier: Are there any reasons intrinsic to the structure of mathematical physics which might favor one or the other of the points of view described at the outset of this paper? A possible answer is contained in the investigation of the relationship between the quantum mechanical model and general physical axiomatics contained[6]. In that paper an *embedding* of a physical system is defined as a mapping E of the set $Q \cup S$ (the set consisting of the physical questions and the physical states of the system) into the subspace structure of some complex Hilbert space in such a way that the computational rule provided by quantum mechanics (i.e., $m(q, \varphi) = \text{trace}(E_q E_\varphi)$) can in principle compute all possible data on the system. The existence of such embeddings for arbitrary physical systems is then demonstrated. Thus on the basis of general physical axiomatics alone we can go far towards guaranteeing the existence of quantum mechanical models for any physical system; the one thing lacking from these models is the requirement of countable dimensionality ordinarily imposed on the Hilbert space structures used in physical theory.[13] Even this requirement is fulfilled if we make the following assumption:

(A) For any physical system there exists an at most countable infinity of states φ which determine all the data $m(q, \varphi)$ for any question q,

since then it follows from the arguments[6] that there always exist embeddings, and hence quantum mechanical models, of a physical system with respect to separable Hilbert space. In understanding these results, it is important to be clear about what they do not imply. First, no uniqueness theorem can be claimed – there may be several or even an infinity of distinct quantum mechanical models for a given physical system.[14] Second, the results guarantee the existence of quantum mechanical models, but do not give any specific instructions for producing useful examples of these models. Finally, the results do not guarantee that a quantum mechanical model for a physical system, once found, will rival other possible models on the grounds, say, of mathematical simplicity or elegance. The results only imply, then, that quantum mechanical mathematical structures in principle cannot be superseded on grounds of incapability to cope with new empirical data – there exists a mathematical model of quantum mechanical type to handle any conceivable empirical data.

A further question, however, concerns the physical motivation for the countability assumption (A) above which guarantees the sufficiency of separable Hilbert space structure. This assumption postulates that the entire data content any mathematical model has to deal with will be determined by an at most countable infinity of states. It should be noted that the assumption places no explicit restrictions on the number or structure of the possible observables or measurements that can be made, nor does it place any restriction on either the number or structure of the set of states in the final mathematical completion of a physical system. Translated physically, what is assumed is simply that there is an intrinsic discreteness involved in our encounter with physical reality in such a way that, granting an indefinite extension of the history of physics into the future, and even being over generous to the point of granting an indefinite extension to the working lifetime of individual physicists, the number of distinct experimental situations on which any conceivable measurements can be made is limited to being an at most countable infinity, and thus the actual data content of any mathematical model will be subject to the limitation expressed by (A). The present author believes that this countability assumption is physically well motivated; it is the type of assumption that guided the early workers in quantum theory in their realistic desire to incorporate into the formal structure of physical theory as much as possible of the actual phenomenology of the observer-observed interaction, but it avoids some of the limitations of the particular arguments they used.

The present author believes that these results constitute a basic confirmation and justification of the program outlined above as viewpoint *B*. It would seem that with the introduction of Hilbert space structure the mathematical language of modern physics has reached a definitive stage of sophistication and power, since arbitrary physical situations can be formulated and studied within this language with the assurance that at least adequate computational models for the situations exist. On the other hand, any *a priori* restriction to abelian models only (such as would be placed in viewpoint *C*) would seem to be as much a regression in modern physics, as, say, a restriction to the effect that only Euclidean geometry be used in formulating models of space-time.[15]

Loyola University (New Orleans)

NOTES

[1] R. D. Mattuck, *A Guide to Feynman Diagrams in the Many-Body Problem*, McGraw-Hill, New York, 1967.

[2] J. S. Bell, D. Bohm and J. Bub, *Reviews of Modern Physics* **38** (1966) (series of three papers).

[3] J. M. Jauch and C. Piron, *Helvetica Physica Acta* **37** (1964) 293.

[4] Cf. Letters to the Editor, *Reviews of Modern Physics* **40** (1968) 228ff.

[5] A. R. Marlow, *Journal of Mathematical Physics* (submitted for publication).

[6] A. R. Marlow, *Journal of Mathematical Physics* (submitted for publication).

[7] G. Birkhoff and J. von Neumann, *Annals of Mathematics* **37** (1936), 823.

[8] G. W. Mackey, *The Mathematical Foundations of Quantum Mechanics*, Benjamin, New York, 1963.

[9] Piron, *Helvetica Physica Acta* **37** (1964), 439.

[10] See note 8, p. 66.

[11] For the further development of the quantum mechanical model, cf. note 8, pp. 71ff.

[12] For a full discussion of this point, see S. P. Gudder, *Journal of Mathematical Physics* **8** (1967), 1848 and our note 5, footnote 1.

[13] Due to a suggestion of Dr. Abner Shimony, the general proof of the existence of embedding is independent of the countability assumption (A).

[14] At least some of this multiplicity is removed in the C*-algebra approach to quantum theory; for an account of this approach, see D. Kastler, pp. 179–191 in the volume edited by W. Martin and I. Segal, *Analysis in Function Space*, MIT Press, Cambridge, Mass., 1964.

[15] This modifies and sharpens somewhat an interesting comparison offered by D. Bohm and J. Bub, *Reviews of Modern Physics* **40** (1968), 235.

MILIČ ČAPEK

TWO TYPES OF CONTINUITY

In this paper I am going to deal with two very different kinds of continuity. One is of *mathematical* kind and it is familiar to every student of calculus; the other was named by Poincaré – not very appropriately, as we shall see – *physical* continuity (*le continu physique*). While the obvious contrast between these two different types of continuity is fairly well known, its deeper philosophical significance is rarely analyzed. This lack of interest in it is not accidental; it is due to the persistent influence of the intellectual tradition generated by the three centuries of classical science (1600–1900). We shall see that a more subtle epistemological approach together with the emergence of some new and quite unexpected problems in contemporary physics requires another fresh look at the contrast between both types of continuity and the way it was interpreted both by classical science and classical philosophy.

As far as mathematical continuity is concerned, one illustration from elementary calculus will make its meaning clear. The function $f(x)$ is defined to be continuous at the point x_0 of its interval when for any arbitrarily small number ε another sufficiently small number ϑ can be found such that the following inequalities hold:

$$|f(x) - f(x_0)| < \varepsilon \quad \text{when} \quad |x - x_0| < \vartheta$$

It is clear that this definition of continuity assumes the continuity of the argument, i.e. of an independent variable which is represented by the horizontal x-axis: to every point of this line corresponds a certain real number, representing the value of the argument, and each such value is correlated with a particular value of the function which is visually represented by the segment of the vertical line erected at the corresponding value-point of the argument. The definition thus presupposes the *continuum of real numbers*, each of which is correlated with one particular point of the x-line. Since it is in the nature of continuum that between any of its terms another intermediate term can be inserted, the continuum of real numbers as well as the continuum of the points on a straight line is 'everywhere dense or compact' ('überall dicht' in Cantor's words). Consequently, there is no smallest number since between zero and an arbitrarily small ε there is always some real number. For the same reason there is no minimum interval since every linear segment and, more generally, every geometrical interval, no matter how small, contains a subinterval and so on, *ad infinitum*. This shows clearly that continuity in the mathematical sense of the word is synonymous with *infinite divisibility*: real numbers as well as geometrical intervals are divisible without any limit for the reasons stated in the previous sentence.

Such 'infinite divisibility' is clearly a conceptual construct and its artificial character will become even more conspicuous when we compare it with what Poincaré called 'physical continuity' and whose more appropriate term is 'perceptual' or 'intuitive' continuum. Let us take the example above of a continuous function represented by a smooth continuous curve. The visual perception of its continuity contrasts by its simplicity with the complexity of the mathematical definition in which there occur two inequalities and which refers to such ideal entities as dimensionless points. Perceptual continuity is an indivisible *Gestalt* whose components can be only artificially isolated by the effort of analytical attention; as Berkeley and, after him, Hume recognized, in our perception of space there are no points, no infinite divisibility, but concrete *minima sensibilia*, indivisible quanta of extension which can be subdivided only in our imagination without being actually divisible. Points, instants and actual infinity are mere figments of mathematical imagination.

Confronted with such objections, the mathematicians have not remained silent, especially when theoretical physicists hurried to their rescue. They concede to their sensualist opponents that there are no points and no instants in our sensory experience. But, they said, this is only because our experience is hazy and confused; the incapacity of our perception and even of our imagination to attain the dimensionless points and durationless instants cannot rule out their existence. In this respect the whole weight of the philosophical tradition or at least of its major part was on the side of the mathematician. No theme in the history of philosophy was more recurrent than the unreliability and haziness of our sensory perception; in truth, man did not have to become a philosopher to be aware of various sensory illusions. An enormous extension of our visual field by telescope and microscope in the seventeenth century still further discredited the authority of our spontaneous perception and showed its original limits. Furthermore, in the same century, Newton's identification of Euclid's infinite space with the objective physical space was used as a powerful argument by the mathematician in his polemic against sensualistic philosophers. For the Euclid-Newtonian space was not only infinite, but continuous in the mathematical sense, that is, divisible without limits. One of the first theorems which Euclid established in his *Elements* is that *every* straight segment can be bisected; and by 'every' is meant *any* distance in infinite space, whether huge or minute, whether it is the distance of the earth from the sun or the radius of the minutest atom. In truth, as Sir Thomas Heath pointed out[1], the continuity of Euclidean space is postulated in the third postulate of Euclid which removes any restriction on the size of the circle; such restriction is clearly incompatible with the continuity of space. This idea *of relativity of magnitude*, according to which any geometrical figure can be constructed on any scale, underlay one of the basic assumptions of classical physics about the basic *similarity* (in a geometrical sense) of the microcosmos and macrocosmos. Hence the repeated attempts to construct intuitive and mechanical models of the atom and ether in the last century; in truth, even the original twentieth century model of the atom as a planetary system on a very minute scale was guided by the same – false, as we know today – assumption.

Such then was essentially the answer of the mathematician, supported by the classical physicist, to the sensualistic objections of Berkeley and Hume. To sum it up in one sentence: infinite divisibility of space as well as the existence of dimensionless points is not *disproved* by the fact that they both are inaccessible to our sensory experience which is notoriously hazy and inadequate. As indicated above, the classical physicist followed the line of the centuries old philosophical tradition from Parmenides to Bradley which posited the logically flawless realm behind the chaotic and 'imperfect' flux of our hazy and confused sensory experience.

It was not only the classical belief in the reality of continuous physical space which encouraged the mathematician in his confidence that his concepts are not mere artificial constructs, but that they have their objective ontological counterparts. Time, which Isaac Newton regarded as objectively existing as space, was endowed with the same property of homogeneity, that is, with infinity and continuity. Galileo affirmed that in any interval of time, no matter how small, there is an infinite number of instants, and Bertrand Russell reaffirmed the same claim at the beginning of this century. This assumption was inherent within the very structure of the infinitesimal calculus and without it no rational dynamics would have existed. For the classical concept of motion presupposed the classical concepts of space and time and it inevitably shared their continuity. Without spatio-temporal continuity, it would be impossible to speak of the identity of the material particle in different points of its trajectory, that is, in different points of space and different instants of time. When the quantum theory indicated the possibility that this may not be so, it came as a real shock and it indeed led to the crisis of the classical concept of particle. For, if the trajectory of a certain corpuscle is not continuous, that is, if the corpuscle does not exist in all the points intermediate between its successive positions, how can we be sure that it is the *same* particle? Thus the persistence of the material particle through time, its very identity – what Kurt Lewin called *Genidentität* – is assured by the possibility of tracing its *continuous* displacement through space and time, i.e. to its existence in *every* point and *every* instant of the corresponding spatio-temporal interval.

But let us not anticipate the discussion which belongs to the second part of this paper. For our immediate purpose let us add that the absence of the instants on the sensory and, more generally, psychological level did not imply their non-existence on the physical level. The fact that the psychological present has a certain duration has been known for a considerable time; but the very name by which it was designated – 'specious present' – indicated that it was not regarded as the true, mathematical present which is strictly durationless and does exist on the physical level. Again it was assumed that only the haziness of our perception prevents us from perceiving the true instants of the physical time in their precise, sharp-edged unambiguity. In this sense the psychological or mental present is indeed only apparent, 'specious', being merely a fuzzy, imperfect representation of what in the physical world is an enormous number of successive, rigorously point-like presents. In truth, this number is infinite since every temporal stretch, no matter how short, must be regarded as an infinite aggregate of durationless instants. Thus the function – or rather malfunction – of our temporal perception is to condense spuriously the infinity of the knife-edged, infinitely thin instants into the deceptive phenomenal unity of the 'specious present'.

This is clearly a *dualistic* view in which objective reality possessing a clear-cut logico-mathematical structure, is opposed to the qualitative realm of appearance, what Kant called a 'rhapsody of sensations', seemingly devoid of any order. It was precisely the haziness of our sensory and introspective experience which led philosophers to regard it as a mere 'appearance'. But on this point we have to be on our guard against the haziness of our own terminology; for *appearance* does not mean *unreality*. In this respect those who uphold the reality of the logico-mathematical structure of nature, independent of our perception and awareness, are guilty of certain haziness of language which then inevitably affects the clarity of their own thought. For, no matter how uncertain and questionable appearance may be, it still *does* exist; it cannot be dismissed as unreal. Otherwise the whole distinction between 'appearance' and 'reality' would be impossible. This was the old problem of Parmenides: if reality is One, where does the illusion of diversity and change come from? If it comes from 'the finite perspective of individual minds', do we not concede *ipso facto* the existence of 'individual minds' or 'finite perspectives', that is, something *alongside* or *outside* the Eleatic One? And by conceding the reality of something else besides One, are we not departing from our original rigorous monism? In a similar way, we can call the quality of yellow color of the sodium spectral line a mere 'appearance' of what in the objective physical world is the electromagnetic wave of the length of 5890 ångströms; but this does not change the fact that the quality of yellow is *actually* and *indubitably* experienced and cannot be dismissed as simply unreal. What is wrong and must be rejected is the *hypostasizing* of the quality of yellow, that is, its projection into the physical realm; there is no question that 'yellow' *qua* quality does not have any physical status. In this respect neorealists and John Dewey were grievously wrong in claiming that "things are what they are experienced as" and it is difficult to understand how some persons with their knowledge of elementary college physics could make such a claim. But it is equally wrong to conclude from the non-existence of qualities on the objective, physical level their non-existence in general; yet, this is the common fallacy of behaviorism and similar epistemologically innocent trends. For the self-evidence of introspective quality cannot be questioned; for if *esse est percipi*, and, while I may be wrong in *interpreting* this quality, I cannot err in *having* it. This should be known since the time of Descartes, in truth since the timeof St. Augustine. To use Professor Feigl's words, who certainly cannot be accused of any sympathy for Descartes or Berkeley, the mental qualia are not *evidenced*; they are *evident*.[2] For, let us repeat it, without the mental qualia, without sensory qualities being *in some sense real*, the whole distinction between 'reality' and 'appearance' loses its meaning.

As mentioned above, it was the so called 'haziness' of the sensory and introspective qualities which arouses the distrust of not only the behaviorists, but also of nearly every quantitatively minded scientist. But again we have to be on guard against our linguistic habits. 'Hazy' is a relative term as much as 'disorder'; as Bergson, incidentally, one of the most misunderstood and unjustly abused thinkers, convincingly pointed out, we speak of 'disorder' only when we do not find the type of order which we expect.[3] We complacently speak of the vagueness of music or poetry and claim that they both are 'less definite' than the 'normal language'. This is a mistake or at least a misunderstanding; there are not a few philosophers and musicians who claim that music is *more definite* than the words since the spoken language cannot express the passing individual nuances of our inner life which possess their specific qualities despite the fact that our object-oriented language does not have the ready-made symbols for designating them[4].

These preliminary remarks lead us to our main topic – that of qualitative or intuitive continuity. The fact that this second type of continuity is radically different from the mathematical continuity does not mean, that it does not have its own structure and its own specificity which can be analyzed. All depends on the tools by which we analyze it.

Let us then analyze as closely as possible what is that second type of continuity. We touched upon it before when we recalled that in our perception of extension there are no points and in our perception of time there are no instants. This was the meaning of Berkeley's and Hume's *minima sensibilia*; on the sensory and introspective level there is no mathematical continuity, no infinite divisibility. Let us focuss our attention on the problem of temporal continuity since temporality is a more pervasive feature of our experience than extension; extensionality or spatiality is characteristic of the sensations of touch and sight and only indirectly, by association with optical and tactile sensations, of other categories of sensations; it is altogether absent in imageless thought and emotive experience whose temporal aspect is still very pronounced. Kant expressed this fact in his own jargon by saying that "space is the form of outer sense while time is the form of both outer and inner sense." There is thus no reason to fear that the results of our analysis will lack in generality if we confine our attention to the analysis of psychological time.

Our analysis will start with our first negative conclusion: that perceptual and introspective continuum does not consist of mathematical durationless instants. On this point the agreement is nearly general; even Bertrand Russell, whom nobody can suspect of any antipathy against mathematical continuity, explicitly conceded it. It is true that C. A. Strong and Alexius Meinong tried by a very artificial way to establish the existence of the point-like psychological instants; I dealt with their views in my other papers and do not intend to repeat my criticism here.[5] A great majority of philosophers and psychologists side with William James and George H. Mead about the non-existence of knife-edge present. Now if we agree on this point, it is very easy and tempting to accept what seems to be the only alternative; that qualitative continuum, being not divisible *in infinitum*, must be atomic. This was the step which Hume and after him all associationistic philosophers took. For associationism is nothing but psychological atomism. Thus the perceptual and psychological time would be a succession of *minima sensibilia* – which were named differently by different thinkers – *impressions*, *ideas*, *sensations*, *Vorstellungen* and finally *elements* by Mach. (He used the term *Empfindungen*, i.e. sensations, too.) Even William James, in spite of his vigorous criticism of psychological atomism, comes very close to it when he speaks of the specious present as "elementary sensation of duration" or as "the unit of composition of our perception of time."[6] Thus the atomism of duration seems to be present on the sensory and psychological level; the second type of continuity would then be no continuum at all, but a *discretum*, the succession of contiguous temporal segments.

But such representation of qualitative continua is only slightly less inadequate than the theory of instants. Everybody acquainted with the development of psychology since about 1890 knows how the inadequacy of psychological atomism were exposed by Gestalt psychologists and holistically oriented philosophers. It has been pointed out that associationism was based on inadequate psychological observation which mistook artificial conceptual constructs for authentic 'elements' of introspective experience. What we call 'sensations', 'impressions', 'images' etc. are nothing but artificial entities carved out from the continuity of 'stream of thought'. William James in the passages of unsurpassed subtlety in his *Principles of Psychology* showed how psychological atomism focussed its attention on the sensory nuclei of the psychological self and ignored almost entirely more elusive imageless fringes and relations, although it is precisely in these dynamic non-sensory links that the true continuity of consciousness is located.[7] Today we would say that our object-oriented language is ill suited to express adequately the structure of the processes to which the category of 'thinghood' clearly does not apply. Bergson's and Dilthey's criticism of atomistic psychology were pointing in the same direction.

But it was not only psychological experience which militated against the atomistic theory of the intuitive continuum. Equally serious was the intrinsic, logical difficulty which was inherent in the very concept of *contiguum theory*, as we may call it. This difficulty will become immediately obvious when we approach this theory on a more abstract level. For the main reason for postulating the atomistic structure of psychological time was the rejection of mathematical continuity, that is, the rejection of instants. But a very cursory glance at the doctrine of temporal atomicity will show that it *covertly reintroduces the very concept which it purportedly rejects.* For in assuming that time consists of contiguous, successive segments, it implicitly postulates the boundaries by which these segments are separated; and such boundaries must be instant-like, if we want to avoid an infinite regress. For were they not, were they constituted by shorter segments interposed between the successive atoms, the same question concerning their boundaries would re-emerge. It is rather interesting to observe that even such an outstanding and serious thinker as Lovejoy did not perceive this difficulty.[8]

This difficulty is far more serious since the only alternative to the atomicity of time seems to be its mathematical continuity. Both alternatives are unacceptable since the intuitive, qualitative continuum consists neither of durationless instants nor of sharply delimited, contiguous segments. Are we not facing a contradiction here? Or is the dilemma 'atomism versus continuity' falsely stated in the sense that it is *not* logically exhaustive?

Before considering this possibility, we must take a look at two serious attempts which have been made to conceptualize the structure of the intuitive continuum. The first one was made by Henry Poincaré in his book *Science et hypothèse* in 1902. It was he who coined the rather misleading and narrow term 'physical continuum'. (What he was really dealing with was *the perceptual continuum* which is one of the cases of qualitative continua.) He showed in a precise way the paradoxical and in a sense 'logically scandalous' structure of such continuum when we try to express it in a rigorous way. Its most paradoxical feature is that *the relation of equality does not seem to be transitive in it*; while its two contiguous terms are indistinguishable from each other, the non-contiguous are:

$$A = B, \quad B = C, \quad A \neq C$$

Poincaré illustrated it by the observation known since the time of Gustav Theodore Fechner: while we do not perceive the difference of weight between 10 grams and 11 grams or between 11 grams and 12 grams, we do perceive it between 10 and 12 grams. Similar examples can be found in different sensory or introspective continua, such as the scale of different shades of color, of different intensities of the same pitch etc.[9]

Poincaré obviously dealt with that type of sensory continuum whose terms are given simultaneously; it is therefore even more striking to see that Bertrand Russell arrived at the same results when in 1915 he tried to conceptualize the temporal qualitative continuum whose terms are successive. (This is the second attempt in this direction which I have in mind.) Russell showed that the relation of psychological simultaneity or 'temporal togetherness' is *not* transitive. Let us hear his own *ipsissima verba*:

Suppose, for example, the sounds *A*, *B*, *C*, *D*, *E* occur in succession, and three of them can be experienced together. The *C* will belong to a total experience containing *A*, *B*, *C*, to one containing *B*, *C*, *D*, and to one containing *C*, *D*, *E*. ... In the above instance, *C* is at the end of the specious present *A*, *B*, *C*, in the middle of that *B*, *C*, *D*, and at the beginning of that *C*, *D*, *E*.

Evidently, when we designate the relation of 'being in the same specious present' as 'psychological simultaneity', then this relation is not transitive. This is what Russell explicitly states when he considers a slightly different concrete example:

Suppose that I see a given object *A* continuously while I am hearing two successive sounds *B* and *C*. The *B* is simultaneous with *A* and *A* with *C*, but *B* is not simultaneous with *C*.[10]

There seems to be an obvious way out of both Poincaré's and Russell's paradox and this way was already indicated before. All we have to assume is that qualitative continuum, whether its terms are simultaneous or successive, is merely 'apparent' or 'illusory', and that the alleged logical difficulty stems from its 'haziness'. In other words, this difficulty disappears when we focuss our attention on the underlying physico-mathematical continuum which is allegedly 'the only real'. Illustrated by a concrete example: when we gradually increase weight, pressing on our hand, from 10 to 12 grams, the only thing we must consider is the continuous range of magnitudes through which the physical stimulus passes; within this continuum each term is sharply distinguished from any other and the logical absurdity of the non-transitive equality can then never arise. The whole difficulty is removed, if the scheme above $a=b$, $b=c$, $a\neq c$, is replaced by the following one: '*a* is *indistinguishable* from *b*, *b* is *indistinguishable* from *c*, but *c* is *distinguishable* from *a*'. In other words, the paradox arises merely out of the limited capacity of consciousness to differentiate the minutely different stimuli. The same is true of the intuited temporal continuum: the underlying *mathematical* continuum is *the only real* and in it the transitivity of simultaneity is fully preserved. The apparent intransitivity of 'temporal togetherness' or 'psychological simultaneity' is due to the haziness of our temporal experience, more specifically, to the fact that our psychological present is merely 'specious', without definite boundaries. The only true present is the mathematical, instantaneous, 'knife-edge' present in the physico-mathematical continuum of events.

I hope that I was not unfair to the convential explanation of both paradoxes which both Poincaré and Russell suggested. It remains to be shown that this explanation simply will not do.

In the first place, nobody claims that the intransitivity of equality exists on the level of physical stimuli. Neither does anybody deny that two stimuli whose difference is imperceptible are physically different, though indistinguishable psychologically. But it is clearly meaningless to call two sensations *qua* sensations 'different, though indistinguishable'. For the sensations resulting from two minutely different stimuli are *qualitatively the same*; to affirm their difference despite their imperceivability, does not make sense. A difference which is neither sensed or felt, simply does not exist *psychologically*; if we continue to say that two indistinguishable sensations are 'really' different, we are not speaking of sensations *qua* sensations, but of their *external stimuli.* In other words, we are unconsciously slipping from the language of perceptual data to that of physical stimuli. The paradox of the intuited continuum cannot be dismissed when we insist that it does not exist on the physical level; it continues to exist on the psychological level whose paradoxical structure it reveals. And this level can be called 'appearance' only with respect to the physical stimuli, but not with respect to itself; for as William James stated unanswerably: "A material fact may indeed be different from what we feel it to be, but what sense is there in saying that a feeling, which has not other nature than to be felt, is not as it *is* felt?"[11]

Are we then back to the dualism of two continua, one mathematico-physical, other qualitative-psychological – to what Whitehead called "bifurcation of nature"? This seemed to be true still at the beginning of this century, but it is rather doubtful today. Today we cannot assert dogmatically that physical nature down to its deepest microphysical level is describable adequately in the terms of mathematical continuum. In truth, there is a considerable circumstantial evidence that the applicability of the concept of spatio-temporal continuity to the quantum level is very questionable. Not only a number of physicists, but even some outstanding mathematicians admit that the belief in infinite divisibility of space and time is nothing but 'an enormous extrapolation' of our experience to which nature has no obligation to conform.[12] Various theories postulating the minimum intervals of time (chronons) and motion (hodons) are symptoms of the growing distrust of physicists to the traditional concept of continuity. It is therefore quite possible, indeed probable, that the physical world, at least in its deepest microphysical – and microchronical – strata does not possess such sharp edges and clear cut contours as the last century physics hopefully expected; in other words, that there is, to use Professor Margenau terms, an "elementary diffusion"[13] in nature. This is perfectly compatible with the fact that nature on the macroscopic level of the physiological stimuli is *for all practical purposes* continuous in the mathematical sense. The main reason why we find it so difficult to give up the applicability of this concept on the microphysical level is that this would be tantamount of the admission of the irreducible *qualitative* element in nature. We are still very strongly committed to the dogma of bifurcation of nature which divides reality into two completely different realms – the homogeneous and quantitative realm of matter and the mental or phenomenal realm of qualities. If we give up this dogma, as Whitehead did in his organic philosophy of nature, our reluctance will disappear together with other difficulties which Cartesian dualism created.

What does this mean with respect to our problem? It means nothing more, but also nothing less than the possibility that the physical and intuitive continuum are intrinsically not so different as it was believed; more specifically, that the microchronic structure of the physical continuum is not *essentially* different from that which we call with Weyl "intuitive continuum."[14] As stated above, it is neither the dense continuum of instants nor the contiguum of the atomic segments. This is not as irrational as it appears. The main reason why the dilemma 'atomism versus continuity' appears to us as logically exhaustive, is that we are habitually inclined to think in the visual terms; as long as we symbolize time by a geometrical line, there are clearly only two possibilities: either this line is divisible *in infinitum*, and then the only indivisible elements are zero-intervals, that is, extensionless points-instants; or it is not – and then it must consist of finite segments. Needless to say that our natural tendency will be to prefer the first alternative. But if we consider some other continuum, for instance the succession of sounds in melody, we see clearly that it transcends the opposition between atomism and continuity. This was already touched upon when I briefly discussed the inapplicability of both the concept of instant and of atomization in psychology. A more detailed discussion would require a more extensive digression into Gestalt psychology and phenomenology which would be beyond the scope of this paper.

It must be further added that the above mentioned anti-thesis 'atomism versus continuity' is *spurious*, as long as we mean by the latter term the continuity in a mathematical sense. Again, there is a considerable agreement among philosophers and mathematicians that the so-called 'mathematical continuity' is a *disguised discontinuity*, that is, discontinuity, so to speak, infinitely repeated. In Bergson's words, the "intellectual representation of continuity is negative, being, at bottom, only the refusal of our mind, before any actually given system of decomposition, to regard it as the only possible one."[15] In the words of Bertrand Russell quoting approvingly Henry Poincaré, "the continuum thus conceived is nothing but a collection of individuals arranged in a certain order, infinite in number, it is true, but *external to each other*... Of the famous formula 'the continuum is unity in multiplicity', the multiplicity alone subsists, the unity has disappeared."[16] In other words, the points and instants in the mathematico-physical continuum are *as external* to each other as are discontinuous finite segments.

But such discontinuity is clearly absent in intuitive continua. For instance, the temporal continuum is not the succession of the well defined atomic segments. Wittgenstein observed correctly that "visual field has no limits"; similarly, as Dr. Efron noted, we do not perceive the onset of perception, we are not aware of its incipient temporal edge; therefore such mathematical instantaneous edge does not *psychologically* exist.[17] Several thinkers then hastily concluded – Lovejoy, for instance, – that by eliminating the boundaries between successive specious presents we eliminate also their differences. Such claim is based on the assumptions that every qualitative difference implies the *externality* of the qualities in the question – the assumption whose falsity is shown by a simple perception of melody or by any temporal *Gestalt*. In truth, the simplest perception of difference exhibits in its very structure the co-presence of two different, though not mutually separated qualities. But we have to be on guard against our own habitual language: this 'co-presence' is not 'co-instantaneity' nor should 'two' be understood in the usual arithmetical sense of two mutually external units. Qualitative continua are heterogeneous, qualitatively differentiated, and not consisting of additively distinct elements. For this reason the paradox of intransitivity of equality is merely apparent, a mere pseudo-paradox since it arises only when we try to translate the structure of intuitive continua into the language of distinct, atomic elements. This atomistic language is conspicuous in both Poincaré and Russell; they both treat the sensory continuum as a *contiguum* of distinct, atomic elements, each of which can be designated by a distinct and self-identical symbol: A, B, C,..., etc. The paradox which crops up merely shows the inapplicability of such atomistic language to the experienced continua.[18]

Let us say, in conclusion, that mathematical continuity is *very approximately* applicable on the macroscopic level since on that level time and space are *practically* continuous in the mathematical sense. Very probably it is inapplicable on the microscopic, i.e. quantum level and it is manifestly inapplicable on the sensory and introspective level. In other words, on either of these levels neither time nor space are infinitely divisible; extensionless points and durationless instants have neither physical nor psychological existence. This suggests that the structure of the microphysical time-space is perhaps not *essentially* different from that of the directly experienced qualitative continua.

Boston University

NOTES

[1] Thomas L. Heath, *The Thirteen Books of Euclid's Elements*, Dover, New York, 1956, pp. 199–200.
[2] Herbert Feigl, 'The 'Mental' and the 'Physical'', *Minnesota Studies in Philosophy of Science* II (1958), p. 466.
[3] H. Bergson, *Creative Evolution*, Random House, New York, 1944, pp. 240–44.
[4] William James, *The Principles of Psychology*, Dover, New York, 1950, pp. 251–2: "Which is to say that our psychological language is wholly inadequate to name the differences that exist .... *But namelessness is compatible with existence*...", p. 254: "It is, in short, the re-instatement of the vague to its proper place in our mental life which I am so anxious to press on your attention." (Italics added.)
[5] Cf. my articles 'The Elusive Nature of the Past', in *Experience, Existence and the Good*, Essays in Honor of Paul Weiss (ed. by Irwin C. Lieb), Southern Illinois Univ., Carbondale, 1961, pp. 130–33; 'Memini ergo Fui?, in *Memorias del XIII Congresso Internacional de Filosofia*, Mexico 1963, esp.pp. 422–26.
[6] W. James, *op. cit.*, pp. 609–10.
[7] W. James, *op. cit.*, pp. 243ff.
[8] A. O. Lovejoy, 'The Problem of Time in Recent French Philosophy', *Phil. Review* **21**, 532–33.
[9] *Science and Hyphothesis* in *The Foundations of Science* (tr. by G. B. Halsted), The Science Press, Lancaster, 1946, p. 46. Cf. also *The Value of Science*, *ibid.*, pp. 240f.
[10] B. Russell, 'On the Experience of Time', *The Monist* **25** (1915), 218–228.
[11] W. James, *Some Problems of Philosophy*, Longmans & Green, London, 1940, p .151; *The Principles of Psychology* I, p. 63.
[12] D. Hilbert and P. Bernays, *Grundlagen der Mathematik*, Jena 1931, pp. 15–17; E. Schrödinger, *Science and Humanism*, Cambridge Univ. Press 1955, pp. 30–31.
[13] Cf. H. Margenau, 'Methodology of Modern Physics', *Phil. of Science* **2** (1935), 164–187, esp. 174–75.
[14] H. Weyl, *Das Kontinuum und andere Monographien*, Chelsea Publishing Company, New York, n.d., pp. 65–74.
[15] H. Bergson, *Creative Evolution*, Random House, New York, 1944, p. 170.
[16] B. Russell, *The Principles of Mathematics*, The Norton Co., New York, 1964, pp. 346–347; H. Poincaré, *The Foundations of Science*, p. 46.
[17] L. Wittgenstein, *Tractatus Logico-Philosophicus*, 6.4311; R. Efron, 'The Duration of the Present', *Annals of the New York Academy of Science* **138** (February 1967), 714: "The onset of a perception cannot be perceived for it is not an object of perception.... Analogously, we do not *perceive* the edge of our visual field or the 'borders' of our blind spot."

[18] F. B. Fitch in his penetrating article 'Physical Continuity', *Philosophy of Science* **3** (Oct. 1936), 486–493, rejects the alleged contradiction between atomism and continuity: "The reputed contradiction arises only if we try to employ a Democritean theory of solid, unbreakable, definitely determined atoms of material substance." (p. 490). When he says that "indistinctness may well be a fundamental clue to the nature of physical continuity" (p. 487), he is very close to David Bohm who in his *Quantum Theory* (Prentice Hall, 1951) pointed out the striking similarity between the mental processes and quantum phenomena, especially their indivisibility: "In any thought process, the component ideas are not separate but flow steadily and indivisibly, An attempt to analyze them into separate parts destroys or changes their meaning." (p. 170.)

A. PAPAPETROU

GENERAL RELATIVITY – SOME PUZZLING QUESTIONS

The formulation of the special theory of relativity is doubtlessly an important event in the history of physical sciences. We talk of *prerelativistic physics* to denote physical sciences as they were before that event.

Prerelativistic physics is based on the Newtonian concepts of *absolute space* and *absolute time*, absolute in the sense that these concepts are assumed to be given '*a priori*' and in particular to be unaffected by any presence of matter here or there. The absolute space is the 3-dimensional Euclidean space and the transformations allowed for the Cartesian-like coordinates are the Galilei transformations. The absolute time flows uniformly everywhere; one assumes the existence of signals propagated with arbitrarily large speeds – up to an infinite speed – and one can set synchronously running clocks at distant points of the space, thus establishing absolute simultaneity.

After a long period of unquestioned validity, supported by numerous and important successes, the first difficulties appeared towards the end of the 19th century. To mention two of them: From microphysics the increase of the inertial mass of particles moving with speeds approaching the speed of light; from field theory the incompatibility of the Maxwell equations with the Galilei transformations.

The special theory of relativity has been formulated in order to overcome an increasing number of difficulties of this type. The basic concept in special relativity is the Minkowski 4-*dimensional space* or space-time, which profoundly modifies the prerelativistic concepts of space and time. Indeed, the Minkowski space-time is the result of an organic interrelation of the previously totally independent concepts of absolute 3-dimensional space and absolute time. This interrelation is a direct consequence of the *principle of special relativity* which can be put in the following form: *All non-accelerated observers are equivalent to each other completely*, i.e. with respect to all kinds of physical phenomena. In other words there is no privileged non-accelerated observer like the 'observer at absolute rest' of prerelativistic physics.

Boston Studies in the Philosophy of Science, XIII.

When we demand the equivalence of non-accelerated observers with respect to the propagation of light in vacuum we conclude that this propagation must be isotropic and have the same speed for any non-accelerated observer. This has then the following important consequence: *The speed of the propagation of light in vacuum is the maximal velocity obtainable in nature.*

The existence of a maximal velocity has as an immediate consequence the impossibility of defining an absolute simultaneity and it also leads directly to the impossibility of attributing an absolute meaning to the 3-dimensional space and to the time separately: In the Minkowski space-time we can again talk of the 3-dimensional space and of the time, but only in a relative sense; i.e. of the 3-dimensional space and time which are the products of the decomposition of the space-time by a given non-accelerated observer.

The 4-dimensional Minkowski space has two basic qualitative properties in common with the 3-dimensional Euclidean space:

(i) it is a *flat* (non-curved) space and

(ii) it is *absolute*, in the sense that it is given '*a priori*', being unaffected by the presence of matter here or there.

It differs from the Euclidean space in one important aspect: It has an *indefinite* (i.e. non-positive-definite) metric. The last property leads to the existence of the light-cones which describe in a simple as well as elegant way the demands we have formulated on the propagation of light signals.

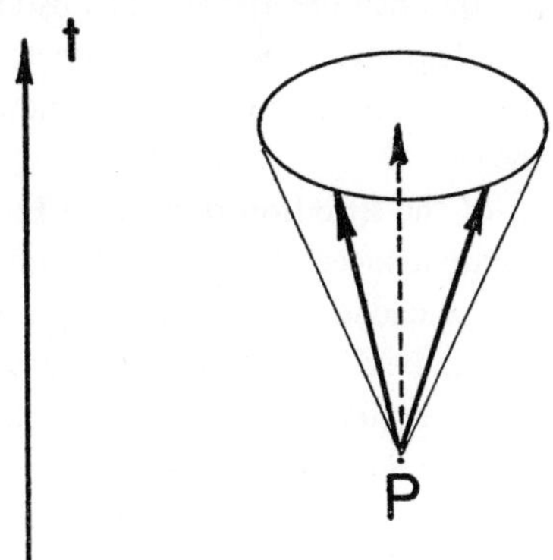

The diagram on Figure 1 represents these demands in a somewhat intuitive way. The arrow in the figure indicates the direction of the time axis and the *future* light cone of the point P will have the form shown schematically on this figure. Light signals emitted from (or passing through) P travel on the *surface* of the cone. On the contrary, any particle having non-zero rest mass must move in the *interior* of the cone in order to have a speed which is smaller than the speed of light.

Summarizing we can say that Minkowski space is certainly not more complicated than the prerelativistic concepts of absolute space and time; it may even be considered, from a certain point of view, as conceptually simpler. There is an impressive list of successes of the special theory of relativity in explaining or correctly predicting new phenomena. For all these reasons special relativity is considered today as a basic part of contemporary physics.

General relativity was formulated only about 10 years after special relativity and from an entirely different motivation. There was not any pressing necessity for its formulation because of experimental facts which could not be explained in the existing theoretical frame. On the contrary, it was the time of frequent new successful applications of the special theory of relativity. In fact general relativity came out as the result of a deep-going analysis of the question how to formulate a *relativistic* theory of gravitation. It is true that the Newtonian theory of gravitation had only successes and practically no failure at all; a minute residual in the motion of the perihelion of Mercury, which remained unexplained, was not a sufficient reason for changing the theory. Still the theory was based on the prerelativistic concepts of absolute space and absolute time, which we already found to be inadequate. It was therefore logically necessary to reformulate the theory of gravitation using the new concept of the 4-dimensional space-time.

A thorough analysis of the situation persuaded Einstein that it should be impossible to formulate a successful theory of gravitation in the frame of special relativity. He decided then to modify again the concept of space-time and thus he arrived at the general theory of relativity. The new modification is, in a certain sense, less drastic than the one made in special relativity. The space-time is again of the same type as the Minkowski space; it is a Riemannian space with the same local properties: The same signature and therefore the same structure of the light cones. But there is an important quantitative difference: *The curvature of the space is non-zero*. Moreover the non-vanishing curvature is not given '*a priori*' but depends on the actual distribution of matter. Indeed, the field equations established by Einstein for the gravitational field constitute a quantitative (differential) relation between the curvature of the space-time and the distribution of matter.

Thus in general relativity the gravitational field has been reduced to the geometry: it has been *geometrized*. We can still talk of the gravitational field in the usual way by comparing the actual curved space with the flat Minkowski space and by simply considering the difference of the metrics of the two spaces; but this is possible only in the case of *weak* gravitational fields. This case, in which the actual space-time is nearly Minkowskian, is realized in almost all physical problems. In particular the gravitational field of the sun and of all ordinary stars is weak. In the case of weak gravitational fields no fundamentally new questions arise. But there is now at least one case in which the gravitational field will not be weak: We shall have a *strong* gravitational field in the case of a neutron star, the recently observed *pulsars* being almost certainly neutron stars.

In the case of a strong gravitational field the old way of looking at weak fields is not applicable. Entirely new situations do arise and in extreme cases we face certain puzzling questions, which concern me now.

Strong gravitational fields were encountered for the first time in the *cosmological* solutions of the field equations of general relativity. From the physicist's point of view the cosmological problem is of an *asymptotic* nature. Scientific research gradually broadens the limits of the already explored region of the Universe. We have also to remember that the best working hypothesis is certainly the one according to which the Universe is infinite, in extension as well as in variety. It follows that the knowledge of the Universe as a whole – which would constitute the complete solution of the cosmological problem – should be considered as possible only asymptotically for $t \to \infty$.

However, partial cosmological questions have been asked already in prerelativistic physics. All these questions are related to a simplified model of the Universe based on the following two hypotheses, usually called the *cosmological principle*: The general features of the average distribution and properties of matter are the same

(i) at all points (homogeneity) and

(ii) in all directions (isotropy).

In prerelativistic physics cosmological questions have always led to difficulties. The origin of the difficulties has been essentially the infinite extension of the 3-dimensional space combined with the use of a time-independent model of the Universe, which may at that time be currently accepted.

A different situation presents itself in general relativity: Using the Einstein field equations one obtains a solution of the dynamical cosmological problem, again simplified by the assumption of homogeneity and isotropy, which determines a well defined evolution. Thus general relativity has given for the first time the possibility to formulate and discuss in detail the dynamical problem of the simplified cosmological model. I shall try to describe in some detail the results obtained from this discussion and the more general consequences, to which these results lead.

The assumptions of homogeneity and isotropy have the consequence that the 4-dimensional space can be considered as the product of the (1-dimensional) time-line and of the 3-dimensional space, this space being of *constant curvature*. It follows then that the metric of the 4-dimensional space is, in appropriate coordinates, of a simple form which makes the problem of solving the field equations of general relativity rather easy. The solutions obtained are the following:

(i) The Einstein solution (1917) which is time-independent. It has been obtained from the field equations which contain the so-called cosmological term.

(ii) The de-Sitter solution (1917) which is time-dependent. This solution does not take into account the distribution of matter in the Universe, being a solution of the field equations (with cosmological term) for empty space.

(iii) The Friedmann solution (1922) is also time-dependent and can be derived from the field equations with or without a cosmological term. I may recall that there are 3 possible cases: The 3-dimensional space can be (a) of positive constant curvature and then it is closed, or (b) of zero curvature or (c) of negative curvature. In the last two cases it is an open space.

The Friedmann solution has been found to be better adapted to the experimental facts since the discovery of the all-important *Hubble-effect* (1936): Observations of the spectra of distant galaxies revealed the existence of a systematic red-shift which increases nearly linearly with the distance of the source. The only possible interpretation of this effect seems to be the following: The distribution of galaxies in space is in a state of *expansion* with a nearly constant rate of expansion, the observed red-shift being simply the corresponding Doppler-effect. The Einstein solution of the cosmological dynamical problem cannot account for such an expansion, while the Friedmann solution was already predicting it. I shall therefore discuss the Friedmann solution in some detail.

The space-time described by the Friedmann solution is, as we have just said, in a state of expansion *now*. Looking backwards in the past we find from the mathematical description of the solution that the corresponding 3-dimensional space was smaller and smaller, so that at a certain moment it was *just one point*. Prior to that moment there was no geometrically meaningful space-time, i.e. there was no 'prior to that moment'. At that moment the space-time and its material content has been 'created' in a kind of an explosive act, the now famous 'big bang'.

This was the past. What will be the future?

We mentioned already that in the Friedmann solution are contained 3 different cases corresponding to positive or zero or negative curvature of the 3-dimensional space, which is closed in the first and open in the last two cases. In the first case the mathematical description of the solution shows that the expansion slows down with increasing time and changes into a contraction at a certain moment; the contraction continues then until the 3-dimensional space finally shrinks to a new point-singularity. This second singularity, which is of the same type as the one we had in the beginning, will be the end of the space-time with its content. In the two cases with an open 3-dimensional space the expansion is simply slowing down continuously; no subsequent phase of contraction and consequently no second singularity exists.

More detailed discussion of the Friedmann solution has shown that we shall have a closed or an open 3-dimensional space if the average density of matter is larger or smaller than a certain critical value. Present estimates of this density give a value smaller than the critical one: We may be in an open space and so escape the second singularity!

The cosmological singularity is the first of the strange consequences of general relativity, on which I had to report in this survey. It is certainly remarkable that from a physically meaningful equation an answer has been obtained to a question which was considered until then as a metaphysical one. The difficulty is to find out how truthful this answer may be, in view of the strong simplifying assumptions we have been obliged to introduce.

The second question, which I am going to discuss now, is related to a *physical* and not to the cosmological problem. Again we shall have to do with a singularity, which however is of a very different type, as I shall explain in the following.

The gravitational field of the sun and more generally of ordinary stars is a *weak* field: The metric of the space-time surrounding the sun differs from the metric of the Minkowski space-time of special relativity at most by an amount of the order 10^{-6}.

This result is described in some more detail as follows. The mass of the sun is measured in 'gravitational units' by the quantity

$$\frac{GM}{c^2} = m;$$

G is the gravitational constant, c the speed of light and M the mass in ordinary units (cgs-system). The quantity m has the physical dimensions of a length and the value of m for the sun is $m \approx 1$ km. The radius of the sun is $a \approx 10^6$ km. It is the ratio m/a which gives the order of magnitude of the difference between the metric describing the sun's field and the Minkowski metric. I have already mentioned that a weak gravitational field does not lead to any fundamentally new questions or conceptual difficulties.

A star which remains isolated will in the course of time be cooling off and contracting. Finally its thermonuclear energy content will be totally exhausted and we shall then have a *cold star*. Cold stars with spherical symmetry have been discussed on the basis of the Einstein field equations. This discussion led to the following important result: If the total mass of a spherical cold star is larger than a certain limit – which, incidentally, is of the order of the sun's mass – then the internal stresses cannot balance the gravitational forces and consequently at the end of the cooling period all matter contained in the star will have to contract until it reaches finally the centre of the star; this is the *gravitational collapse*.

The reason for this collapse can be explained in a descriptive manner: In general relativity the stresses also act as sources of the gravitational field. Thus the increased stresses needed to balance the gravitational forces which increased because of the contraction will cause a further increase of the gravitational forces and finally no equilibrium will be possible.

From the Einstein field equations it follows that when the total mass of the star has been concentrated at its centre this point will become a geometrical *singularity*: At this point the curvature will be infinite and the metric degenerate. Because of this it will be impossible to determine at this point the Christoffel symbols and also the velocity of matter, i.e. the quantities entering into the equation of motion. Consequently, it is impossible to write mathematically meaningful equations of motion for the matter which has fallen into the singularity. From the moment of collapse all matter which has reached the singularity will remain there permanently. It is as if matter falling in the singularity is henceforth of a very special kind, for which the concept of motion exists no more.

Note an important difference between this singularity and the cosmological one. In the cosmological case the whole space-time shrinks into the singularity. In the present case we have a singular point – more exactly a singular half-world line – in an otherwise regular space. If the singularity has been formed at the point *A* (Figure 2), we have the singular half-line

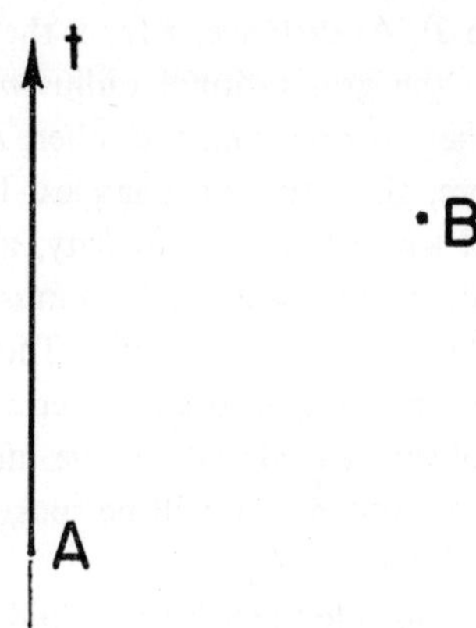

At. (The arrow in Figure 2 indicates the direction of the time-axis.) All points outside *At* are regular and an observer like ourselves might well be at a certain moment e.g. at the point *B*. This observer is then running an important risk which I shall try to explain now.

Outside a spherically symmetric star the gravitational field is given by the solution of the Einstein equations which has been found by Schwarz-

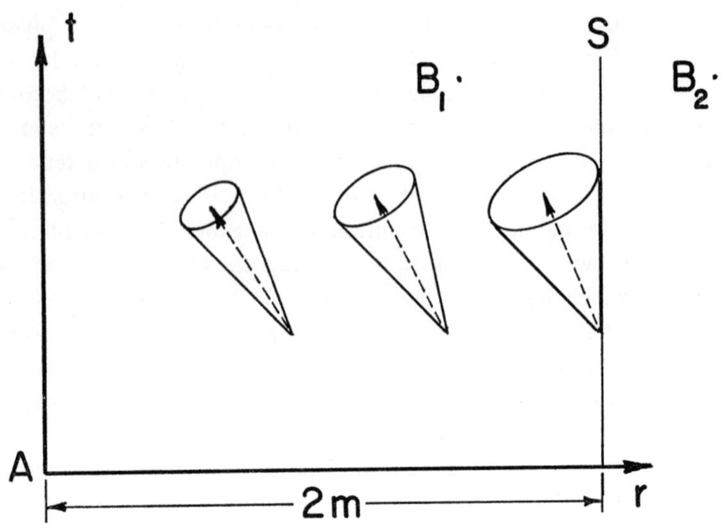

schild, the now famous *Schwarzschild solution.* After the collapse of the star we shall have the Schwarzschild metric everywhere outside the singular line At (Figure 3). At distances r from the singularity which are large in comparison to the gravitational radius m of the star nothing strange happens. But when we approach the sphere $r = 2m$, which is called the *Schwarzschild* sphere, the situation changes. Inside this sphere the light cones are inclined towards the singularity, as shown in Figure 3. Now we know that ordinary physical particles must move in the interior or at most on the surface of the light cone. (Tachyons should behave differently; but their existence has not been proved and is rather doubtful.) It follows that if some observer is already in the interior of the Schwarzschild sphere, e.g. at the point B_1, he will be sucked into the singularity and this in a very short interval.

An observer being just outside the Schwarzschild sphere, e.g. at B_2, can still escape but only if he has a very high speed in an appropriate direction. On the contrary, an observer far away from the singularity – e.g. as far as we are from the centre of the sun, at $r \approx 10^8$ m – will have no difficulty to stay permanently outside the collapsed star by circling around it: It will be sufficient that he has the appropriate peripheral velocity, as in the case of the motion of the earth around the sun. Such an observer can then try to observe the collapsed star. According to the preceding description of the interior of the Schwarzschild sphere the results of the observation will be as follows.

Nothing can come out of the Schwarzschild sphere S (Figure 3) from any point inside this sphere and consequently the observer will receive no signals of any kind – no light signals and no particles of any type – originating at points inside the sphere S. The surface S will therefore appear to the observer as an *horizon*. In particular the collapsed star being inside the sphere S will send no signals of any kind outwards and will therefore appear to any such observer as an absolutely *black* object. One has proposed the name *black hole*, in order to stress the fact that particles moving inwards can cross the surface S and will then fall on the collapsed body; but nothing will come out again.

Note that a collapsed star is not strictly unobservable: There is the gravitational field of this star and the observer can determine this field outside S in detail, e.g. by using an appropriate system of test particles. In this way the observer can arrive at a definite proof of the existence of the collapsed star and even determine its mass. In the opinion of a number of research workers black holes may well exist in nature and methods are now under discussion which would permit us to observe them, of course in an indirect way.

I should also mention that the Einstein field equations have another solution which would be represented by Figure 3 after a time reversal. The future light cones are then inclined away from the singularity, so that there can be only particles moving outwards. This new singularity can eventually blow up at a certain moment, to form a non-singular stellar object. This is the case of the *white hole*, still more hypothetical than that of the black hole, but nevertheless considered seriously by some astrophysicists.

I spoke of the risk of an observer falling in a black hole. However there is also a different aspect of the situation which might be considered as rather optimistic, as I shall try to explain now.

The case of a black hole existing for all times, and not created at a certain moment by the collapse of a 'cold star', constitutes a very interesting geometrical problem which has been discussed in detail. Several types of coordinates have been used for its mathematical description. The coordinates used initially for the solution of the field equations and some other coordinates obtained by rather simple transformations have the following common defect: They determine manifolds which are, in the geometrical language, *incomplete*.

A more complicated transformation has then been found leading to coordinates which extend the Schwarzschild space-time to a complete manifold (Kruskal coordinates). But this manifold presents to us an extremely interesting new feature: In the corresponding diagrams there are *two singular worldlines* and *two space-times*; the one in which observers like ourselves are located and a second one which is, in a certain sense, symmetrical to the first. The two singular worldlines and the two space-times are interrelated in a complicated, non-intuitive way.

The question then arises at once, whether we should consider this second space-time as existing independently. A first discussion has been made of the possibility to send signals from the one space into the other. The results have been negative because of the existence of the singularities, on which must fall every material particle which has crossed the horizon.

Later it has been found that there are other types of black holes, with angular momentum (axial symmetry only) or with electric charge. For these types of black holes the situation is in a certain sense simpler: Some types of particles do not necessarily fall on the singularity after they have crossed the horizon; they move inwards only down to a certain depth after which they move again outwards. They then cross the horizon again but they are going out *into another space*! Thus a black hole of e.g. the electric type may be considered as a bridge from *our* space-time into another one! As human beings complain often about wrong things happening predominantly in our space-time, it might be worth while to try to pass, with the help of such a black hole, into another space-time where things might be more happily arranged. In this sense a black hole might constitute not a threat, but a real hope....

To sum up, the field equations of general relativity lead necessarily to solutions with singularities. We have described here the cosmological singularities and the physical singularities of the type of black holes. The solutions with physical singularities lead to two important consequences which constitute profound modifications of until now generally accepted ideas:

(i) existence of a state of matter to which the concept of motion is not applicable and

(ii) existence of several space-times interrelated through black holes.

Scientists in general do not change fundamental concepts and ideas without compelling reasons and physicists are, maybe, even more conservative in this respect. Is there a compelling reason to accept the more than revolutionary changes dictated by the existence of singularities in general relativity?

It is impossible to answer this question in an objective manner now. A number of possibilities has been suggested which might constitute an efficient obstacle to the formation of singularities. I shall mention two of them:

(i) There may be limitations in the applicability of the classical form of general relativity, analogous to limitations we know to exist for Maxwell's theory. An example of such a limitation in the Maxwell theory is the following. Classically we can have a head-on collision of a proton with an electron, assumed to be charged point-particles, accompanied by the release of an infinite amount of energy. However such a collision is not allowed by the laws of quantum physics.

(ii) The assumption that we may have a theory of unlimited validity is equivalent to the assumption that the Universe is not infinite in variety. Assuming that we have such a theory *now* would mean that we already know *all* phenomena existing in nature – a point of view which is most probably incorrect. There can be still entirely unknown phenomena and some of these phenomena may result in the impossibility of the formation of gravitational singularities.

We shall have to wait until it will be possible to prove either the impossibility of the formation of singularities or the actual existence of black holes.

Boston University and
Institut Henri Poincaré, Paris

H. BERGMAN

PERSONAL REMEMBRANCES OF ALBERT EINSTEIN

The editor has kindly invited me to preface the English translation of my book on the Law of Causality – to which Albert Einstein wrote a foreword 41 years ago – with memories of my personal contacts with Einstein.

I made his acquaintance in 1910 when he was appointed professor of theoretical physics at the German University of Prague. To characterize the spiritual atmosphere that surrounded him when he took up his appointment, I must say a few words about the Prague of those days; for the Prague of 1910 and the Prague of 1970 are not only two different cities but two different worlds. Prague today is a purely Czech city, the capital of the Republic of Czechoslovakia; at that time it was a provincial town in the Austro-Hungarian Monarchy governed from Vienna by Emperor Franz Josef. In those days it was a tri-national town, inhabited by a Czech majority and the German and Jewish minorities. The minorities, however, were the intellectual leaders. The German University was of international renown – I need only mention the name of Ernst Mach who during four years occupied the chair of physics at the German University. And Prague's poets who wrote in German – Rilke, Kafka, Werfel – have conquered the world.[1] The German Theater, under the Jew Angelo Neumann, played a leading role in the German cultural life of the time. Much has been written about the peculiar intermediate position between the Czechs and the Germans that was held by the Jews. They formed a solid unit but spoke German among themselves, thus becoming supporters of the German minority in its struggle against the Czechs. This peculiar position was aptly illustrated by a contemporary anti-Semitic Czech witticism about the German Theater in Prague: "Director Jewish, actors Jewish, audience Jewish – and that's what is called 'German National Theater' (Deutsches Landestheater)." It is equally true, though, that the Jews as intermediaries and translators greatly benefited Czech culture – notably Max Brod.

It was necessary to outline, though briefly, this complicated intellectual background, to enable an understanding of Einstein's position in the

German-Jewish society of Prague, when in 1910 he was called to the German University to become the successor of (if I remember rightly) Lippich.

As for me – born in Prague in 1883 – I had been studying at the German University, philosophy with Marty and Ehrenfels, mathematics with Georg Pick (who was subsequently murdered by the Nazis in a concentration camp), and physics with Lescher and Lippich. I had obtained my doctorate a few years earlier, and it was, therefore, a matter of course that I participated in Einstein's seminary. The physics and the mathematics seminaries were housed in the same building – physics on the upper floor – and I remember vividly a moment when Einstein came excitedly rushing downstairs to the mathematician Pick: he was in urgent need of help by a mathematician (for the solution of an integral, I think) and asked Pick to come up. His assistant at that time was L. Hopf (born in Aachen, I believe), a man torn between physics and psychoanalysis – a new discipline that was just coming into fashion; once he said to his friends: "when I am reborn, I should like to be born a psychoanalyst."

When I am talking here about friends, I mean a philosophical circle of disciples of my aforementioned teacher, the Swiss Anton Marty. His field was the philosophy of language, and his principal work bore the title 'Basic Research for General Grammar and Linguistic Philosophy' (Halle, 1908). He was an orthodox pupil of the philosopher Franz Brentano. Initially our circle called itself the 'Louvre Circle' after the Louvre coffeehouse where we used to gather. Later on, we got together in the drawing-room of my then mother-in-law, Berta Sohr-Fanta, where Einstein was a frequent visitor when we were reading Hegel's 'Phenomenology of the Spirit'. I scarcely remember whether Einstein took part in these readings. Yet I well recall a very popular lecture he held before this score of non-physicists on the special theory of relativity; the basic principles had been published in 1905 under the title of 'Electrodynamics of Moving Bodies'.

The above-mentioned physics seminary was situated out of town, near the hospital, whereas Einstein with his first (Yugoslav) wife lived in the suburb of Smichov, and I frequently had the privilege of accompanying him evenings after the seminary on his half-hour walk home. In Mrs. Fanta's or in my house he met the pianist Ottilie Nagel, with whom he often played duets – she at the piano and he playing his violin.

Of the many talks I had with Einstein in those days, one in particular has lived on in my memory. It was in the spring of 1911, when we were living at Podbaba, a suburb about an hour's railway journey distant from Prague. Einstein came to visit us, and on the road from the station to our house he told me about the great amount of work he was just about to undertake (he had in mind his general theory of relativity) and about the lectures he gave to his students on elementary theoretical physics. I remarked that it must be rather hard to expound elementary theories at a time when he himself was preoccupied with equations of a cosmic nature. He agreed at first but, after further reflection he said: "Actually I am glad to be allowed to lecture on these elementary matters, for in my own research it often happens that I pursue and explore some thought, only to realize in the end that I have been wandering in a maze; all those weeks of hunting a phantom would have been lost, had I not been giving my lectures and thus done something useful during all that time."

In those months Rudolf *Steiner* came to Prague and lectured on anthroposophy. Then (as now) I was interested in the scientific aspect of anthroposophy, and when I once told Einstein that I was going to a lecture, he proposed to come with me. He listened to the lecture, but there was nothing less than a spiritual dialogue between the two. Einstein came out laughing. Rudolf Steiner's mystical way of thinking went counter to Einstein's physical trend of mind. Yet later on the realm of the occult must have occupied him also. When I visited him in Princeton in 1953 I found him immersed in a book on parapsychology that had just been published, and he said to me – as if interested and repelled at the same time – "It can't be true!"

I do not recall having ever discussed *Judaism* with Einstein during those years of 1910–1912, although I took an active part in Zionism. I do not think Einstein was interested in Judaism at that time. So long as he lived in Switzerland he was, I think, not even a member of the Jewish community – though he himself has told me that at the age of 12 he had gone through a short phase of deep religiousness. However, when he was called to Prague, as a professor he automatically became an Austrian civil servant, and in Franz Josef's empire no civil servant could be an atheist. Even Einstein had to belong to a religion of his choice, and he joined the Jewish congregation. In those days this was, however, a pure formality without any inner significance. It was only ten years later that Einstein, under the influence of the Zionist leader Kurt Blumenfeld, came closer to Judaism, and in particular to Zionism whose faithful servant he was to become later.

As to myself – I served in the Austrian army during the first world war and lost contact with Einstein. He had left Prague two years before the outbreak of war and was succeeded by Philipp Frank whom he had recommended. After the war I spent a year in London as secretary of the recently founded Cultural Department of the Zionist World Organization. Thus in 1919 I was an eye-witness to the great excitement that gripped the whole of the British people when it became known that the expeditions sent out to Brazil and Africa to observe a solar eclipse had proved the correctness of Einstein's theory of relativity. Banner headlines in the London press screamed: "Scientific Revolution! Newton's Theory Confuted by the Research of a German Physicist!" On November 6, 1919, the famous meeting of the Royal Society took place, so well described by the philosopher Whitehead. Einstein, who hitherto had been unknown to the public at large, became world famous overnight, and remained so for the rest of his life. "I have become sort of a newspaper celebrity" he said to me in 1953 with a laugh.

It seems to me that Einstein's sudden fame, based upon a discovery that most people did not and could not understand, beautifully expresses the chivalrous character of the British people. This scientific event had, after all, taken place during the armistice period, before the peace treaty between England and Germany was signed, and the British scholars were honoring a scientist from an enemy country who had triumphed over their own Newton. At that time in London I had a feeling that the victors' longing for peace played no insignificant part in the public acclaim accorded to the unassuming Einstein all over the world.

In 1920 I emigrated to Palestine from London and in 1923 saw Einstein again, when he visited Jerusalem on his way back to Berlin after a lecture tour of Japan. Asked about his impressions of Japan, he mentioned one thing that had impressed him most: that the Japanese employed human beings where we would use machines. Not that machinery was unknown to them, but in Japan humans were cheaper than machines. Whether this remark is true or not does not concern us here. I quote it only because it shows his social interest and his concern with the fate of the working man. During Einstein's stay in Palestine, there was a meeting of the Palestinian

Workers' Organization – nowadays known under the name of 'Histadrut' as a powerful political and economic factor – which at that time was still in its infancy. Einstein was invited to attend the meeting in Tel Aviv and delivered an address to the workers who received it enthusiastically.

Another event of those days might be worth mentioning: it was of a humorous nature but was characteristic of the Palestinian people of the time and their encounter with Einstein. It happened two years prior to the opening of the Hebrew University on Mount Scopus above the slopes of the Mount of Olives. The University did not yet exist, but Mount Scopus had already been purchased from the English family who had been living there and whose house was the only building on the site. The leader of the Zionist Executive was Menachem Ussishkin, a man with a great sense of historical drama. He hurriedly had the main hall of the house converted into a kind of lecture hall, and Einstein was invited to "deliver the first lecture on Mount Scopus." For the Jerusalem of those days it was a big social event, attended by the elite of the city (the majority then consisting of the heads of the various religious congregations). Ussishkin asked Einstein to step onto the platform, saying: "Mount the rostrum that has been waiting for you these past 2000 years." With a slightly ironical smile Einstein played his role in this theatrical performance and, after Ussishkin's bombastic words, began to lecture on the theory of relativity in his simple, unadorned German. That was the 'first lecture of the Hebrew University' held two years before its inauguration.

Of great seriousness was Einstein's attempt, six years later, to allay the Jewish-Arab conflict. It was 1929, the year of the first organized Arab uprising against the Jews, and in the course of the riots the Arabs founded an English language weekly 'Falestin' with an Indian Moslem editor named Achmed Achtar. Einstein's fame at the time was so great, even among the Arabs, that the editor turned to Einstein with an invitation to suggest a solution of the conflict. Einstein's reply, written on February 25, 1930, is a document characteristic for his manner of tackling political problems. He took a direct, straightforward attitude, without political or diplomatic deviousness. He wrote: "Our (the Jews' and Arabs') position is bad because Jews and Arabs are opposing each other as warring parties in the face of a third factor – the British Mandatory power. Such a position is beneath the dignity of both peoples, and it can only be improved through an understanding between ourselves." And then he proposed the formation of a council composed of Jews and Arabs in equal numbers. Both Jews and Arabs should elect four representatives each: a doctor, a lawyer, a worker and a religious minister. It should be a strict condition that none of the eight was dependent in any way on a political institution. The discussions of this council should be secret, and resolutions could be published in the name of the whole council only if they were carried by at least three votes from each side. Although this council would have no legislative powers, it should gradually earn the respect of the mandatory power and thus gain political influence.

Achtar, the editor, accepted the proposal enthusiastically. "We are happy to see the name of Professor Einstein connected with a plan which – had it been conceived by anybody else – would surely have been suspect. Now a round table for discussions will be possible." (The text of this letter was published in Hebrew in 'Sheifoteinu', the periodical of 'Brith Shalom', the Jewish peace movement of the time.)

Einstein's proposal bore no practical fruit. But his active concern and his straightforwardness in approaching difficult social problems surely are characteristic traits of the picture we may draw of Einstein's personality. Einstein had failed to take into account (and this seems to me the basic grounds for his failure) the fact that in times when nationalism is rampant, it is impossible to find four men who are independent of the political powers-that-be to such a degree that they could freely voice their views. The failure of the Jewish 'Brith Shalom' in those years may also have been due to the impossibility of finding a partner on the Arab side with whom such a dialogue could have been held.

The last time I saw Einstein was when in 1953 I was permitted to visit him in Princeton. Our conversation lasted one and a half hours and dealt almost exclusively with actual problems of physics. Einstein explained to me the tragic tension between himself and the younger generation of physicists: his pupils had adopted the statistical method that had developed in physics as a consequence of the quantum theory, whereas he himself refused to go along with them and to admit that "God casts the die". Although he himself had revolutionized physics, his own thinking was too deeply influenced by classical conceptions for him to be prepared to acknowledge that *chance* had a place in physics. During the past years he had been working on the 'World Formula'. The World Formula was there, but there was no way to prove or to disprove it. "Thus my World

Formula is like a locked box that cannot be opened" he said, laughing all the time, and told me with amusement what appeared to me, the listener, a tale of loss of many years of his life.

This humor in such a tragic story at the same time revealed to me the religious basis of Einstein's life. The world admired, even adored him, whilst he saw himself and his work objectively, as part of the universe. "My internal and external life," he once said, "depend so much on the work of *others* that I must make an extreme effort to *give* as much as I have received."

NOTE

[1] An excellent description of the milieu of early twentieth century Prague is given in Hans Tramer's article 'The Tri-National City of Prague' in the volume published in honour of Robert Weltsch's 70th birthday (Tel Aviv, 1961, pp. 138–203).

On April 18, 1969, the central organ of the Czech youth federation 'Mladá Fronta' (The Young Front) published an article under the headline 'Prague Forgets Einstein', complaining that there is no Einstein memorial tablet in Prague, nor has a postage stamp been issued in his memory. Only in the entrance hall of the Institute of Physics a commemorative plaque was unveiled in 1966; after the renovation of the building this was to be affixed at its front – I do not know if this was done.

H. BERGMAN

THE CONTROVERSY CONCERNING THE LAW OF CAUSALITY IN CONTEMPORARY PHYSICS

In memory of my teacher Anton Marty (1847–1914)

FOREWORD

It is well known that present-day physics, influenced by the facts of atomic physics, places in serious doubt the feasibility of a rigorous causality. This difficulty will probably be rectified when a philosophy specialist, praiseworthy for the unusual extent of his knowledge of philosophical literature, his independence of thought, and his knowledge of the actual relevant physical facts along with attempted explanations of those facts, analyzes the problem. May this little book contribute to promoting the best in our present-day attempts at merging physical and philosophical thought.

Berlin, March 1929 A. EINSTEIN

PREFACE

Allow me to premise the text with a few personal words. Whoever takes upon himself the effort to compare the present paper with my earlier books published in German would be able to see conflict between my present position and that which I accepted earlier, especially in my book concerning Bolzano. This conflict is concerned above all with the law of causality. At that time I believed – following Brentano – that this law was infinitely probable and therefore certain in practice. Hume's attack on science thus seemed to be dissipated.

I had to recognize, however, that the law of causality itself cannot be proven by the theory of probability; the applicability of this calculus cannot be pertinent to practical experience itself without something further (see sections 22ff of this paper). Its applicability needs a foundation itself because it is not obvious why the laws of probability should be valid in a 'world' given value by chance alone. A new hypothesis is thus needed here. Upon what should this hypothesis be based – this was the question.

It was with this consideration that the strongest pillar of Brentano's theory of knowledge crumbled. I realized that the law of causality cannot be proven, but must be given a foundation not simply for reasons 'of convenience', but as a precondition of practical experience. I assume as legitimate the protest that practical experience itself certainly needs grounding, and that the diversion of the most important assumptions of practical experience from the possibility of practical experience is really, as Pichler said, "to brew a drink from thirst." This was at the limit of demonstrability.

Recent physics now poses for us a new problem: it points out that the law of causality is not necessary for the practical knowledge yielded in some fields of physics; instead, physics can rest content with the hypothesis that the probability calculus itself can be applied to reality and that physics would have to be content [with this application] because the experimental assumptions for using the law of causality cannot in principle be given. The present paper is concerned with this question. I have taken the trouble to merge epistemological trends with the new turn in physics without failing to recognize the enormity of the dangers which threaten [scientific] inquiry if we actually abandon the epistemological goal of rigorous causality.

In my accounts of the works of Heisenberg and Bohr I have not relegated myself to tracing the exact progress of their thought but have interpolated, to a certain extent, in places where I had had to struggle with difficulties in comprehending their lectures. If mistakes have been made in doing this then I am, of course, responsible for them.

Comprehension of the problems in recent physics was made accessible to me by my friends Dr. M. Reiner and Dr. S. Sambursky in Jerusalem, to both of whom I give my heartfelt thanks.

I have dedicated this book to the memory of my teacher who, if he were still with us, would certainly have shaken his head critically over much which is said in it. The difference in [our] views, even concerning the basic questions of philosophy, cannot lessen the respect which I have for Anton Marty, a paragon as a human being, teacher, and philosopher.

Jerusalem (Hebrew University), August 9, 1928

TABLE OF CONTENTS

18. Born's polemic against Lasker.
19. Heisenberg's doctrine of the impossibility, in principle, of a simultaneous and exact determination of position and impulse.
20. Bohr's theory of complementarity.
21. Historical remarks concerning the assumption of contingency in nature.
22. Does the use of probability calculus presuppose causal law?
23. Acceptance of causal law is not necessary for the probability calculus, but is to be postulated separately for its applicability to reality.
24. Spontaneity theory. Phenomena as play with dice.
25. Can the postulate replace the functions of the law of causality?
26. The synthetic character of that postulate.
27. By what means is the minus purchased through presuppositions?
28. Miracles have become possible according to natural law. Mere pedagogical significance of such arguments.

III. TELEOLOGY IN PHYSICS?

29. Influence of the future upon the past?
30. Disagreement with the hypothesis of the probability connections.
31. Hamilton's Principle.
32. Influence of the future would destroy the determinateness of the present given.
33. Penetration of teleology into physics.

IV. PROBABILITY AND FREE WILL

34. Does recent physics allow room for freedom? Misunderstanding in interpretation.
35. Gaps in causality do not imply freedom.
36. The postulational character of causal law, and the law of the applicability of probability theory.
37. Abstract character of physics. Compatibility of methodological determinism with freedom of the will.

I. CAUSALITY

1. In 1900 Planck initiated the first thrust of the quantum theory and in 1905 Einstein published his first paper concerning the theory of relativity. A quarter-century has thus passed since physics developed its new complex theories, the theory of relativity and the quantum theory. Both theories have aroused an unusually strong philosophical interest and, as hardly any other scientific theories before this time, contributed to clarifying the conflicting views concerning the meaning of fundamental philosophical concepts and concerning scientific method as used in general and as used in physics in particular.

The following is, I believe, the result of this development for the theory of knowledge: empiricism and positivism, which supported all science, and especially natural science, upon practical experience alone, believed they could show we sort out physical concepts from practical experience and read off physical laws from practical experience, crumbled as under. Never as now has it been so clear what prominent role is played by the significance of facts in physics. Although the Einsteinian view of the world has, as is known, received strong support from empirical data, it is in no sense [my purpose] here to point out the prominence of that support, but instead to attempt to point out the new significance which this view of the world has given to empirical data. The facts have not been shown to be the crucial value of this view of the world, but instead the crucial value – to use Kantian terminology – has been shown to be the transcendental assumptions with which the scientific inquirer approaches nature. It is probable that the theory of relativity has taught us more clearly than earlier to know how to distinguish, in the structure of physics, the empirical elements from the *a priori* elements; proof through practical knowledge had to leave much to *a priori* content or show itself as empty in regard to the empirical and to our methods and means of empirical observation.

On the other hand, however, the role played by definition in scientific structure was revealed in its entire weight. "In to the chaos of the flowing world a system of concepts is defined to such an extent that this chaos seems like an ordered flow. Geometry is thus that framework of relations which carries the chaos ordered by it," as is stated by Hans Reichenbach in regard to the philosophical product of the theory of relativity (*Philo-*

sophie der Raum-Zeit-Lehre, 1928, p. 204). And Planck says in very general terms "that all arrives at the chosen point of departure; for from nothing nothing can arise, [and] without certain assumptions nothing can really follow (*Kausalität und Willensfreiheit*, p. 12).

It is known that a conventional interpretation of science was advocated by prominent physicists such as Poincaré. This conventional interpretation was, with reason, rejected from the philosophical point of view since it was noticed that such interpretations lead all too easily to the erroneous path of fictionalism. The transcendental assumptions of science are no fictions, no conventions. Fiction marks the end of science if it is accepted as a basis for science. But if such an expressed [interpretation] was incorrect and could yield false keys, the objection to empiricism, which was rooted in conventionalism, was correct.

Science is the meaning of reality, not the photograph of reality. As meaning, it is bound to certain presuppositions which do not themselves originate from the empirical. These presuppositions can be of various types and science organizes itself according to the type. These presuppositions are comparable to the tools of workers. If we now only partially see that, in certain fields of physics, conflicting hypotheses have been assumed for years – for example, the wave theory and the emission theory of light enduring next to each other, or the same with classical physics and quantum theory – then such development, which indeed will always and only be temporary and provisional, is suitable for sharpening our awareness of the difference between transcendental presuppositions and the results of observation – between tools and the product of work.

To write off these different and mutually conflicting assumptions as fictions would be mistaken because fiction means a false statement provisionally accepted as true. Fiction therefore has its own inherent contradiction. Thus, if one physicist proceeds with the wave interpretation of light and another with the corpuscular interpretation of light, then both assumptions are tools. There are, if one pleases, different languages, and it cannot be said they are 'fictitious' because they have two different vocabularies or are 'true' for the same reason; the same applies for the categorical presuppositions which the scientific inquirer stipulates when he goes to work.

2. As an example, let us take physics as a whole and its delineation from

other sciences. Planck designated in an excellent way (*Physikalische Rundblicke*, 1922, p. 34) the goal of physics as the complete separation of the physical structure of the world from the individuality of the structuring mind, i.e. the emancipation of anthropomorphic elements. That means: it is the task of physics to build a world which is foreign to consciousness and in which consciousness is obliterated. This is the transcendental presupposition of physics (see my essay 'Über einige philosophische Argumente gegen die Relativitätstheorie' in *Kantstudien*, 1928).

Now, is this presupposition true, is the world foreign to consciousness? This is to ask if the significance of the transcendental presupposition is entirely mistaken. The physicist himself, who thinks epistemologically and understands his own work, will not want to make such an assertion. On the contrary, this is the language with which he has to speak as physicist. The world of physics, however, is a world of meaning, i.e. of abstraction,[1] and we must require of physics that the internal conceptual web of the abstraction be pure throughout. If indeed the conceptual tool should become dogma, then we would remember its transcendental origin. We must guard ever so carefully against allowing transcendental presuppositions becoming dogma[2] as well as against the positivist assertion that they be robbed of the empirical and on this account be declared empty because they *correctly seem to be empty* of a certain empirical content. An example:

The "nonchronologically determined events" (as Reichenbach calls them, *Philosophie der Raum-Zeit-Lehre*, p. 168) originate through the parallelism of two time lines at different spatial points. The fact that our fastest messenger, the light signal, always has only a finite speed even though it does have such a great speed, and therefore that we can stop clocks at different places in the world one after another only by signals which themselves need time to traverse their paths, implies that if we regulate the time from P to P', an event at P seems to be coordinated with a point event at P', all points of which are nonchronologically determined in relation to that point event [P]. If we imagine our quickest message moving ever faster, then the indeterminateness becomes ever smaller and in the case of infinite speed it seems, in return, that one point event is coordinated with another point event and the classical interpretation of time seems to be reestablished again.

This means however: the old interpretation of time has become a limiting case which we are actually unable to attain because we measure time with objects. The limitation to which we are committed here is thus rooted in the method of physics. Within this method the concept of absolute simultaneity is 'empty', i.e. it is not physically realizable with objects. That concept has not on that account become meaningless, however, because it specifies a limiting case which is not meaningless even if that limiting case remains imaginary and unattainable from the standpoint of physics.

The philosopher, who is not tied to the methods of physics and the goal of whom is not to structure an objective world of objects unrelated to consciousness, but whose world is indeed only an artful, methodological abstraction and whose job it is to point out its place in the entire realm of knowledge and to place, to a certain degree, the world unrelated to consciousness back into the world of consciousness, is not bound to the limitation of the structure of physical concepts. The philosopher will thus be able to prevent the discarding of concepts which have become empty for the physicist in the sense that they are not physical, i.e. realizable through objective concepts. Such concepts, even though they are only limiting concepts to the physicist, are of the highest importance methodologically for the ideal of science.

The 'pure' concept thus does not become empty even though it seems empty to the physicist, who knows and uses it only in its empirical dress. It is precisely the assignment of the philosopher to retain the legitimacy of the pure concept and thereby save it, not the least important reason for its preservation being for physics itself, which will need it tomorrow if the empirical will have progressed further in devising, by reflection upon the pure concept, new conceptual tools for itself. In an unsurpassed manner Natorp, in his *Logischen Grundlagen der exakten Wissenschaften*, pp. 392ff, which appeared immediately after the first publications of Einstein and Minkowski, clarified the pure sense of space and time and confirmed the relative right of physics along with defending the absolute right of philosophy. He thereby rendered a greater service to both philosophy and physics than those philosophers who, identifying or recognizing as the same both pure and empirical-physical concepts, fought against the theory of relativity because it did not accept absolute space and absolute time for making its measurements. It is the proper combination of the pure and the empirical, of the Choresis and Metexis of idea and reality, that has the key to the solution of the philosophical difficulties of recent physics.

An American experimental physicist who recently wrote a book concerning the logic of modern physics, P. W. Bridgman (*The Logic of Modern Physics*, New York, 1927), argues that the highest merit of Einstein is that he taught the physicist to define his concepts through the physical operations by which they can be actualized instead of defining them through their conceptual content. The philosopher will gladly recognize this service. But when Bridgman goes further by declaring that the concept is synonymous with its corresponding operations (therefore, perhaps, time is synonymous with the measurement of time), the philosopher will point out that this identification of the concept with its physical operations would represent a new type of nominalism, and this nominalism would have to hinder physical progress severely because it would destroy the significant development of a concept promising progress in empirical possibilities. [Why?] Because this development, which always replaces one set of operations serving momentarily as the definition of a physical concept with another set of operations, occurs by retrospection upon the pure concept lying at the base of each empirical concept and thus propelling concrete inquiry. It is the job of the philosopher to portray that pure concept and yet, at the same time, give physics freedom to have that elasticity and looseness for the empirical concept, which is required for the momentary state of the empirical.

3. The contemporary development of physics requires such a relaxation of the concept of causality. How far can such a freedom be allowed to extend? This is the question with which we will have to occupy ourselves. Before we approach this question, the pure meaning of the concept of causality and the law of causality should be discussed in the sense of the foregoing statements. Once we have clarified these notions, we will be able to answer the question as to what extent concessions can be made to the empirical.

Brentano defined the law of causality in the following way (*Versuch über die Erkenntnis*, p. 108): "With each becoming, that which becomes is linked with something which comes before as the effect with the cause; in other words, something which comes before determines that which becomes." A double formulation of the law of causality has repeatedly been asked for (for example, Benno Kohn, *Untersuchungen über das Causalproblem*, 1881, p. 35, as well as Brunschvicg, *L'expérience humaine*, p. 526). Thus Kohn adds to the law of causality, which he formulates as vaguely as Brentano, a principle of causality which states that equivalent causes always draw equivalent effects to themselves (the same with Maxwell, *Matter and Motion*). It seems to me that such an addition is unnecessary because the unequivocal relation is already present in the concept of determination. If after an *a* a *b* follows one time and a *c* follows at another time, as the supporters of the statistical law of nature now accept, then one is no longer allowed to designate with confidence that *a* is the cause of *b*.

Kohn's assertion, called the 'Causal Principle', is of course less comprehensive than the general law of causality and thus one could retain it without accepting the general law of causality. This is the position of Reichenbach, who calls the general law of causality the 'determination form of the causal hypothesis' and rejects it while he accepts the 'implication form of the causal hypothesis' (if there is *A*, then there is *B*). ('Die Kausalstruktur der Welt', in *Sitzungsberichte der Bayr. Akad., Math.-Nat.*, 1925, pp. 133ff.)

The problem with which philosophy has grappled since Hume is to produce a proof for the validity of the law of causality. These endeavors have not been successful. There has been no success in refuting Hume and establishing the law of causality. But in the process of attempting to refute Hume, the concept of 'proof' of the most prominent assumptions has been clarified for science and, more important than the unsuccessful formal refutation of the Humean objections, is the production, by philosophy, of an understanding of the nature of the most prominent scientific principles. It has been recognized that the problem of giving a "proof" of the most important assumptions has itself been posed improperly. Only from the perspective of pre-critical philosophy was posing such a problem possible: science then passed as the image of reality and the question posed by Hume was thus justified at that time. Is it the case, it was asked, that reality is actually such that the law of causality is inherent in it? And to this question there was no answer possible. It was recognized, first through the Copernican revolution of Kant, that the process of science is a meaning of that *X* which is the basis of all meaning and which could only be identified figuratively with reality or the world, and thus that these concepts are themselves already meaningful. This meaning is bound, however, to certain presuppositions, 'synthetic *a priori* judgments', which the meaning itself offers and which determine the framework in which appearances should be determined and interpreted; stated otherwise, 'synthetic *a priori* judgments' constitute the categorical tools with which we approach appearances. The general validity of the law of causality is therefore not a statement demonstrable by logic, but expresses our resolution 'to make use of all means to comprehend each given fact as subjected to regularity and, as long as this inquiry is not successful in respect to regularity, it is not to be viewed as proof for the refutation of the law of causality but as an indication of the presence of an uncompleted assignment' (H. Gomperz, *Problem der Willensfreiheit*, pp. 15ff.).

The law of causality, as such a supreme categorical presupposition, can neither be proven nor refuted. Practical knowledge is not capable of that because each non-proven causal relationship only signifies an incomplete assignment and not a refutation. It is therefore with validity that Ehrhardt speaks (*Zeitschrift für Philosophie*, 1909, p. 223) of the methodological generality of causality, which therefore survives even if we are not somehow in a position to distinguish a change from its causes; we always conclude only that to this time we have not been able to detect the cause. We are able to do this because we can never prove exceptions to the law of causality.

On the other hand, and *a priori* proof of the law of causality would only be possible by reducing it to the purely analytic judgments of logic; such a reduction, as Hume saw, cannot be effected. Here we are clearly concerned with a hypothesis which is laid at the base of practical knowledge.

4. Kant collectively named as pure science all synthetic *a priori* judgments which lie at the base of natural science and first made it possible "to establish the *a priori* conditions of the possibility of practical knowledge," which "are at the same time the sources from which the all-encompassing laws of nature derive." (*Prolegomena*, #17). "The laws of nature which lie at the base of science are known *a priori* and are not mere laws of practical knowledge" (*Metaph. Anfangsgr. d. Naturwissenschaft*). The law of causality is in this sense a proposition of pure science. Now,

when in recent times attacks against the law of causality were raised from the camp of natural science, when it was said the law of causality has ceased to be rigorous and unequivocal in its meaning for the natural sciences and that it must be replaced by a much more slack law which no longer demands unequivocal application in particular instances but instead is satisfied with a regularity of a statistical type, a lawfulness of the average, then we must pose to ourselves the general question: if a proposition of pure science (or, in general, a synthetic *a priori* judgment) cannot be refuted or proven by either formal logic or practieal knowledge, what further way is there to give substitute support for (Del Negro: "supplementary 'amendment' to") such a law?

From the standpoint of the Kantian philosophy interpreted strictly, this question makes no sense. Synthetic *a priori* judgments are fixed for all times according to Kant. The following alone seems to be the essential progress which philosophy has made in this area: that the rigid system of synthetic *a priori* judgments has been made flexible without sacrificing the basic idea of Kant – that science is the meaning of reality, the latter being qualified by certain categorical presuppositions. It was recognized that scientific development consists not only in collecting more and more *a posteriori* facts in a framework already determined forever, but that this framework itself is capable of expansion and development and that it is precisely this development of the basic synthetic *a priori* principles which signifies essential progress in science.

Now, how is such an *a priori* axiom changed? I do not say 'refuted' because a refutation in the strong sense is as little possible as a proof. I see two possibilities available by which a synthetic *a priori* axiom can be given up or modified: (a) if it can be shown that empirical data cannot permit the use of the axioms. If we were to lose all our ability to move ourselves, we would probably drop certain principles which we place at the basis of our present physical superstructure, which would probably become useless to us as a result. If the physical apparatus were not granted us to grasp or determine the identity of certain physical entities, then there would be no sense in applying to them the law of causality, which has to do with the determinateness of phenomena when there is repetition of identical events, etc.

We will later see how Heisenberg and Bohr, using this line of thinking, sought to prove the emptiness of the law of causality for atomic physics.

(b) One could also imagine science placing at its base, above all for ordering the data of practical knowledge, an axiom which exacts more than is necessary and thus that science, as soon as this is recognized, will give up the far-reaching demands of such an axiom and, instead of requiring an axiom *entia non sunt multiplicanda praeter necessitatem* will be content with an *a priori* axiom less postulated than realized spontaneously. Thus an interpretation of nature according to which all natural facts can be declared causal accepts less in regard to the lawfulness of nature than an interpretation which includes both causality and teleology in explaining natural facts. One will thus have to be content with the former axiom – with mere causality at its base – presupposing, of course, that this axiom is actually satisfactory for explaining natural facts. Further on we will see, using the same considerations, that we can accept in place of the clearly rigorous relationship of necessity between cause and effect a relationship of probability between cause and effect for the reason that this latter relationship should be extendible to explain other areas of nature or even to explain the entirety of nature.

Using both the above criteria, therefore, we will proceed to test the objections brought forth against the hypothesis of causality in contemporary physics.

5. It is now our job to examine which functions this hypothesis fulfills within scientific work. Only if we know these functions are we really in the position to ask if the hypothesis of rigorous causality can be replaced with a looser hypothesis. Not until then can we determine which functions of the earlier foundation can be fulfilled by the new foundation and whether and how far it is integral to the establishment of science.

In science itself causality has two functions to fill: (1) the function of determining the reciprocal time-junctures of events and (2) the function of directing the process of scientific observation such that prediction of the future is possible.

Leibniz was probably the first to refer to the function of causality in determining time. In 'Metaphysischen Anfangsgründen der Mathematik' (*Hauptschriften zur Grundlegung der Philosophie*, tr. by Buchenau, 1904, Vol. 1, p. 53) he writes: "If, of two elements which are not simultaneous, the one comprises the basis for the other, the former is considered the precedent and the latter the consequent. My earlier condition comprises the basis for the being of the later [condition]. And there, on account of the connexion of all things, the earlier condition in myself also comprises the earlier condition of other things, and thus it likewise contains the basis for the later condition of other things and is there with earlier than them. All existing elements are therefore ordered in relationship to each other according to the relationship of simultaneity, or before and after each other."

In commenting on this position of Leibniz [3] Cassirer notes: "These attempts to define space and time are characteristic for the general methodological ideal that Leibniz has before his eyes. If most determinations of space and time hold good as immediate and obviously certain and concepts, on the other hand, are held to be derived elements, then the reciprocal relationship is a factor here. Especially emphasized is the decision concerning whether two subjects follow or precede each other objectively, presupposing the collation of these subjects from the point of view of cause and effect. We first make the decision with the help of a purely conceptual principle concerning the factual order of the individual elements in the order of time."

Cassirer's elucidation of these few words quoted here from Leibniz brings into correct perspective what was hinted at by Leibniz but what was later analyzed more clearly and completely by Kant in the second analogy of the *Critique of Pure Reason.*

For Kant, causality is, for the following reason, the presupposition for the objective statement of chronological order: The manifold of appearances is observed by us as truly successive at all times without retrospection as to whether, in an actual situation, there is a succession or simultaneity. Whether we now observe the parts of a house one after another or, per chance, the different positions of a ship moving down the river, the observations always follow each other subjectively. It is the job of science to determine the objective juncture of time from this subjective succession. Now, however, time cannot itself be observed, and therefore one cannot determine the position of events in time by their relationship to absolute time, because absolute time is not part of the observation. It must be the case, therefore, that "appearances themselves determine their positions in time and make these (positions) necessary in the order of time, i.e. the same that therefore follows or happens must follow, according to a general rule, upon that which was included in the previous states-of-affairs by a succession of appearances which in turn brings forth and makes necessary, using the understanding, precisely the same order and continuous connection in the order of all possible observations." It must, therefore, "be thought that the relationship between states-of-affairs [which follow each other in the observation] necessarily determine that which precedes, that which follows, and that neither could be placed in reverse [order]." The concept which makes it possible for us to determine in the objective progression of time one event as earlier and another as later is the concept of cause and effect. We thus take as the basis of practical knowledge a fixed special law of causality, according to which *A* is the cause of *B*, *B* is the cause of *C*, etc., [doing the same] first by experiment and then with ever greater certainty, and this hypothetical basis for the law of causality *ABC*... makes it possible for us to determine the objective succession of events *ABC*.

Through multiplication of the single laws of causality, which we take as the basis of practical knowledge, we strive to attain the establishment, for each given natural phenomenon, of its preceding *X* and its following *Z*, so that no single natural phenomenon remains isolated. The logical sequence in scientific work is thus the following: first comes the hypothetically accepted law of causality and then, upon the basis of this law of causality, the determination of sequence in the succession of appearances. This determination would not follow if the universal causal connection of appearances were not assumed (see Cassirer, *Leibniz' System*, p. 282).

6. Still remaining is mention of the post-Kantian author Lotze who, in the chapter devoted to time in his *Metaphysik* (edited by von Misch, p. 284), realizes that: "Only in the content of the phenomenon itself, not in a form existent outside of the content in which the phenomenon occurs, can the basis of its chronological order lie and, at the same time, the basis of its succession.... That which we can call the past we first consider the condition without which the present is not, and in [the present] we see the necessary condition for the future; this one-sided relationship of dependency, separated from the content which is in it, and expanded to cover all cases which are permitted in this sense, leads to the idea of infinite time, in which each point of the past shapes the transition point to the present and the future even though the future does not shape the transition point to the former."

We will maintain, therefore, that the relationship of causality is, in Kant's sense, set at the basis of events by us for the purpose of making their order in time possible; [the relationship of causality] is an attribute of the events themselves. In no way can we remove from events their chronological order, neither by assuming absolute time, which we could use for dating, nor by disclosing for the events which is earlier and which is later, which is cause and which is effect. The subjective succession of observations cannot, in respect to the scientific ideal of physics – which excludes consciousness – be brought into play; when such [subjective succession] is called upon, it is a sign of retrogressive development (see above and Section IV).

A more precise description of the process is that we determine, or one might almost say we designate, a perceptible substitute for absolute chronological order. For example in the movement of the fixed stars in the sky about the earth, the sky could be the 'clock' (not for measurement of time, but for determination of chronological order) from which we read off the chronological order of all other events and hence also their interconnection as cause and effect.

7. The relation of cause and effect is itself fully symmetrical. That which applies to cause also applies to effect, with the single exception that the cause is earlier and the effect is later. Hence the relationship of cause and effect can be used by us as a measure of the objective determination of time provided that that which we want to consider as cause and that which we want to consider as effect be established by its basis, the law of causality. On the other hand there is no topological structure which could enable cause to be distinguished from effect. In other words: If we have before us a causal relation *U-W*, then we have no way of distinguishing the direction of *U* to *W* from its reverse if we do not already have as its basis a 'clock' which tells us what is earlier (and therefore is cause) and what is later (and therefore is effect). This 'clock' could subsist independent [of the events], so that we determine specifically, for each particular causal relationship, that which is cause and that which is effect and likewise point out the direction of time in each particular causal relationship. More simply, it does indeed happen that we establish a procedure, for a large group of phenomena, by which we determine the sequence of cause and effect: *U-W-X-Y-Z*. This procedure then serves as the determinant of

chronological order within the group of phenomena. A natural sequence is thus designated as irreversible for the respective groups of phenomena and finally for the entirety of nature, which means nothing other than that what is cause and what is effect is determined for all phenomena. We then measure all other relationships of causality against the [same] sequence and hence determine for them what is cause and what is effect: that is the cause which, measured by the basic irreversible sequence, is the earlier.

For clarification we can imagine before us a film of a rock falling to earth, a stream flowing down a mountain, etc. We should not know how we should project the film. If the film were projected in one direction, then the rock would seem to rise and the stream to flow up the mountain; if it is projected in the other direction, we would see the opposite. In order to determine how to project the film, i.e. to determine the direction of the past → future, we must establish a law of causality as a basis. Indeed, we do not need a special law for the rock and a special law for the water; rather, it is sufficient to establish voluntarily a law of causality, as for example: water flows downhill and thus the direction of all other relationships of causality upon the film is given empirically from the film. We note empirically what is cause and what is effect. But a procedure for causality must voluntarily be established. In the beginning, therefore, there is the voluntary establishment [of a procedure for causality]. We establish an irreversible procedure and define what is cause and what is effect. We define for the world that the total amount of entropy is increasing. With this [definition] the direction of past – future is definitively established and thus it is given empirically from thence on what is cause and what is effect for each particular causal relationship. The voluntary definition of a sequence makes it possible for us to comprehend all other scientific sequences. This is indeed the way of the hypothesis, which neo-Kantian philosophy has put in a special light. (See the parallel exposition concerning the law of inertia in Natorp's book *Die logischen Grundlagen der exakten Wissenschaften*.)

The sequence established by us as irreversible for a group of phenomena must not be irreversible in the absolute sense. What does it mean, however, to say that the sequence could be regarded as reversible and that, therefore, the causal relationship *B-A* could be recognized as possible in addition to the causal relationship *A-B*?

It means that, in the former case, *B* is earlier than *A*. Previously *B* was later than *A* by definition. Now, then, another physical procedure must be established for defining time as irreversible.[4] If all world processes are reversible, as is accepted by mechanical physics, then all physical determination of the earlier from the later is relative and there is no absolute distinction between the direction of past and future. On the other hand, if there is an absolutely irreversible physical process (as there was, for example, with the increase of entropy before it was shown reversible by Boltzmann's explanation), then there is an absolute sense of past and future. If this is not the case, i.e. if we have no irreversible natural processes by which we could distinguish the earlier from the later, then we would have no means of physically distinguishing one sequence in the world from another if, per chance, the world process were also to be reversed and hence consist of two possible orders, *A-Z* and *Z-A*.

8. In contrast to the view presented here, according to which events themselves exhibit no topological structure enabling cause to be distinguished from effect since cause is distinguished from effect only through its position in time, Hans Reichenbach repeatedly attempted to show that cause and effect are distinguishable from each other because the cause – effect relationship is asymmetrical. In his very noteworthy new book, *Philosophie der Raum-Zeit-Lehre*, Reichenbach traces time and space back to causality.

In tracing space and time back to causality, he candidly sees the essential philosophical consequence of relativity theory and that that consequence could not be affected by new physical discoveries, i.e. by new facts. Reichenbach accomplishes this by tracing space back to time and then time, for its topological structure, back to causality. *W* is called later than *U* if *W* is the effect of *U*. Therefore it is originally through the law of causality that we succeed in establishing an objective chronological order. In this sense Reichenbach is in agreement with Leibniz and Kant. He himself objects to the circular definition. Is it not the case, he asks, that such an effect is precisely the reverse of what is later? Are we able to know, in general, what is cause and what is effect if we do not already know the order of time? To this self-generated objection Reichenbach gives an answer which, it seems to me, is incongruous with the truth he himself believes, i.e. that one could infer the relation of causality from the facts themselves. We are of the view that the causal relation is symmetrical in itself and that between cause and effect there exists no other distinction than that of their respective position in time and that one cannot therefore look to the phenomena themselves to determine which is cause and which is effect. According to our view, one first posits a relatively irreversible progression as 'clock'. Reichenbach believes, on the other hand, that the relationship of cause–effect is asymmetrical and that this asymmetry has nothing to do with temporal order. On the contrary, temporal order lies in the fact that different causes have different effects following them, while reversible but different effects must not be traced back to different causes. 'If E_1 is the cause of E_2, then small variations in E_1 will be connected with small variations in E_2. On the other hand, small variations in E_2 are not connected with variations in E_1'. Thus cause would be distinguishable from the effect without having to make reference to chronological order.

9. In a paper entitled 'Der Begriff der Verursachung und das Problem der individuellen Kausalität', which appeared in *Logos* (1914), I attempted to show that the asymmetry of causal structure cannot be successfully explained in this way. All that can be said regarding cause can just as well be said concerning effect with the single exception that effect is simply later than cause. Only temporal position distinguishes cause and effect.

In the well-known scholastic rule 'ab effectu conclusio non datur' is expressed the view which counters the opposed view of Reichenbach; namely, the counter view that it would probably be impossible for similar causes to bring about different effects, but possible (which Reichenbach denies) that different causes produce similar effects. I believe both interpretations are false and that the connection of cause and effect to be one-to-one. (Concerning this scholastic rule and the 'multiplicity of causes' in general see B. Kohn, *Untersuchungen über das Kausalproblem*, Chapt. 3. Many of Kohn's statements are, *mutatis mutandis*, statements countering Reichenbach's view.)

The counter-instance brought forth by Reichenbach – that a variation in the effect is linked with no variation in the cause – can only be a valid counter-instance if he considers the partial cause rather than the total cause. Were it possible to have a variation in effect E_2 without variation in the cause E_1, this would indicate that the same cause brought forth different effects (once the effect E_2 and once the varied effect E^*_2). This would indicate, however, a completely causeless variation of the cause–effect relationship and therefore strictly contradicts the law of causality. It is impossible to observe the combinations E_1E_2 and $E_1E_2{}^*$, because this would contradict the law of causality. "Because if the one sign followed another sign *as* the other sign, then there would be, using the universally valid concept of law, certainly no law; or there would be a breach of the law (Fechner)[5]."

I attempted to show in the cited discussion that only confusion of partial cause with cause, partial effect with effect, leads to acceptance of asymmetry in cause and effect, and I would like to expand further on this in light of the example used by Reichenbach. "We send a light signal from *A* to *B*. If we hold red glass in the light path at *A*, then the light at *B* will certainly be red. If we hold the red glass in the light path at *B*, then the light at *A* will not be colored." (p. 163). Response: The light at *A* is not, in the latter case, the total cause of the process, but the partial cause. The placing of the red glass at *B* is a new partial cause which changes the effect. The example used by Reichenbach only shows that the changing of one partial cause does not change the other partial cause, and that an effect can be realized by using two partial causes which are not dependent upon each other (Light at *A*, glass at *B*). It is therefore shown that a variation of the partial cause does not vary the other partial cause, but it is not shown that the effect could be varied without variation of the cause.

In a way analogous to Reichenbach's variation of partial causes without varying other partial causes, partial effects could be varied without varying other partial effects and thereby could lead one to the false conclusion that effect could be varied without cause being varied. I have already brought this out on pp. 86-87 in the *Logos* article already mentioned.

I believe, therefore, that the former explanation of causality, according to which chronological order is self-contained and according to which cause and effect are differentiated by their position in chronological order alone, must be upheld.

10. In his article 'Kausalstruktur der Welt und der Unterschied von Vergangenheit und Zukunft' Reichenbach pursued in detail his view that

there is a topological structure permitting distinction of cause and effect. And indeed he traces this distinction back to the appearance of nodal points in the causal chain; points, that is, at which one effect has two causes or one cause has two effects.

We will also have to show that the attempt to differentiate cause from effect other than by temporal means is unsuccessful even in this detailed explanation by Reichenbach. Reichenbach admits that in the simple causal chain *A-B-C...* there is no direction demarcated, thus leaving the continuous chain with no indication of time direction. Therefore he distinguishes cause from effect by the characteristics of a structural net. There should be a complete difference between two causes having one effect and two effects having one cause.

Let us give his argument (in which we will replace Reichenbach's implication of probability, which we have not yet discussed but will later discuss, by simple implication; this replacement will change nothing in his argument). Reichenbach argues as follows:

(a) Two causes, *A* and *B*, produce an effect *C*.

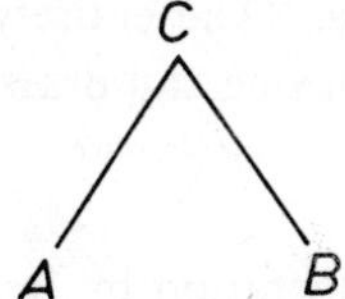

Example: At *A* and *B* two different billiard balls are thrown free; *C* is their collision. *C* occurs only when both occurrences *A* and *B* take place. If the ball at *B* does not move away from that point, then *C* will also not occur. Therefore: if *C* takes place, one can just as well conclude that *A* as well as *B* occurred. If, on the other hand, merely *A* occurred, then one cannot draw a conclusion about *C* because *C* follows only when *A* and *B* occur together. Therefore Reichenbach believes one can extract a definition of effect, or the later occurrence; the implication is intransitive, i.e. one can draw a conclusion concerning *A* if *C* occurs, but one cannot draw a conclusion about *C* by knowing only the occurrence of *A*. If the implication goes only in one direction, then the preceding occurrence is temporally later. "The intransitive fork has a topologically distinguished corner *C*."

(b) According to Reichenbach's view, the situation is entirely different if one cause has two effects:

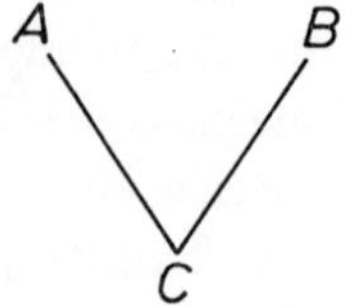

The above means the following: From the occurrence of *A* and *B* together one can certainly conclude that *C* occurred. *C* implies *A*, but *A* certainly does imply *C*. But this means that from *B* as well as from *A* one can conclude *C*. *A* implies *C*, *C* implies *A*. The relation is transitive. The fork of the previous example, directed to the future, was intransitive while the fork in this example, directed towards the past, is transitive. Thus one cannot draw conclusions concerning the effect from the partial cause; on the other hand, one can indeed draw a conclusion concerning the cause from partial effects. "The entirety of all causes permits conclusions about the future, but one can draw conclusions about the past from a partial effect."

11. I believe this entire deliberation by Reichenbach is unsupportable and rests upon a confusion of partial cause with cause, partial effect with effect. Namely, one can give example (b) a formulation suiting example (a) precisely and thus rendering the parallelism of cause and effect fully transparent. Thus, just as Reichenbach says in the first example that from a partial cause *A* one could not draw a conclusion concerning the total effect *C*, one could just as well say: from a partial effect one cannot draw a conclusion concerning the total cause. Because – this is the case which Reichenbach does not bring up – if the cause *C* has the effects *A* and *B*, and *A* together with *B* has *C* as cause, and if *A* occurs but not by *C*, *A* having *D* as its cause, then *C* would occur as well as *B* because, indeed, *C* has *A* and *B* following it. So as in the first example *A* does not imply *C* i f *A* alone occurs, but instead implies *D*, in the latter example *A* alone does not have *C*, but *D*, as cause, and we must thus expand both schema in the following manner:

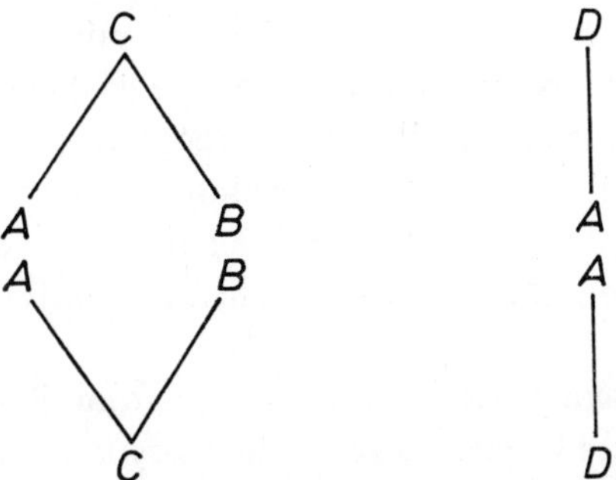

And now we can formulate case (b) in precisely the same way as case (a): if *C* takes place, one can draw a conclusion about *A* as well as *B*; if, on the other hand, *A* took place, then one cannot draw a conclusion about the occurrence of *C* because only *A* and *B* together have *C* as a cause. Case (b) is therefore also intransitive. The formulation of case (a) can be transformed verbally to the formulation of case (b); between cause and effect, past and future, there is no topological difference to be found.

Reichenbach concludes the following from the alleged transitivity of the second fork and intransitivity of the first: If two events have a common cause, then from the occurrence of *A* one can conclude that *B* also occurred (case b); if, on the other hand, both events have a common effect *C*, then one cannot conclude from the occurrence of *A* that there was a simultaneous occurrence of *B* (case a) because *A* could occur without *C* and therefore also without *B*. The conclusion, on the contrary, follows immediately: in the second case the occurrence of the effects *A* and *B* together has another cause presupposed as the cause of *A*, which appears alone. If one accepts in the second case, however, that *A* occurs only together with *B*, then one must, if he wishes to have a parallel schema, also accept in case (a) that *A* occurs only together with *B*, therefore leaving out, in both cases, our two valid schemata *A-D*.

For case (b) Reichenbach uses the following example: "*A* and *B* may again signify putting a billiard ball into motion, but here *C* indicates the complete cause, perhaps the signal by which both balls are put into motion. If I observe only *A*, I can conclude immediately that the signal was given." I answer: No! – because it could indeed be the case that, through some sort of oversight, billiard ball *A* was put into motion alone, without *B* and without *C*. Case (a) has analogous instances.

12. Reichenbach, however, uses his schema to portray a difference in the conclusions that can be drawn concerning the past and the future. From the partial effect one can draw conclusions concerning the cause, but from the partial cause one cannot draw conclusions concerning the effect. It could therefore be the case that the past is objectively determined, but not the future. Therefore there could be a science of history, but only of the past.

Let us recall here what we cited above from Reichenbach's book concerning space and time. There he discusses the situation in which a change in effect brings about no change in cause. It is now interesting to note that Poincaré draws, from this view, precisely the opposite conclusion concerning judgments about the past than the conclusion drawn by Reichenbach in his essay concerning the causal structure of the world. While Reichenbach maintains that a conclusion about the past is more certain than a conclusion about the future, Poincaré shows in his polemic against Lalande (*Dernières Pensées*, p. 18; *Science et Méthode III*) that if this view were valid it would be possible to have a change in effect without a change in cause, and that it would be precisely the past which one could not unravel, although it would be easy to explore the future.

How in truth does the matter stand regarding the drawing of conclusions about the past in relation to drawing conclusions about the future? First and foremost it should be said here that in inanimate nature there is no priority of the past over the future. To the physicist, a conclusion concerning the past is not easier and no more certain than a conclusion concerning the future.

The situation is first changed with the presence of living beings, especially with the presence of men. Now, Poincaré pointed out in the passage of his 'last thoughts' already cited that a conclusion about the past is more certain the more the case under consideration represents an exception from natural phenomena. "If a rolling flint stone were accidentally to remain at rest on a mountain, it will finally fall down to the valley; if we were to run across one already fallen down completely, it would be a trivial effect and would teach us nothing concerning the entire history of the flintstone. We would not be able to know at which point of the mountain it had been lying. If we, however, were per chance able to find a stone in the vicinity of the summit above, we would be able to confirm that it had always remained there because if it had been upon the slope it would have rolled down into the valley; and our guarantee will be the more certain the more the case is an exception and the more chance it has not to become like the norm."

Now what is special as relates to the science of history, with which Reichenbach is concerned, is that conclusions about the past are causal only in a most insignificant way. If we find somewhere a historical document and grant it our belief [as true], this conclusion is not causal in the sense that we could draw conclusions from the historical document concerning its cause, the author, and then further draw a conclusion concerning the events which caused the author to write about them. The crucial fact is, rather, that we grant our belief as to the meaning of the historical document. Here commences the moment of meaning. It is not that we would know more about the past of this blackboard before us than about its future. Indeed, we do not know from which quarry it comes. The material event is as unknown concerning its past as its future. And neither from the stone nor from its causal past do we infer its history.

13. Above we have represented the second function of causality as a guide to the scientific process of perception and as that which makes a prediction of the future possible. We need not add much new material to our previous statements clarifying this second function. We saw that past and future are equivalent directions within the relationship of causality: from the cause to the effect and from the effect to the cause. They are equivalent to each other in regard to scientific research. Assuming the world is governed by rigorous causality, the Laplacean spirit, which knows the constellation of the world and all of its laws of causality in a glance, can calculate the future just as well as the past and between the two there is no difference other than direction. Past and future in physics have no absolute sense; instead, merely a sense of direction. The same which is the future extending from one event is the past extending back from another event. Here past and future are only relative because the concept of the now, which separates past and future for our psychic life into two sharply divided directions, does not exist for physics. That this is the case is of the highest significance for understanding the methodological sense of physics. It is the job of physics to construct a world of objects in which consciousness plays absolutely no role; in which we deal only with purely objective concepts having no relationship whatsoever to our

consciousness. Whether the goal of this ideal of physics is realizable is another (negative) question. This scientific ideal makes it impossible to introduce into physics concepts which, objective and independent of living consciousness, are not comprehensible. And to this class of incomprehensible concepts belongs, above all, the concept of the Now. Lotze (*loc. cit.*, p. 287), says: "This characteristic [the Now] does not depend on the content of events; on the contrary, what should be said as regards this characteristic is first rendered clear by the expression of the Present, in which our language auspiciously fulfills its need for a subject." Therefore this purely subjective concept of the Now, the Present, has no place in physics.

Reichenbach defends an opposing view in the Academy essay already cited. He believes he can distinguish the now-point as the border between past and future by means of a new interpretation of the causal relationship. Reichenbach himself correctly avoids identifying the concept of the Now with the concept of simultaneity. The expression 'I live now' is not identical with the expression of the form 'I live simultaneously with such-and-such event'. Because if there were such identity, there would be no special 'now', the meaning of the word 'now' would be traceable back to the concepts 'earlier', 'later', and 'simultaneity'. But this does not exhaust the sense of Now.

Reichenbach attempts another approach: he traces the physical uselessness of the 'now' back to the determinism which earlier dominated physics. He shows that determinism is unable to give an account of the fact that "my existence is a reality; Plato's life, however, only throws its shadow on reality now." Reichenbach replaces determinism with a certain indeterminism, by which the effect is not produced from the cause with certainty, but with probability. The difference between past and future should now be that the past stands fast while the future is undetermined. "The present is that wave upon which the world passes from a condition of indeterminateness to a condition of determinateness." Reichenbach renews in this way a Greek metaphor according to which the invariable past is in competition with an indeterminate future. Upon this metaphor Diodorus of Megara based his sophism: "From something possible can result nothing impossible. Now, however, it is impossible that something past be other than it is. If it would have been possible that something past was different than it was, then from the possible would have resulted something impossible. It is therefore not possible to have something different in the past than it was." (Translated from Zeller. See Brunschvicg, *L'Expérience humaine et la causalité*, p. 510.) In our own time Jonas Cohn (in his book *Voraussetzungen und Ziele des Erkennens*, 1908, Chapter XII), clarified the capital position of the present by describing it as that which stands between the demands for complete explanation by determinism and for complete explanation by freedom.

According to Reichenbach, then, the Now is a cross-section in the state of the world; the Now has an objective meaning. Even if no man were to be alive, there would be a Now. 'The present state of the planet system' would thus be a statement just as determinate as 'the state of the planet system in the year 1000'.

In opposition to this definition one must ask: which Now is intended when it is said 'the present state of the planet system?' That of the year 1800 or 2000 or otherwise? Reichenbach answers thus: the Now is the wave of the passage from the state of indeterminateness to that of determinateness. This passage from indeterminateness to determinateness has occurred at all times (if Reichenbach's indeterminism is correct) and will continue to occur. If, however, the question were answered in the following way: the indeterminateness of the year 1800 has already become determinate, then it is to be asked: for whom? Obviously for us; for the present, for our Now. And so this definition of Reichenbach seems to be related to the Now, which it should, above all, be defining. What is the objective difference between the Now of the year 1800 and the Now of this moment? Upon this one can only answer: the Now is the moment of transition from the determined to the indeterminate; that is, one comprehends the Now of this moment through another Now, much more than through it itself.

14. For us a confrontation with this position of Reichenbach is even more important when we accept as permissible his substitution of determinism by a limited indeterminism of probability, which we will discuss later. We will not on this account be able to conclude, however, that the concept of the Now makes no sense from the standpoint of determinism. Reichenbach writes: the problem is formulated as the question concerning the distinction of past from future. For determinism there is no such difference... the passing (of time) brings nothing new; that which will take

place in a hundred years is given to me in the same sense as the events of the past war, and I could discuss in basically the same contemplative way the wars of Napoleon VII and the battles at Verdun.

There exists, then, in regard to the 'now', no difference between Plato and me; I can just as well say that Plato lives now and that I am yet to be in the future. Indeed, I was able to say that Plato lived earlier than myself because there is an 'earlier' and 'later' for determinism. But there is no 'now', there is no designated point of time, and the feeling that my existence is a reality, but that Plato's life only throws its shadow on reality, must be erroneous.

On the other hand it is probably correct to say that within determinism all aspects of the world, viewed physically, have the same reality. But what Reichenbach puts forth here as determinism is not that, but a physical interpretation of the world which recognizes no psychological categories and for which there is no 'I'; therefore there is no 'my life' even though the concept of the 'now' is very closely bound to that of the 'I'. Whoever, like ourselves, accepts as permissible the substitution of physical determinism by indeterminism will not want to grant that the concept of the 'now' has a legitimate place in physics interpreted via indeterminism. If one also accepts that the future is not entirely determined by a temporal cross-section – as we want to do along with Reichenbach – one will only be able to say that this indeterminateness exists as much for Plato as for myself and that I cannot distinguish physically who lives 'now'. The difference is clearly a psychological difference.

In reverse: for once let this psychological category of the 'I-experience' be granted, so that the 'I-experience' is as permissible from the standpoint of determinism as from the standpoint of indeterminism. The question as to what the 'now' is has nothing to do with the question as to whether the future is unequivocally determined by a cross-section of the Now or not. 'Now' is the mode of time of the living 'I'. Then, – Reichenbach will ask – does Plato live now or Napoleon?

Answer: I live now. And this 'I' is neither Plato nor Napoleon, but clearly – I, one experiencing subjective reality not restored to objectivity by a name. On this account physics does not recognize the I-reality.

Therefore: The concept of the 'now' is not to be grasped physically. We will be content with the distinction between past and future as a distinction in the direction of time, and that there is no more a physical boundary between past and future than there is a boundary and transition between the plus-direction and minus-direction of a straight line.

Thus prediction of the future is related to the concept of causality in precisely the same way that determinateness of the past is related to the concept of causality; both relationships are identical and are different only relative to direction.

While we thus maintain that the functions of causality in knowledge are: (1) to make possible an objective determination of time and (2) to make possible a calculation of past and future, we want to examine to what extent these functions can be carried over to natural laws not as interpreted by a rigorous causality, but as interpreted by a statistical conception of laws of nature.

II. RELEVANCE OF PROBABILITY

15. The significance of statistical laws of nature has become greater and greater during the course of the last decade, but last year for the first time it was believed that an antithesis could be maintained between the statistical law of nature and the principle of causality. When the statistical interpretation of the theory of gases celebrated its great triumph, no one conceived of it constituting an antithesis to rigorous causality governing particular phenomena; physicists considered such a possibility as little as the statistician of population, while he gives us certain laws concerning the average behavior of the population, considers the possibility that individual cases are not guided by a rigorous causality even when those cases do not square with the law of averages. Only the following should be said concerning the statistical laws of gases: that certain characteristics of gases can be better grasped by considering averages. "Just as the insurance statistician is satisfied to ascertain the year of birth and not the exact time of birth of his human statistical objects, the physicist working with statistics can renounce a precise measurement of the momentary phase values of the partial system." (Smekal, *Handbuch der Physik IX*, 'Theorie der Wärme', p. 199). It is a question, therefore, of using the theory of probability in gas theory and discovering statistical laws of averages. Discovery was crowned with such great success not because the law of causality was breached, but because it was found that "in a complicated complex of similar causes (!) a certain uniformity develops, in the long run, according to the law of large numbers." (Gatterer, *Das Problem des*

statistischen Naturgesetzes, 1924, p. 17). It is not a question therefore of contingency, of a breach in the law of causality in the absolute sense, but of the laws of averages of the complicated complexes, in which each particular phenomenon is governed thoroughly by a rigorous causality.

"Research has shown that if we assume the validity of dynamical laws for each single and separate occurrence (of molecular movement), and therefore assume rigorous causality, then the laws of probability, determined via observation, result directly" (Planck).

16. Physicists first changed over to a radical interpretation of the laws of probability when they learned to apply the laws of probability to the disintegration of radioactive elements and to the transitions of stationary states in quantum theory. There was no success earlier in determining the causes behind disintegration of a uranium atom and the development of enormous energy therewith. There has, nevertheless, been success in establishing the lawfulness of this decomposition, but in doing so the search for laws governing particular instances of disintegration was abandoned and what is even more, acceptance of a description of the average chance for decay amongst a number of atoms using only the theory of probability postponed all too quickly attempts to explain why there was an average chance for decay – it was as if the decay were completely causeless and occurred by chance. I quote W. Bothe, "Der radioaktive Zerfall" (*Handbuch der Physik, Volume XXII*): "The probability that an atom will disintegrate in a given amount of time is independent of the length of time during which the atom has already existed; one can represent this probability as $\lambda\, dt$ where λ is a characteristic constant, the 'disintegration constant', for the radioactive element in question. In brief it can be said: the process of decomposition is characterized by chance events. The question is still completely open as to whether the chance here is to be interpreted as 'chance' in the sense of a real suspension of causality or whether the 'chance' is primarily a result of the complexity of conditions which must be integral to the constantly changing constellation of atomic components. (See the chance distribution of numbers in the 10's decimal column of a logarithmic table.) S. Rosseland (*Nature*, 1923) does not regard it as conclusive that those electrons in the energy levels whose paths come very close to the nucleus influence decomposition.

"People have become accustomed to this uncertainty today, especially since it is known there is not a question here of a peculiarity of the atom nucleus, but that entirely similar situations exist in the outer electron shells. The analogy between the fundamental process of light transmission and radioactive decomposition was first brought forth by Einstein" (*Physikalische Zeitschrift*, 1917).

The analogy about which Bothe speaks here is the following: according to Einstein a stationary state of an atom has a certain probability of changing to a lower energy state by transmitting radiation from within itself without need of an external cause. Similar probability factors led Einstein to recognize that an atom changes to a higher energy state upon absorption of energy. The question now arises as to whether these quantum jumps are causally regulated by some yet unknown mechanism or whether the law of probability is predicated upon nothing except itself – in which case pure chance governs the disintegration of radioactive elements. I quote again the *Handbuch der Physik*, Vol. XXIII (Pauli: 'Quantentheorie', p. 11): "With this problem it is above all a question of introducing general hypotheses concerning the frequency of the start of the transmission process and having these hypotheses square with the emission and absorption of radiation. For the occurrence of spontaneous processes, which are bound up with the emission of a quantum $h\nu$, Einstein accepted a statistical law that is analogous to the law of radioactive disintegration. If an atom is in an excited state n, then there should be a certain probability that an emission process will occur in the period of time dt... The Einsteinian law indeed does determine the mean behavior of many atoms, but says nothing about the moment of time of the emission process for a single excited atom. A single excited atom seems governed only by chance according to our present state of knowledge. It is much discussed, but still an unresolved question, as to whether we have discovered a basic failure of the causal description of nature or have discovered only a provisional imperfection in the earlier theoretical formulation."

17. In order to fill out Pauli's discussion, three typical views of physicists will be mentioned next. Planck repeatedly asserted (see, for example, *Rundblicke*, p. 97) that the presupposition for the existence of statistical laws in the large is the validity of rigorous dynamic laws for particular events, even if knowledge of them eludes observation because of our

crude senses. Nernst sees in the description of nature by statistical laws more a sign of the weakness of human comprehension than something inherent in nature itself (see his essay: 'Zum Gutigkeitsbereich der Naturgesetze', *Naturwissenschaften*, 1922). According to Nernst we must take account of the possibility that, for the problem of quantitative calculation of individual processes, our reasoning power breaks down. "It is admitted that the previous common version of the principle of causality as an absolutely rigorous law of nature strangulated the spirit like the Spanish Boot, and thus indeed the present duty of natural science is to loosen these chains enough so that the free steps of philosophical thought will no longer be hindered." Nernst refers to an analogy between theological and physical interpretations because theologians formerly would have asserted that all events occur strictly according to logic in God's comprehension, even though the human mind would be unable to grasp the same. So Nernst probably believes that nature itself is ruled by a rigorous causality, but that the knowledge of men is limited. "The human mind is unable to see these natural processes in their finest detail."

Exner (*Vorlesungen über die physikalischen Grundlagen der Naturwissenschaft*, pp. 657ff.) advocates a more extreme view than either Planck or Nernst: "Were we to have the opportunity to investigate precisely the falling of a body in empty space, we would undoubtedly find the acceleration constant and the traversed path corresponding to the laws of falling objects. Does it follow from this, however, that the same would prove true using time not counted in seconds, but in billionths of seconds or even less? Perhaps the acceleration is not constant, but fluctuates very quickly about a mean value, and perhaps the movement of the falling object is not uniform in the smallest time spans, but accelerates irregularly?

In the course of conversation Boltzmann agreed completely with this view and held it not only possible but even very probable that the falling object moved backwards, perhaps not in a straight line, but in a zigzag line."

We have recounted these three positions here in order to illustrate the multiplicity of views of research scientists concerning the question of statistical natural laws. We will later return to the views of others concerning the same subject and to a theoretical discussion of the question itself. The debate itself has its roots not only in theory, but also in the area of experimental testing, such that in recent times a number of researchers, themselves having played a prominent part in the development of quantum theory, attempted to show that the question of a causal determinateness for particular occurrences is in principle undecidable because our physical means of testing particular cases is fundamentally inadequate.

18. Next we will present the popular statement (*Vossische Zeitung*, 12 April 1928) which Max Born gives of this problem in a polemic against Emanuel Lasker, whose assertions we will discuss at a later time: "Physics maintains the notion (of causal law) expressed somewhat sharply as follows: If the state of a closed system is known precisely at one instant, then natural laws determine the state at every later point in time. The laws of physics accepted earlier always had claimed this characteristic. Such an interpretation of nature is deterministic and mechanistic. For freedom of any type, of the will or of a higher power, there is no place. It is probably for the latter reason that all 'good rationalists' value this intuition so highly.

But recent physics has found, and verified with a huge amount of empirical data, laws which do not square with this deterministic schema. May they now assert that the causal principle formulated above is false? In any case they do not say it is false; they only say it is 'empty', meaningless. It is, namely, a conditional sentence beginning with 'if'.

What if the assumption 'if, etc.' were never realizable? We have already seen the claim that to know the precise state at a given instant is practically impossible; but it has also been accepted that this is only a technical defect and thus that, with the art of experimentation progressing, we will get closer and closer to knowing the precise state at a given instant.

Recent discoveries in atomic physics have led a few physicists to the opinion, however, that that claim is in principle unrealizable: *the laws of nature are themselves so constituted that they prevent the precise determination of a momentary state.* In order to understand this noteworthy assertion, one should realize that for determining the state, primarily the state of each minute particle, some type of probe is needed, be it only an electron packet or a fine light beam. But each probe destroys the state, indeed destroys it in fact. Electron and light probes are crude if the object of study is either atoms or electrons. Old Lichtenberg already saw the problem when he posed the question: can young ladies blush in the dark? The physiologist and the doctor know the difficulty in making a diagnosis without entirely changing the object of study by the research conditions, for example, by killing a cell. It is precisely the art, or method, of the experimentalist to choose a probe so fine that the object of study is itself not altered. But the physicist cannot arbitrarily push this method as far as he wishes. In the realm of atoms there is a natural limit for the fineness of the probe. The belief in the existence of this limit is based on deep physical knowledge, which collectively falls under the label of quantum mechanics.

This knowledge reaches its peak in the assertion that matter, exactly like light, can be conceived as waves; and this assertion has recently been confirmed by wonderful experiments. Indeed, according to the type of process, the corpuscular side or wave side of phenomena steps to the foreground.

If one directs his attention to space-time distribution, then matter in motion and light behave like waves: we see movement and interference. On the other hand, if one directs attention to energy effects using the 'causal' model, then matter in motion and light behave like swarms of particles. Therefore there is a close relationship, confirmed by many phenomena, between the frequency of the waves and the energy of the corpuscle; they are proportional to one another. This basic law of quantum theory brings about the mutual exclusion, up to a certain point, of spatio-temporal and energy descriptive models. [Why?] Because in order to represent a phenomenon with limited space-time extension by the wave model, one necessarily needs countless waves of different frequencies; and to these correspond particles with the most varied energies. The closer one confines the model spatially and temporally, the more diffuse becomes the energy and vice-versa. Even with the most precise measurements only 'half' of all the physical data can be known accurately; for example, if position and time are measured precisely, then energy and velocity remain entirely indeterminate, or if energy and velocity are measured exactly, then position and time remain indeterminate or finally, one can measure something of the one and the remainder of the other. In the latter case an exact formulation is made by use of expressions of probability theory; one knows, for example, that a particle is to be found within a certain region of space, but it does not follow that one can also determine its velocity precisely even though a probability for each possible velocity can be given.

... Therefore even the most perfected art of experimentation will not enable one to describe the state of a system precisely, in all details, because nature itself has piled up impassable barriers to it. This is why the principle of causality has become 'empty' in its customary sense, at least for the physicist. The principle of causality contains no testable statement of the form of natural laws. One can accept or reject the principle of causality – to the natural scientist it makes no difference. But his philosophic need is so strong that he troubles himself to replace the rejected formulation of the principle of causality with a better formulation, one which possesses a realistic statement. Perhaps that need is also at the base of recent investigations into quantum theory. [Directing such investigations] are statements of the following type: If one determines the state of a system empirically as precisely as possible using natural laws, then the further course of events of the system is limited just as each later state is empirically determinable in its turn.

Such a 'more moderate' formulation also seems basically more satisfactory than the older formulation, which assumes a heavenly observer. This new view of nature also leaves those who wish it the privilege of imagining the world as clockwork, even though he cannot get very far with this privilege. He must certainly satisfy himself with a limited perceptibility of mechanism and a limited application of determinism.

19. Heisenberg claimed the following in an article in the *Zeitschrift für Physik*, Vol. 43, 1927:

Position and impulse of an electron are not determinable at the same time. In order to establish the position of an electron, one must illuminate it. This assumes a light-electric effect which can therefore be interpreted as a quantum of light meeting the electron and being bent or reflected by the electron. As a consequence, however, the electron changes its impulse in a variable manner at this instant. The more precise the determination of position, i.e. the smaller the wavelength of light used, the greater the change in the impulse of the electron. Therefore: The more precise the determination of position, the less precise the impulse. One could also determine the position of the electron by impact, which is carried out by use of very fast particles, but this in turn leads to a variable change in the impulse. On the other hand the impulse of a particle can be measured as precisely as one wishes by measuring its velocity (Doppler effect), but then it must be assumed that the light is long-wave so that recoil can be ignored. But then the determination of position becomes correspondingly imprecise.

Just as questionable is the occasional use of the phrase 'path of electrons'. In order to measure the path, one would have to illuminate the atom, but the quantum of light would knock the electron out of the path. The word 'path' thus has no rational sense here. Every experiment to determine phase destroys or else changes the atom.

All concepts used in classical theory for description of a mechanical system can also be precisely defined for atomic models. The experiments which use such definition, however, contain an empirical indeterminateness if we demand of them simultaneous measurement of two parameerst such as position and impulse, where precise determination of the one is made at the expense of precise determination of the other. Here we can suggest a comparison with the theory of relativity. If there were, in the theory of relativity, signals which propagate at infinite speed, then that theory would be superfluous. The situation is similar with the definition of the concepts of position, velocity, etc., in quantum theory, where an indeterminacy, not foreseeable in the theory itself, is introduced by the experiments.

"Concerning the strong formulation of the law of causality – if we know the present precisely, we can predict the future – the premise, not the conclusion, is false. We cannot in principle determine all levels of the present. Therefore all observation is a choice from a totality of possibilities and a delimitation of the future possibilities. Now since the statistical character of quantum theory is so closely bound to the imprecision in all observation, one could be misled into presuming that behind the statistical world, which is accepted as true, there is concealed another 'real' world to which the law of causality applies. But such speculations seem unfruitful and senseless to us. Physics should formally describe only the relationship of observations. Indeed, one can characterize the true situation much better as follows: since all experiments are subjected to the laws of quantum mechanics, the invalidity of the law of causality is definitively established by quantum mechanics" (p. 97).

To these claims by Heisenberg the following is to be noted: Heisenberg expresses himself more pointedly than Born. Born speaks of the emptiness of the law of causality, Heisenberg of the definitive establishment of the invalidity of the law of causality. Only there is a simple logical error on his part. One cannot claim a sentence of the form 'if–then' invalid because one claims its presupposition is not realizable or, as he himself vaguely says, its presupposition is false. The falsehood of the premise by no means implies the falsehood of the conclusion, not to mention the falsehood of the conditional relationship which alone is the content of the law of causality. There can thus be no talk of a definitive establishment of the invalidity of the law of causality by means of the quantum theory; at most one can speak of its inapplicability.

20. Bohr gave Heisenberg's thoughts an essential foundation when he worked out a so-called theory of complementarity. ('Das Quantenpostulat und die neuere Entwicklung der Atomistik', *Naturwissenschaften*, 1928.) The mutual 'complements' are the causal viewpoint and the spatial-temporal viewpoint. That we can establish, using ordinary observational perceptions, events as spatial-temporal as well as causal, implies that the Planck quantum of action is small in its dimensions in comparison with ordinary observational perceptions. On the other hand, the classical physical concepts must be modified if they are to be applied on the atomic scale. Our description of natural phenomena is based on the assumption that phenomena can be observed without being essentially influenced by the observational process. Observations of atomic processes, however, evoke a reciprocal response due to the means of measurement that cannot be ignored; thus an autonomous physical reality cannot be assigned to either the observed phenomena or the means of observation. For each individual instance there is a question of suitability or purpose for which one wants to introduce the notion of observation and the irrational descriptive characteristic brought about by the quantum postulate, which means that each atomic process has a characteristic inconsistency which is foreign to classical theories. With the exclusion of all external influences, however, goes the exclusion of every possibility of observation and, above all, the concepts of space and time lose their direct sense. Nevertheless, if we concede that reciprocal of effects are brought about by means of measurement, then a unique definition of the state of the system is no longer possible, and although there can now probably be talk of space and time, talk of causality in the usual sense would not be possible. Spatial-temporal

descriptions and the demands of causality are therefore complementary, mutually exclusive characteristics of describing the content of experience symbolizing the idealization of the possibilities of observation or definition.

This complementarity finds expression in both theories of light – wave theory and emission (light quantum) theory – which have fought each other anew during recent years. Both theories are correct according to Bohr, but the wave theory requires spatial-temporal comprehension whereas the quantum theory requires calculation assuming causal comprehension. Both requirements cannot be satisfied at the same time according to the theory of complementarity, and thus there cannot be a unified theory of light. I quote Bohr: "The spatial-temporal propagation of light is described, as is known, in an analogous way by the electromagnetic theory of light. In particular, not only the interference phenomena in empty space but also the optical characteristics of material media are completely expressed by the superposition principle of wave theory. Nevertheless, the principle of conservation of energy and impulse finds its obvious expression in exchange of radiation and matter, just as this effect is evident in the photoelectric effect and the Compton effect, precisely according to the conception of light quanta developed by Einstein."

On the one hand the idea of wave action has been reproached because it cannot account for energy transfer in discontinuous quanta; hence there has been doubt concerning the unqualified preservation of the principle of superposition. On the other hand it has been believed that the idea of quanta requires that energy and impulse be exchangeable from one quantum to another during a certain period of time and not instantaneously, while investigation indicates that the exchange occurs instantaneously at a spatial point. So both theories of light have run into difficulties. Nevertheless, these "doubts – on the one hand concerning the unqualified preservation of the superposition principle and on the other hand concerning the general validity of the conservation principles, the validity of which seems inconsistent – are known to be refuted decisively by direct tests."

Thus it seems that both theories, although inconsistent with each other, are supportable and a decision between them seems impossible. "This state of affairs might explain the infeasibility of a causal space-time description of light phenomena. As long as we wish to maintain the laws of space-time light propagation of light effects, we are committed to statistical ways of thinking which use the quantum postulate. On the contrary, to maintain a causal requirement for particular events, by maintaining that light processes are characterized by quantum effects, means a retrospective renunciation of spatial-temporal relations. Naturally there can never be talk of a fully independent use of spatial-temporal description and of a causal concept. On the contrary: each interpretation of the nature of light exhibits a different attempt to adapt experimental facts to our usual way of thinking, meaning that limitations in our classical concepts are expressed as complements. Thinking in terms of the characteristics of material particles leads to an analogous conclusion."

Here investigations of the most recent data support de Broglie's interpretation of matter as a wave phenomenon. Here also is an interpretation of matter as waves on the one hand and as elementary particles on the other. Bohr again comments on this situation: "Just as with light we must continue to face, as long as we restrict ourselves to classical concepts, the question of an inevitable dilemma regarding the nature of matter; this dilemma might even be viewed as a way of analyzing the material of practical experience. Actually, the question here is not of mutually conflicting interpretations of phenomena but of complementary interpretations of phenomena, these interpretations together offering a natural generalization of the classical mode of description."

These two interpretations of light and matter, spatial-temporal on the one hand but causal on the other, shed light immediately "upon the simple formulas which structure the mutual basis of the light quantum and wave theories of material particles. If we designate Planck's constant as h, then we have the known formulas $E\tau = I\lambda = h$, where E and I are energy and impulse, and τ and λ are the co-ordinated frequency and wave length, respectively. In these formulas the two different interpretations of light and matter face each other bluntly. While energy and impulse belong to the concept of particles and therefore can be characterized by the classical interpretation using space-time co-ordinates, the frequency and wave length are based on an even, harmonic wave characteristic which is unlimited by spatial-temporal considerations." "In the language of the theory of relativity, complementarity can be stated such that according to quantum theory there exists, in general, a reciprocal relationship between the maximum precision of definitions using spatial-temporal or energy-impulse vectors. This state of affairs might be regarded as a simple symbolic expression for the complementary nature of spacial-temporal description and a causal requirement. At the same time, however, the universal character of this relationship allows reconciliation, to a certain extent, of the conservation laws with the spatial-temporal representation, because instead of assuming events to occur together at a spatial-temporal point, one assumes the concurrence of imprecisely defined individuals within a finite spatial-temporal region," and therefore both aspects are defined imprecisely.

Thus it is that Bohr describes the impossibility, in principle, of a rigorous conception of spatial-temporal causality; this impossibility compels us to be satisfied with imprecise laws of averages in those regions for which h can no longer be considered small.

21. It cannot be our concern to examine whether Bohr's view is valid or whether the impossibility, in principle and not merely contingently, of finding spatial-temporal and causal determinations for the particular phenomena of the atomic realm is meaningful. Here philosophy can only accept the thesis of the natural scientist and must leave the responsibility for it to him. We are concerned with the critical epistemological significance of the fact, accepted by Bohr, that only statistical laws are accessible to us in the atomic realm; or rather, we are concerned with the critical epistemological presuppositions of Heisenberg's and Bohr's theses. Has our science become presuppositionless by exchanging rigorous causality for a causality of averages? It has been asserted from many sides that henceforth, all laws of nature are to be viewed only as statistical laws. This is a complete misconstrual. Just as in logic the laws of induction cannot be proven inductively, the presupposition upon which rests application of natural processes cannot be deduced statistically. Münchhausen cannot pull himself out of the swamp by his own pig tails this time. On the contrary, the use of statistics for comprehending nature rests upon a transcendental *a priori* principle in precisely the same way that use of rigorous causality for comprehending nature was based upon transcendental presuppositions. The number of presuppositions natural science must make and which belong to pure science in Kant's sense (see above) have not diminished, but the transcendental presupposition now used by us assumes less than those before.

Next let us make a few more historical remarks: To us today a rigorous lawfulness in nature seems self-evident and we must fight a strong habit of thinking in order to grant chance any leeway. How little our present-day view was self-evident during antiquity and the Middle Ages is pointed out by Troeltsch in his encyclopedia article concerning contingency in Hastings' *Encyclopedia of Religion and Ethics* ('Die Bedeutung des Kontingenzbegriffs', W.W. II). Heinrich Gomperz (*Willensfreiheit*, p. 154) cites as proof for a philosophy of contingency the ancient Epicurus, "who indeed let atoms swerve (absolutely randomly) from the vertical, even for the tiniest distances, during their fall" (also see Theodor Gomperz, *Griechische Denker III*, p. 81; this example taken from Epicurus is also used by Brentano, *Versuch über die Erkenntnis*, pp. 20, 145).

Comte is to be named above all others in the 19th century; his noteworthy views regarding this subject was pointed out by Émile Meyerson in his book *Identité et Réalité* (p. 6). Comte believed laws of nature would in no way be able to withstand detailed research. He did not believe a particular natural law discovered by us would be shown as merely approximate upon more precise investigation and hence he did not believe that such a natural law could be replaced by a law that is better and more suited to reality. Instead, he believed that investigations into nature, pursued far enough, would teach us to recognize facts which would destroy each confirmation of lawfulness in nature. He thus adumbrated, to a certain degree, the discoveries made in research by Heisenberg and Bohr and indeed came to the strange conclusion that research which goes into too much detail should be forbidden because such research would be condemned to unfruitfulness and originates only in 'a childlike curiosity'. He thus protested against the use of the microscope, which he believed had already become far too prevalent. Levy-Bruhl certainly believed that Comte would have regarded the microscope to be only a human limitation. In this case his position would be the same as that of Nernst (see above).

René Berthelot, who in his book concerning pragmatism (*Un Romantisme Utilitaire*, 1911, pp. 240ff., 409ff) pursued a study of the related streams of scientific development in the 19th century, names Victor Regnault (1810–1878) as especially influential. He influenced not only the French scientists but also other scientists, especially Lord Kelvin, with his views concerning the merely approximate character of natural laws. "One must admit," he taught, "that physical laws are universal truths only within given limits and thus that they cannot be regarded as true in the absolute sense outside these limits. There are approximate relationships and we can measure the degree of approximation. This degree of approximation corresponds to the degree of our observational power and the increasing precision of our measuring instruments. But we do not have the right to assert that physical laws apply as rigorously as a mathematical formula."

In Germany Nietzsche referred to the 'coarseness' of scientific formulas and to the 'vagueness' necessary for both experimental success and for experiments to make sense, in order to support his interpretation of scientific truth as an erroneous idea that is biologically useful. It is amazing that, for our needs (machines, bridges, etc.), the assumptions made by mechanics suffice. These are, indeed, very crude needs and the "little errors" do not become evident. The imprecision of natural laws "is the condition of existence of trade; we would go hungry without it; scepticism and caution are essentially latecomers and are only infrequently allowed." (*Unveröffentlichtes aus der Zeit der Fröhlichen Wissenschaft*, WW. XII, pp. 30–39).

The views of H. Poincaré and Ernst Mach concerning statistical natural law are known, Poincaré repeatedly expressed the view that natural laws are perhaps only a consequence of mean values. In *Wissenschaft und Hypothese* he implies that the imprecision of our measuring apparatus has promoted the discovery of certain laws, and that it would have been unfortunate for science if it had been generated at a time when the instruments of observation had made investigations possible that are too precise.[6] Mach likewise believed that laws of nature would be realized only through a schematization of reality.

22. In the meantime, the number of names of scientists who have shared the statistical interpretation of natural law (and we could add many more to those already named) is inversely proportional to the uniformity of interpretations regarding the significance and philosophical meaning of their positions. We already saw that Comte and Nernst regarded statistical natural law only as a sign of human limitation.

For more highly organized minds, on the other hand, chance would seem to conform objectively to law. In nature itself rigorous causality reigns even though it has not revealed its purity to us. In accordance with this view, chance in nature would constitute only an exterior facade. The view that statistical natural law could be possible only if a law of strict causality reigned in nature (even if unknown to us) was particularly defended because it was said that probability theory itself, upon which statistics is built, would have meaning only within a rigorously-applicable, causal lawfulness.

Because if all happens by chance – it was argued – why should number one not always turn up by chance on dice? We would argue in ordinary life, if something like this were to happen, that the die with which we play is not homogenous and does not have its center of gravity in the middle. That means we would say the following: the regular turning up of number one on a die depends upon laws, primarily the law of gravity and the construction of the die, because we do not admit of chance. If one does admit of chance, however, then one could not ask why one side of a die turns up rather than another. It is indeed the case that if something happens by chance there is no questioning the 'why'. The fact of the matter seems to be that trust in the law of large numbers is possible only by assuming that rigorous causality governs all chance.

In this sense one often finds defended in philosophical literature the view that probability has meaning only in a law-bound world. Thus Leibniz seems to have thought (see Couturat, *La Logique de Leibniz*, p. 275), like Bolzano, that in each single instance (of throwing a die, for example) a rigorous lawfulness (unknown to us) conditions the result, "through which it happens that only one (of the possible cases) occurs and the occurrence of the remainder is collectively impossible." (*Wissenschaftslehre III*, §317, A. 3).

Thus Lotze asserts (*Logik*, Meiner edition, p. 442): "One can predict only those occurrences which are dependent on others within a law-governed world; [one cannot predict], however, occurrences which have existence completely independent of each other. It would only be a meaningless play of wit to maintain, before anything whatsoever, there is equal probability that there is something and that there is nothing at all[7]... it follows that the probability for the existence of something is really $=\frac{1}{2}$.... It would certainly be another case if we wanted to determine the probability of each basic fact from given data. Assuming lawful relations in all reality, these would provide, as epistemological bases, a condition which would necessarily bring about the exclusive acceptance of the one or another form of each occurrence." Cassirer seems to think similarly (*Erkenntnisproblem* II, p. 359: concerning Hume): "In order to designate an occurrence as 'probable', we must imagine the particular conditions upon which the occurrence is dependent and compare them together in our thought with other circumstances which would lead to another result. We cannot, however, carry out this comparison, nor establish any type of priority whatsoever of an occurrence or its opposite, if we do not fix as its basis an order of phenomena which themselves remain constant in the flow of time .... To assert probability, then, is to conclude there is objective certainty."[8]

The views of English scientists concerning this question are found collected together by Lenzen in 'Nature and Contemporary Physics' appearing in *Essays in Metaphysics*, p. 32 (published by the University of California). I further call attention to the view of the German scientist Kurt Riezler ('Über das Wunder gültiger Naturgesetze' in *Dioskuren II*): Without reference to a non-statistical, i.e. a clearly dynamic, regularity, all statistical regularity hovers helplessly in empty space.

Leopold says the following in his book *Sind Naturgesetze veränderlich*? (1926): "it even seems that regularity in the elementary world is also a presupposition for finding statistical laws of average." See also Brunschvicg, *loc. cit.*, p. 368: "The probability calculus is built upon determinism."

23. Nevertheless the justified response to these objections to statistical laws of nature seems to be only this: In a world wholly given over to chaotic chance all statistics, which are founded upon the law of large numbers, would be meaningless. If all is purely chance, there no law rules. There is no 'law of chance' because every law is a curtailment of chance. Indeed there is no way to accept the theory of probability as *a priori*; therefore the law of probability in respect to the law of large numbers laid down by it, is feasible only with reference to something further in a chaotic world. It would in this sense be a great mistake to accept an *a priori* law consistent with chaotic chance. In this regard Brunschvicg says very correctly: "But it would by no means be a question of putting chance in formulas."

Mathematical definition of probability has absolutely nothing to do with its application to nature. If we say (with Schlick, *Naturphilosophie*,

p. 457) that if that which has the greatest mathematical probability is also that which appears frequently in corresponding amounts in nature, then this is in no way an analytic *a priori* truth of the same value as $2 + 2 = 4$; rather, the applicability of the theory of probability to practical experience is a new principle, which we either substantiate or, alternatively, must accept as an axiomatic postulate for our investigations.[9] In this sense one must by all means concede to Lotze and the others that an applied calculus of probability makes sense only by presupposing regular interrelationships in the world.

On the other hand the question remains open as to whether one, in order to guarantee statistical natural laws, must go as far as the law of causality goes in delimiting chance and therefore must presuppose a fully unequivocal necessity in individual phenomena or whether one can be satisfied with accepting as the highest basic postulate "that that which has the greatest mathematical probability also appears frequently in corresponding amounts in nature."

Here it is suggested only to postulate the applicability of the theory of probability to reality. This postulate makes fewer assumptions than the law of causality itself and has in this respect an advantage over the law of causality. (There have been many attempts to base the law of causality upon the theory of probability. This seems to us clearly unrealizable because it is impossible to derive an assumption which makes more presuppositions from an assumption which makes fewer assumptions. At most one could attempt to derive probability – according to Brentano the infinite probability – of the law of causality, but then one cannot avoid the question as to how we can legitimately accept that probability which appears frequently in corresponding amounts in nature.)

In any event, however, and here we are in accord with the research scientists mentioned above, a regularity must be presupposed. In chaos itself no law rules.

24. We must accept the following position concerning the question of statistical regularity upon the basis of what was said earlier: acceptance of such regularity does not absolutely imply domination by chance. Chaos does not break in, but the rigid regularity of causality is loosened up somewhat. The 'straight line' of an unequivocal natural regularity is broadened to a stream; natural law has received some 'elbow room'.

One such interpretation of nature was set forth by Heinrich Gomperz in his book concerning freewill (p. 153) under the title of 'spontaneity theory'. According to this theory material elements would show certain particular, momentary peculiarities in their behavior. It is conceivable that two entirely identical particles of matter or the same particle of matter at different times be excited to somewhat different reactions under identical conditions. The so-called regularity relation would only be an average value of the actual relations, and the latter would not be identical to each other, but would only be approximately like each other.

That natural laws seem to apply precisely would be the case only because the science of inorganic nature deals only with masses containing countless material elements and only because in these masses the individual and momentary deviations from the norm would compensate each other.... Natural laws would be interpreted as such rules of the average behavior of material masses in the theory of spontaneity. Nature thus perhaps seems to be ruled by a lawfulness, even though it opposes our desire for unequivocal causality with 'a resistance of medium strength'. The world behaves, in contrast to our desire for order, like matter with some plasticity (Gomperz, p. 15).

Amongst contemporary physicists Reichenbach expressed most clearly the law of the probabilistic union between cause and effect. And if we were not able to accept his attempt to derive a physically useful definition of the Now, then his position in regard to the law of causality will need consideration also. I quote from the previously mentioned Academy essay *Kausalstruktur der Welt*, p. 138. Reichenbach replaces the law of causality "by the assumption that between cause and effect there is a relationship which conforms to probability... we therefore think of a world in which all dependencies are of the same type, like the turning up of a given side of a die in dice playing. Each step of events is a roll of dice, and only the great probability of particular sequences has led us astray into seeking an absolute regularity hidden in those sequences. With this interpretation we have likewise arrived at a uniform assumption concerning the character of events, only we have omitted the assumption of causality. Such a world has only probable connections between its elements. It is the requirement for a minimum of presuppositions which compels us to renounce rigorous causality."

Hence it is this relaxed assumption which could replace the law of causality and we now need ask ourselves whether this assumption, as a basis for comprehending the world, can fulfill, in regard to the scientific treatment of practical experience, both functions we have attributed to causality: determination of the temporal position of natural events and prediction of the future.

25. Let us assume now that every cause is not united uniquely with its effect, but could be united in two ways, primarily such that *a* could result in either *b* or *c*. Would the purpose of the causal schema in Kant's sense be shattered by this assumption? No. Just as we know the position in time of *n* after *m* through knowledge of the causal law '*m* produces *n*', so we could determine the temporal position of *b* or *c* through knowledge of the law of causality '*a* produces *b* or *c*' and thus the causal schema would fulfill its temporal role. Likewise if the law of causality allowed a multiplicity of causes (*c* can be produced by either *a* or *b*), we could determine the position of *c* in regard to *a* or *b* and vice-versa in spite of multiplicity in the law of causality – always assuming, as was stated before, that a way of determining the earlier from the later in an irreversible physical process remains set for us.

Certainly this probability of an effect following from a given cause or of a cause following from a given effect would become ever smaller the further the determined effect is temporally distant from the given cause (or the reverse). In any case it would be possible, if the one-to-one relation between cause and effect were to be replaced by a relationship of probability, that in a given series of phenomena the temporal order, i.e. the causal order, of the phenomena could be determined and hence Kant's requirement for a causal law would be met.

Now let us turn to the second function of causal law, i.e. the prediction of the future from reconstruction of the past. Such prediction would, of course, be impossible for the particular instance. In spite of this, however, there would be no practical difference if only the average of an infinite number of possibilities is under consideration. Indeed in this case we could not predict the determined effect nor reconstruct the determinant cause, but we no doubt could, with a certainty guaranteed by the theory of probability, estimate that the effect would occur within a certain margin. Sommerfeld requires the following ('Zum gegenwärtigen Stande der Atomphysik', *Phys. Zeitschr.*, 1927, p. 234): "We must require, as long as there is to be natural science, precise prediction of observed phenomena." This requirement is also fulfilled in the case of statistical laws of nature.

So neither by considering one or the other function of causality can one uphold a proof against the replacement of rigorous causal law by a probability function.

The law of statistical average clearly takes over the same functions of prediction and reconstruction earlier fulfilled by the rigorous causal law in the fields of physics under consideration. The only difference is that under the rigorous law of causality the individual case was temporally ordered as well as predicted and reconstructed, whereas under the law of statistical average all this turns out to be based only on averaging.

One could be of the opinion – and this seems to be Reichenbach's view – that the introduction of probability relations would provide a priority of the future over the past and consequently would no longer be valid for our discussions (see above). [Why this view?] Because probability is a probability of future events and not past events. But this would be unjustified. If we say certain physical events hover about a medium value, this value applies in the same way whether we now pursue the sequence of phenomena in the one direction or the other. The 'dice play' of the world has to do with the effect if we view it from the connected cause, and has to do with the cause if we view it from the connected effect. Our reflections above concerning past and future and concerning the physical incomprehensibility of the 'Now' thus remain the same even if we replace rigorous determination with implication by probability.

26. We have already alluded above to the fact that even with this replacement, certain transcendental presuppositions remain intact and that they quite certainly articulate the applicability of probability calculus to practical experience. If nature were so lawless that it would not comply with probability calculus, then practical experience would be impossible. We thus have the law 'that which has the greater mathematical probability also appears frequently in nature in corresponding amounts'; this law is considered the transcendental condition without which practical experience is not possible. It is very important to recognize that this transcendental condition is not at all self-evident. In Kantian terms, it is a synthetic judgment that reality accommodates itself to the laws of chance in the same way that probability calculus formulates those laws. The theory of

probability, as a purely mathematically formulated theory, has no influence on reality. Rather, this influence is a new fact and has, speaking epistemologically, the same dignity as all statements in Kant's 'pure natural science': these are synthetic *a priori* presuppositions accepted by us because without them practical experience would not be possible. By no means do we surrender ourselves to a total empiricism, which would pick natural laws out of an imaginary reality.

If, therefore, transcendental presuppositions are necessary for the new interpretation of nature just as they are for the classical interpretation, then the new interpretation has a great advantage. That theory is plainly advantageous which, with identical goals for knowledge, makes fewer necessary transcendental assumptions. It is no different than with the axioms of mathematics. We must find a minimum number of presuppositions of the nonempirical type sufficient to establish the empirical. "The question now is: what is the absolute minimum number of necessary epistemological presuppositions in order to make possible a quantitative view of the world?" (Sommerfeld, 'Zum gegenwärtigen Stande der Atomphysik', *Phys. Zeitschr.*, 1927, p. 235.) The statistical interpretation of law clearly makes fewer presuppositions than the classical interpretation, because while the classical asserts that no practical knowledge is possible unless each single phenomenon is regulated by rigorous causality and from its beginning through its entire course lets itself be determined by a differential equation, the statistical interpretation asserts much less: individual phenomena are undetermined, i.e. are 'free' and are subjected only to the laws of chance as ascertained by probability calculus. 'In the world strict necessity rules' demands the one interpretation; 'in the world chance rules according to the calculus of probability' demands the other interpretation, and it therefore clearly requires less than the first. Thus it is that the new interpretation has an advantage over the classical theory.

27. Certainly, each reduction of the number of presuppositions must take its toll somehow. Somewhere there must be a minus cropping up. The minus associated with the statistical interpretation is renunciation of an assertion of sufficient explanation of phenomena, and at least a strong habit of thinking has valued such a claim. We further find nothing astonishing in that world processes run their course by necessity, that stars remain in the orbits which natural laws prescribe for them, etc.

We are not tempted to ask how the star knows the natural law which it has to follow; known or unknown we perceive natural law as a power which permeates nature with its will.

And because we are clearly saturated by this habit of thinking it becomes difficult for us to grasp that nature, so to say, plays dice, that the path for the electron cannot be prescribed by natural law, but that the electron can 'choose' within certain limits. We formerly would expect that the electron, instead of deciding in a chancy, groundless manner upon one of several possibilities was, like the donkey of Buridan, doing nothing and was kept to its path. In actuality we find a similar view expressed by Bolzano (*Wissenschaftslehre III*, p. 317, A. 3): "If we wanted to assume that each event (the equally possible events of probability theory) are actually the same as each other, then clearly no man, much less God himself, could decide why one as opposed to another occurs; actually, none would occur because none could take precedence over the others on account of the complete identity of circumstances surrounding each of the events." One remembers that with a similar argument Galileo, Wolf, d'Alembert, Euler, Laplace, etc., attempted to derive the principle of inertia because there was no reason at hand as to why a body should move in one direction rather than in another. Concerning similar argumentation by Archimedes concerning equal weights see the second letter of Leibniz to Clarke (see Kohn, *Untersuchungen über das Kausalproblem*, p. 57).

There is an unusually strong habit of thinking, against accepting individual phenomena as indeterminate. If one really extracts from the concept of law the remainder of animism and voluntarism still adhering to it, then it is nevertheless all the more, or all the less, mysterious that the electron remains in a path prescribed for it by law than that it willfully 'chooses' another path amongst those which are lawfully allowed it. The one phenomenon is, in its essence, as incomprehensible as the other; but it is not the business of physics to assuage this incomprehensibility of universal or partial regularity in nature; physics has described and retained this lawfulness in its form without also being concerned as to whether metaphysical conclusions may be drawn from the 'law-bound' or 'free' interpretations of the cause-effect relationship. (See the attempt to understand statistical regularity teleologically by Th. L. Haering, *Philosophie der Naturwissenschaft*, 1923, pp. 617, 622.)

Once one has surmounted the mental resistance to indeterminacy in material nature, one will realize that, in respect to the comprehensibility of natural phenomena, the one theory is as good as the other.[10]

The physicist demands above all that one must be clear concerning the limits of basing the cause-effect relationship on probability. If we have set forth for this interpretation as an *a priori* presupposition 'that that which has the greatest mathematical probability also appears frequently in nature in corresponding amounts', it is implied that events or series of events with lesser probability can happen. That we throw dice with the same number turning up twenty times in a row has a very small probability, but nevertheless it does have a probability. And if we base nature on the postulate of probability, the improbable event will indeed occur seldom, but yet it will happen. And so Seelinger rightly says ('Über die Anwendung der Naturgesetze auf das Universum', *Sitzungsberichte der Bayer. Akad. d. Wiss.*, 1909, p. 20): "Talk of a probability of event *E* makes sense only if the nonoccurrence of *E* under the same circumstances (therefore the occurrence of its opposite) is assumed possible. If one disavows this possibility, then the use of probability theory is at most mere play." In order to abide by our example, if we throw the dice for an appropriately long time and if the laws of probability calculus apply, then every improbable throw must occur sometime and there is nothing astonishing or in need of explanation when the improbable throw occurs. If we carry this example over to natural processes, it means that improbable combinations would occur if sufficiently long time is allowed for the occurrence – and nature does not lack time!

To use an example of Schlick's: it is certainly an event of very low probability that the unordered movement of particles so order themselves by chance into a stream such that the total flow continues moving under cooling conditions while the unordered movement of the particles change to ordered movement. But since nature can throw dice limitlessly, even this improbable event must occur according to the basic postulate of probability without there being any need, or even ability, to clarify why. Such an occurrence is nothing more than the improbable case actualized. The physicist will have to be ready to concede that, with the admission of these events contrary to natural law, all claim for explanation is abandoned; he must be prepared to be satisfied with the explanation that 'the improbable event has happened'. See Smekal, *Handbuch der Physik IX*, p. 213: "the type of median value expressed need not be applicable to the concrete individual case... arbitrarily long times for realization of 'irregular' distribution conditions are fundamentally possible." And in the same book, Jäger (p. 374): "All theorems produced by Boltzmann are thus not statements of certainty, but statements of probability... and that, a gas complies with the second law of thermo dynamics is not unconditionally certain, but is so colossally probable that we can accept it as practically certain... But deviations, 'fluctuations', are present and can also be authenticated under circumstances."

28. The theory of probability, however, ushered a moment of uncertainty into physics. Winternitz (*Relativitätstheorie und Erkenntnistheorie*, p. 221): "To attribute free decisions to the atom means to renounce the job of natural science in this field." The law of causality in its rigorous form is "the kernel of the *a priori* principles of physics." In 1923, Planck emphasized in his Academy address concerning freewill: "that acceptance of causality without exceptions, a complete determinism, provides the presupposition and precondition for scientific knowledge... it renders scientific thought synonymous with causal thought."

Brunschvicg identifies the law of causality with the following formula: there is a universe (*loc. cit.*, p. 536). (Similar statements are collected together by Gatterer, *Das Problem des statistischen Naturgesetzes*, p. 27ff.) In fact these thoughts are not insignificant. We want to express it crassly but clearly: wonder has become possible within the context of natural law. Refuge is opened for uncertainty, and into this refuge a theory can escape easily if the real phenomena do not comply with the theory. In this sense one can understand why Emanuel Lasker, speaking of modern physics, cries out that 'culture is in danger!' (In the same book mentioned above, 1928.) But this is certainly an expression of doubt in statistical natural laws, a doubt which is not epistemological but only methodological or, so to say, pedagogical. If nature were really created such that it is not regulated by rigorous causal law, but by an elastic law of probability, then one will clearly have to accept these peculiarities in the deal. "We have not constructed a nature adequate for our comprehension; instead we have, as much as we are able, merely satisfied ourselves with the given" says Exner (*Vorlesungen über die physikalischen Grundlagen der Naturwissenschaft*, p. 709). Naturally this applies only *cum grano salis*, because the 'given' of nature is also constructed. Th. L. Haering (*loc. cit.*, pp. 86, 593) introduced the appropriate expression 'stage of resignation' but also notes correctly that there can be as little talk here of a melancholy relinquishing of an epistemological ideal as the relinquishing of an ideal rectilinear and effortless splitting of a mountain by an engineer who must work with actual given rock formations. The philosopher can only make the natural scientist aware of the consequences which such an explanation can have, but here as in other places Nicolai Hartmann's saying applies: the philosopher will again have to learn to be satisfied with confrontation of the problem and its paradoxes.

III. TELEOLOGY IN PHYSICS?

29. Especially entangled in the battle against causal law from the standpoint of statistical natural law is another assault on causal law as the predominate form of natural explanation in physics: worthy attempts have been made to supplement causal explanation with teleological explanation; to permit, indeed to require, that along with a determination of the future by the past there also be determination of the past by the future. Certain facts of quantum theory seem to make such a formulation of natural law necessary. In Bohr's atomic model the electron begins emitting rays just as the jump begins; the type of radiation, however, is determined by the final orbital level as well as the initial orbital level. The atom must know, to a certain extent, the final state into which it will transform before it radiates because the radiation frequency is dependent upon the difference in energy between the initial state and final state. (See the problems collected together by von Schottky in his essay 'Das Kausalproblem der Quantentheorie als eine Grundfrage der modernen Naturforschung überhaupt', *Naturwissenschaften*, 1921. It indeed seems that many problems brought forth by him are dependent on atomic models rather than on factual difficulties. The question related to the atomic model may have lost much of their biting edge with the subsequent development of wave mechanics.)

These physical facts or theoretical considerations have brought physicists to the point of expressing views concerning the influence of the future on the past which, if one draws all the consequences of such views for our comprehension of the world, would be sufficient to turn our entire world structure on its head.

To be sure, physicists have not drawn and believed the consequences; they have only dealt with formal changes resulting from this new meaning of causal law. Thus Planck writes ('Die physikalische Realität der Lichtquanten', *Naturwissenschaften*. 1927): "The form of many very general axioms of general mechanics and atomic physics approaches interpretation of the outcome of a process as dependent not only on the beginning state but also on the end state, and thus introduces a certain direct and reciprocal effect of two temporally different states. The principle of causality would thus not be influenced in its essence, but only in its form." And Sommerfeld ('Grundlagen der Quantentheorie', *Naturwissenschaften*, 1924): "It seems as if initial position and final position had equal right in determining the phenomenon. This would conflict to a certain extent with our inherited feeling for causality, according to which we readily believe the course of a process established by its initial situation. It does not seem to me out of the question that quantum phenomena could remodel our ideas here.... In any case we must require, as long as there is to be a natural science, univocal determination of observed phenomena, i.e. the mathematical certainty of natural laws. How this univocal correspondence comes about, i.e. whether it is given only by the initial state or is given through the initial and final state jointly, we cannot know *a priori* but must learn from nature."

Now it seems to me, by all means, that the revolution in the mode of thought required by accepting that "the pushing power of the past is replaced by the suction power of the future" (Riezler) is far more extensive than both quoted physicists are inclined to accept.

30. Next it must be said that this acceptance conflicts with the assumption to which the previous chapter was dedicated, and according to which physical events were not rigorously determined, but occurred by chance within a space with a certain latitude. Only from the standpoint of phenomena completely determined does talk of a 'reciprocal action of past and future' really have a (relative) sense.

The facility with which theories of that type can be installed today is traceable to the theory of relativity, which effaced the difference between time levels to a certain extent. But one must not forget that the theory of relativity grew upon a rigorously deterministic base. It was only possible for Einstein and Minkowski to let temporal phenomena solidify somewhat in a temporal region which does not lapse, but stands still, by clearly accepting that the future was considered determined on a univocal basis by the past. If we accept, on the contrary, that the future is not univocally established by the present and thus that a certain region of flexibility be allowed for phenomena, then the future is (and not only for us) in principle in determinate by the present moment; yet something indeterminate cannot exist and cannot have an effect. Here, then, one hypothesis of quantum theory is contrary to another.

But if we also use our imagination more, the influence of the future upon the past seems contradictory for a rigorously deterministic interpretation. It has already been seen above that the law of causality is our only evidence by which the temporal position of facts in the outer world can be designated. W is later than U if W is the effect of U. This is the basic schema for temporal determination. We must assume a basic law of causality $U \rightarrow W$ in order to determine the temporal position of both U and W. However, if W were the cause of U and should also be later than U, we would lose all evidence by which to establish the temporal position of physical facts.

31. It is often noted that many very general laws of mechanics, especially Hamilton's Principle, are also supported by the conception that the future influences the past. "Physics," says Schlick (*Naturphilosophie*, p. 433) "often finds it expedient to express natural laws such that they treat processes as dependent on the future as well as the past."

One law of this type is Hamilton's Principle – 'the principle of least action'. It is of greater significance for the formal structure of physics because "it is precisely a most universally applicable principle that can be used to comprehend all newly discovered natural laws, including those of relativity theory, as consequences of a principle of least action; it thus seems to possess the highest rank of formal generality. It can hold that rank because its formulation requires the least number of presuppositions concerning the special sort of mutual dependence amongst natural processes."

According to Hamilton's Principle, each system moves naturally to reach a given state in a given amount of time so that the mean difference in kinetic and potential energy over the entire period is the least possible. (Formulation according to Haering, *loc. cit.*, p. 618.) Hamilton's Principle is similar in its structure to the principle which Fermat established in the

17th century as the principle of the shortest light time. That principle stated that a light ray passing from a given point A to another point B uses a shorter time to cover its actual path than it would need to cover the distance of every other path between A and B. "It is as if the light possessed a certain intelligence and followed the commendable plan of arriving at its preordained goal as quickly as possible. In doing this, light does not have time to probe different possible paths, but must decide immediately upon the correct path." (Planck, *Kausalgesetz und Willensfreiheit*, p. 11). (See, concerning the relationship of this principle with that put forth by Hero of Alexandria: Haas, *Einführung in die theoretische Physik*, p. 97, and Haas, *Materiewellen und Quantenmechanik*, p. 13.)

Next: we have already mentioned that the claim – or concession – that future influence on the past has a relative sense only from a purely deterministic standpoint and therefore that this aspect of quantum theory is opposed to another aspect, which regards natural phenomena as indeterminate in respect to particulars and determinate only in respect to average value.

This conflict makes reference to Hamilton's Principle especially clear because it is precisely this principle which is set forth as the true expression of the claim that all natural phenomena are unequivocal. Thus Mach says (*Die Mechanik in ihrer Entwicklung*, Chapter 3): "One sees that the principle of least action and thus all other minimum principles in mechanics express nothing other than that, in the cases under consideration, exactly as much happens as can happen under the given conditions and that those cases are determined, and determined unequivocally, by those conditions." And Mach quotes favorably a saying of Petzoldt: that minimum principles are nothing other than analytic expressions for the fact that, in practical experience, natural processes are determined unequivocally.

32. The absolutely deterministic interpretation of the relationship between future and past, expressed in Hamilton's and related principles, makes comprehensible the use of integral for differential equations, and that the initial and final points of phenomena can be used for the initial point and the direction at the initial point. (See concerning the history of the principles of Hamilton, Maupertuis, etc.: Couturat, *La Logique de Leibniz*, Note XVI; Winter, *Revue de metaphysique et de morale*, 1924, pp. 92ff; Whitehead, *Science and the modern world*, p. 77; Cassirer,

Erkenntnisproblem II, p. 426.) It seems to me that reference to minimum principles and to their parallels with the assertion that the future influences the past, as is maintained in contemporary quantum theory, fails. It may be that, historically and psychologically, physicists who established those basic principles let themselves think in terms of expediency and thus viewed mechanical movements as if a self-moving body would resolve to meet the requirement of minima. Maupertuis asked himself, when he wanted to determine the path of a particle moving through a force field, how a path would have to be created to comply in its perfection with the perfection of God – and thus discovered his principle of least action.

So another scientist may formulate more valid laws of nature by beginning with the assumption that natural laws are simple or with the assumption of economy in thought or nature. Nevertheless, the question is indeed not whether a law could ever be discovered via some sort of teleological considerations, but whether teleology is integral to the actual expression of the law, whether an influence of the future on the present has been accepted in the sense that unequivocal determination of the present is given over to such influence. Such however, is clearly not the case with the cited minimal principles. "These laws," says Weyl (*Philosophie der Naturwissenschaft*), "are mathematically equivalent to differential laws, connecting only infinitely continuous elements with each other, and hence we view them today only as another mathematical form of the law of causality..." One does the same by defining the differential law such that the direction of progress from point to point undergoes an infinitesimal parallel displacement, or by defining the integral law such that the direction of progress is the shortest line connecting two of its points. As regards the actual law itself, it makes no difference whether we view it as determined by initial point and direction, or by final point and direction, or finally by initial point and final point. Between 'causality' and 'finality' there is no difference that can be established as regards regularity; regularity is not at all concerned with scientific knowledge, but with metaphysical significance. (Similarly Zilsel, 'Über die Asymmetrie der Kausalität und die Einsinnigkeit der Zeit', *Naturwissenschaften*, 1927: It makes no difference whether one says process is determined by parameter values at two points of time or through twice as many parameter values at one temporal point.) We can therefore understand the integral principle, which calculates, in a convenient way, the state of a system at a given time with the help of its later state without drawing conclusions about a highly dubious and special future influence on the past from its metaphysical implications. (See Strauss, *Annalen der Philosophie*, 1928, p. 76; Riezler, *Kantstudien* XXXIII, p. 377.)

We see therefore: formulation of a law like Hamilton's Principle does not imply acceptance of a retroactive future power. Dingler (*Der Zusammenbruch der Wissenschaft*, p. 117) correctly pointed to the fact that admission of such a law would render present and past indeterminate in principle. Since the future is not yet (now) determined, it can always become something else. If the present depends on the future, then the present is also indeterminate even though previous physics always held to the axiom that the given is determinate. This determinateness is a presupposition of scientific work because only that which is itself determinate is determinable. (And indeed as much determinable as it is determinate. In the case of statistical natural law, the particular phenomenon would be in principle indeterminate and hence also indeterminable.)

"Determinability in the unending progression of knowledge requires that propositions be determinate. The subject matter to be determined must be thought of as completely determinate, such that determinateness is attributed to the conception of the final point towards which thought strives in an unending process." (Winternitz, *Relativitätstheorie und Erkenntnistheorie*, p. 221.)

33. Now, clearly the retroactive action of the future on the past can be understood in two senses: either as a causal effect of later events upon earlier events, as already explained, or as a purely teleological action of the future as a goal. As is known, Aristotle distinguished between two sorts of causes – efficient cause and final cause – and it can be understood that the effect of the future is the effect of a purpose towards which there is striving. While we said above that the influence of the future upon the past is relatively meaningful only according to the deterministic conception, because only the determinate can act, it is here being said that the teleological meaning of the influence of the future certainly goes hand in hand with the contingency and indeterminateness of phenomena. One could say: there exists no sufficient reason which explains natural phenomena. The particular event is subject to chance within its realm of free play. It is precisely on this account that the particular event can be

understood teleologically. Because natural law leaves room for chance and permits a state of suspense between different possibilities, alternatives under consideration can only be resolved at the end of the state of suspense. (See Whitehead, *Science and the modern world*, p. 134.) Thus the statistical account of causality goes hand in hand with the reintroduction of finality into physics. In actuality there exists a frequent tendency in physics to introduce considerations of finality in this sense. Medicus, in his book *Die Freiheit des Willens und ihre Grenzen*, pp. 88ff., expressed and explained the facts and theories of atomic physics such that the atom searches its goal; he thus sees in contemporary physics a confirmation of Schelling's interpretation of nature. A book soon to appear in England by L. L. Whyte (*Archimedes or the Future of Physics*) attempts to link physics to biology by explaining physical laws as purposeful laws and to present physical processes as irreversible, just like biological processes.[11] Therefore Whyte rejects both the Maxwellian requirement that physical laws should not explicitly incorporate a temporal factor and Boltzmann's attempts to show that entropy processes are reversible; Whyte requires that the atom be viewed as an organism. In such endeavors this English natural scientist shows the influence of Whitehead. (Concerning Whitehead, see my article 'Der Physiker Whitehead', in *Die Kreatur*, 1928, pp. 356–363.)

Above we have attempted to allow possible validity to an interpretation of nature permitting a realm of freedom and to explain this interpretation as physically admissible. Here we must also, however, point out that such an interpretation, which does allow arbitrariness in nature, has the danger of permitting physics to regress into its womb by requiring teleological explanations – and such an interpretation does, of course, focus upon such explanations. One must remember that the basis of modern physics was laid down precisely in the battle against teleology.

"The modern concept of force – writes Cassirer in his introduction to Leibniz, *Hauptschriften zur Grundlegung der Philosophie* **2**, 7 – originates in the use of enticing analogies presented to us by events in life... Because the mathematical method of calculation teaches us to determine the effect completely and uniquely from the cause, such a method eliminates all foreign forces – which are not allowed to appear and actively control this realm as significant factors.... The mathematical concept of law eradicates (according to Kepler) the biological concept of form and withdraws from it every use in explaining natural phenomena." In this day and age, however, the feeling threatens to be lost that teleological ways of speaking should only be provisional ways of expression that require causal explanation in addition. (See Haering, *Philosophie der Naturwissenschaft*, p. 618.) Within physics, biological methods are always foreign. Whether physical explanation suffices for understanding nonliving nature, or whether physics will sometime be forced to give way to biology in its own domain (in the way that it was previously hoped the fields of biology, concerned with life itself, would, on the contrary, give way to physics and chemistry), or whether it will sometime be necessary to introduce biological principles to explain atomic processes, the facts and the facts alone must decide. In no way, however, should the methods be confused.

Hence it should never be concluded from some scientific supraphysical, or metaphysical, standpoint that time could be thought to be transcended that one may accept mutual dependence amongst all temporal phenomena. Such views have no place in physics. We must recognize that the given can be given meaning by different categorical presuppositions in different ways. But these different ways of giving meaning may not be confused with each other. Clarity of method is to be maintained for only in this way can the success of different sciences be guaranteed.

IV. PROBABILITY AND FREE WILL

34. In this concluding chapter we shall deal with the question concerning the repercussions upon the problem of free will brought about by replacing causality with probability. Since Kant, philosophy exhibits an unceasing battle against the mechanical view of the world. Philosophers, desiring to preserve the moral teaching entrusted to them, seek to evade the noose of mechanism. It is a desparate struggle of philosophy with physics.

At the present time, however, the leaf seems to be turning color. Physics itself gives up mechanism and speaks of contingency in its laws. What influence will this new turn have upon philosophy?

Before beginning I would like to caution against the belief that physics can in any way solve philosophical problems. There was a time when physical determinism was uncontested and the fear of philosophers unwarranted because philosophical considerations had already shown the limitations of this determinism. So all the less need, at this point, to tie extravagant hopes to physical contingency.

Naturally the new interpretation of physical laws, if it can be upheld, facilitates comprehension of the influence of nonphysical factors upon physical phenomena. In this respect the new turn in physics has sanctioned the conception due to Boutroux in his book *Die Kontingenz der Naturgesetze* (Geman tr., 1911). He proceeds – I cite N. Hartmann, *Ethik*, p. 621 – by giving a classification of regularities. But in doing this no law is absolute or free from exception. All laws determine only in part the structure of the layers of being to which they belong, leaving those layers partially undetermined and thus open to determination by other sources. All laws thus contain a certain contingency – they are not rigorously necessary. According to this interpretation the higher regularity certainly takes precedence over the lower – but only on so far as the latter is incomplete.

N. Hartmann, *loc. cit.*, opposes Boutroux: "The contingency of natural law is not demonstrable. We see no 'incompleteness' of validity in the entire range of recognizable natural regularity. Science repeatedly makes us aware that laws not continuously proven true are seen as false upon deeper investigation of the situation, or are shown to be false; i.e. they are certainly not real natural laws. As soon as the essence of the situation prevails successfully, complete necessity and universality result immediately." Hartmann's protest can no longer be supported and in this sense, therefore, it must be said that the new view of physics actually reveals new perspectives for understanding the influence of higher guiding forces upon inorganic phenomena. Medicus discussed such in his investigation *Über die Freiheit des Willens und ihre Grenzen* (1926), partly in relation to Weyl's views. However, here we must also be careful about drawing conclusions that are too hasty. In no way is room created for freedom, but only for a new type of causality. So long as one believed the electron path from *a* to *b* was prescribed according to rigorous physical lawfulness, there was no room for intervention of nonphysical forces. Now, if one accepts that the electron can make its way from *a* to either *b* or *c* or *d*, room is created for a new direction-giving force *K*, a force not conceivable in physical terms. This supervening force *K* would be able to reestablish necessity in the arbitrariness of *bcd*; for example, it could describe the path as *a* to *b* and thus reestablish the determinism given up by physics for all phenomena. Whether such direction-giving forces can be discovered remains to be decided – but in any case I do not see the door remaining open for freedom. Therefore I do not understand how e.g. the English physicist J. H. Jeans (*Nature* of February 27, 1926), could write that quantum theory opens up new possibilities for freedom.

The most one could say is that as long as physical determinism is conceived of as enclosed within itself, there remains no possibility for intervention by the human will into the physically determined mechanism; now, however, such a possibility is opened up because there is an area of freedom accorded by the law of probability. This is correct. But the 'freedom' with which we are concerned is only freedom of [human] action (*actus imperatus voluntatis*); such freedom thus already assumes the will, including its ability to determine direction. This freedom concerns only that which is realized by the will, not that which gives direction itself, and hence this freedom would concern not freedom of the will itself, but only freedom of the actualization of the will (see N. Hartmann, *Ethik*, p. 581). Concerning this type of freedom, however – concerning certain limits which will continue to persist – there has never been any doubt; there have always been attempts to explain this influence. The influence of the will upon the inorganic world, or generally the influence of nonphysical factors on the physical world, now only becomes easier to understand. But this has nothing to do with the freedom of these nonphysical factors of the will.

35. Freedom of the will itself could first be recognized at that moment when it could be shown there is "a positive lawfulness of the will along with lawfulness in nature; a determinant which itself is not included in causal world processes, but makes its appearance throught the will of men" (Hartmann, p. 590). On that point, however, i.e. whether there is such a new determinant not caused by a previous event, the law of probability, which should take the place of causal law in physics, says nothing. We are only told there is a gap in natural causality or, more precisely, in physical causality. The physicist gives us no word, however, as to whether this gap is filled by a higher type of natural causality (for example an organic causality) which is itself causally if not physically-causally determined, or whether the gap is filled by a causality of the will that is itself undetermined, or finally whether the gap simply remains unfilled. Further, we cannot

find any kind of arguments for one or another of these possibilities from the assertions of physicists.

The question of freedom is the question of the *actus elicitus voluntatis* of the volition itself and not of the action. This question can only be answered, however, by considering the general grounds which justify the universality of causal law, because accepting freedom is inconsistent with accepting complete predominance of causality.

(This is disputed by Hartmann, pp. 591, 622. He believes "freedom in its positive meaning" is not a minus to determinacy, but a plus. The causal nexus would have to allow this plus, because the law of the causal nexus does not guarantee that other determining elements could not intervene in a process. "If one cuts an ideal cross-section through the bundle of causal fibers, the determining elements in this slice certainly always yield total determination of all following stages in a process and clearly build, in this sense, a totality in themselves. But this totality is never absolutely closed; it does not resist the arrival of new determining elements – if there are such - and the process is not broken apart by such arrivals, but only diverted by them. That is precisely the characteristic of the causal nexus: it does not cancel or discontinue itself, but indeed lets itself be diverted. The further course of a process is thus different than it would have been without the new determinant; but nothing in the process influenced by the original causal factors is curtailed on that account; the new determinant becomes part of the diverted process in a manner as unrestrained as if it had not done any diverting at all." I myself cannot agree with this vindication of freedom. Each causal nexus certainly allows a new determinant to enter its ranks, but the question is: from whence this determinant? The causal law says that all that is produced has a cause, and therefore this determinant would also have to be caused.

It is not the arrival of a new determinant to supplement those already extant, but the causeless origin of the new determinant which conflicts strictly with causal law. If there is – as Lotze correctly formulated – an "unconditional beginning," are there events which are not effects [of a cause]? That is the question and here we have only the choice between general validity of the causal law and admissibility of a causeless phenomenon. One cannot make his peace with either of these viewpoints.)

36. Now, has the new turn in physics verified that causal law is invalid? It could not, because this question concerning the law of causality cannot be decided within physics. Physics can show that the law of causality is empty and inapplicable for certain of its fields, and that another, less demanding requirement could suffice for physics, which shows all the more that physics is not paramount concerning this question. Indeed, however, physics again brought into focus the epistemological character of causal law – and therein lies the great significance of the new turn in physics for philosophy. Physics again compels us to recognize that the law of causality is a postulate, and not more – a realization which is not new but which was easily lost with the all too blindly trusting belief in mechanical physics.

Anton Marty fostered, in his lectures on ethics, a strong resistance to the interpretation of the law of causality as a [presupposed] requirement. Logic could not put up with that, which would certainly be a noteworthy advance: to postulate something which one cannot prove. Whoever then requires that, and with what right? If reason is to prevail, then statements must be evident or proven. Science would be unable to build upon propositions having no other basis than the wish that they be so; if such were allowed, indeterminists would have easy play: they would simply say they do not acknowledge such a postulate.

These objections of Marty could only be upheld, however, if the postulates discussed here were arbitrary requirements which one could accept or reject according to his wishes. But not so. It is rather an ultimate requirement constituting the essence of that affair which we call science. This requirement should be heeded because only with its help does inquiry progress; it is not there because of its own self-will, but serves as an organ for constructing something else. (B. Kohn, *Untersuchungen über das Causalproblem.*) In this sense Planck calls the hypothesis of causality the "hypothesis of hypotheses." Certainly these ultimate transcendental axioms or requirements which we lay as a foundation of inquiry "are not astonishing Palladium, but earthly creations of our thought which we should mold daily as we use them" (Kohn, p. 127). And so it can happen that the development of physics replaces the requirement of causality with a more elastic requirement of probability. As we have seen, the new requirement is not arbitrary; rather, the new requirement originates in the essence of science and is adapted continually in scientific progress.

In this respect Brunschvicg is correct in calling the law of causality the imperative of intellectual activity.

Now, however, as concerns the problem of indeterminism, Marty was correct in so far as the imperative character of the law of causality makes it possible for indeterminists to maintain a sphere of freedom next to the sphere of necessary determinism and therefore to justify indeterminism without giving up meaningful determinism – this alone is the great advantage of this interpretation of causality.

37. So are both determinism and indeterminism valid and are we stuck with a double truth? In actuality anyone must come to this conclusion who, mistaking the significance of scientific inquiry, "transforms the practical imperative into a speculative truth" (Brunschvicg, p. 526) and installs as a proposition that which can only be installed as a requirement. We have repeatedly alluded to the fact that science is not a picture but a meaning, worked out with the help of certain 'intellectual symbols'. In these 'symbols' however "lies the moment of freedom" (Cassirer, *Jahrbuch der Phil.* III, p. 35).

How is this 'moment of freedom' to be understood? With the law of causality it is not a question of a proposition – which would have to be true or false – but of a tool. With the aid of this tool we establish a certain order within phenomena to orient ourselves to them. The order established from events in this way, however, is a product of abstraction. Whoever fails to recognize this methodological significance of categorical presuppositions for science and who accepts symbol for reality, then believing there would be something in itself like the world of physics, naturally runs into contradictions. But the reason for this lies, as Brunschvicg says (pp. 533ff.), in that he has reified science into an absolute that is foreign to the scientific character of knowledge. He has simply extrapolated that which his definition contradicts after the extrapolation. He has taken an abstraction, which was created out of the need for scientific orientation, for reality. (See in this regard the difference between realization and orientation in the philosophy of Martin Buber and also my article 'Begriff und Wirklichkeit, ein Beitrag zur Philosophie Martin Bubers und J. G. Fichtes', in *Juden*, 1928.)

This abstract character of science and especially of physics has not eluded the physicists of our time. I choose here only three weighty voices.

Whitehead (*Science and the Modern World*, p. 190) describes the abstract character of physics by the following two traits: (a) it conjectures the essence of the world only in regard to its outer references, in regard to the references of one thing on others; (b) it also conjectures these aspects only in so far as they are expressible in spatial-temporal terms. Indeed, the inner, psychic reality of the observer, for example the fact that he can see red, enters into the conjecture. But it is not these facts themselves that constitute the conjecture; only the dissimilarity of these facts with other observations by the observer enters into the conjecture. The inner life of the observer is only used to grasp the identity of physical facts, but does not itself enter into physics. Therefore physics has no 'intrinsic reality'. Dingler articulated in an exceedingly clear manner in his book *Die Grundlagen der Physik* the decisive roles conjecture, the pure synthesis as he calls it, plays in physics. "The system of mechanics remains completely preserved ... but ... no longer, as with Descartes and the later mechanists, are the smallest particles, which are integral to mechanical laws, the world; instead this entire mechanism is merely an explanation, a way of representing irrational reality. This does not amount to a world that is, as for Descartes, a single large machine; instead, only that aspect of the world which is provisionally created by us with the laws of mechanics can be represented in this way. This aspect of the world is always a finite aspect, however, and will remain finite ... our entire mechanics is only a way of representing the irrational world." "All that is conceptual is the work of men, flowing finally into pure synthesis and recognized as conjecture." From this a clear light shines on Dingler's exhaustion principle, which is very instructive for understanding methods in physics.

And now a third voice. Weyl says in his book *Raum, Zeit, Materie* (3rd ed., p. 263): "I believe physics deals only with that which could be designated in a precisely analogous way as the formal composition of reality. Its laws are violated just as infrequently in reality as there are truths that are disharmonious with logic. Those laws, however, have no bearing on the inner substance of this reality; the basis for reality is not to be captured in them."

Whitehead, Dingler, and Weyl are physicists whose positions concerning the teachings of recent physics (theory of relativity!) are far from each other, and hence their agreement concerning the abstract character of physics seems all the more important to me. The new turn in physics, with which we have been dealing in this essay, has again put in bright light this nature of the free, *a priori* synthesis in physics.

The presupposition of determinism – may it now be the rigorous law of causality, or may it be the looser law of probability – has lucidly shown itself to be a requirement we attribute to the world construed by us from physical appearances for the purpose of its construction. We certainly do not want to undervalue this construction and abstraction, but while we recognize the limits of them as the work of men and do not confuse the artful reality of physics with the immediately experienced reality, this reality discloses to us its true significance, the universally true significance of science and culture: to raise the intellect to consciousness of itself in its own works. "All progress in knowledge is only clarification of its basis in our own intellect."[12]

NOTES

* Originally printed and published by Friedrich Vieweg and Son, Inc., Braunschweig, 1929. Editor of this paper: Government Councillor, Professor, Doctor of Engineering Science, c.h., Karl Scheel, Berlin.

1 Concerning this see the explanations in Chapter IV.

2 See the excellent expositions of Brunschvicg, *L'expérience humaine et la causalité*, p. 533ff.

3 Also see Posch, 'Theorie der Zeitvorstellung' (*Viert. f. w. Phil.* **23**, 308); Bergmann, *Das phil. Werk B. Bolzanos*, p. 148; Reichenbach, 'Die Bewegungslehre bei Newton, Leibniz und Huyghens' (*Kantstudien* **29**, 421ff.).

4 See the analogous way of thinking as regards the measurement of time and, for this purpose, its foundation in uniform movement in Leibniz, *Nouv. Essais*, 2. Buch, No. 16: "Time is measured by means of uniform changes; but even if there were nothing uniform in nature time would still be determined. The reason is that if one knows the laws of the irregular movements, one can always reduce the same to conceivable uniform movements.... In this sense, then time is also the measure of movement, i.e. uniform movement is the measure of irregular movement." (Concerning this problem see Brunschvicg, *L'Expérience humaine et la causalité physique*, 1922, pp. 496ff. Concerning the necessity of a basic irreversible process, see *loc. cit.*, p. 503.)

5 Quoted by Wentscher, *Geschichte des Kausalproblems*. The objections raised against the law of causality brought forth by quantum theory do not yet come into play in our considerations because we are now dealing with the universal law of causality and only later will we explore how far this law can be 'loosened'.

6 The opposing interpretation of Planck (*Kausalität und Willensfreiheit*, pp. 33, 42). According to him it would be precisely the microscopic observer, not the macroscopic observer, who could grasp rigorous laws.

7 Nevertheless see the derivation of the law of causality by Brentano, (*Versuch über die Erkenntnis*, p. 121).

[8] Poisson, in his investigations concerning probability, bases his formulation of the law of large numbers upon causality in that he speaks of occurrences of the same kind, which depend upon invariable causes and upon such other causes which are erratically variable.'' E. von Hartmann also establishes probability on causality, *Kategorienlehre II*, p. 214 (Meiner edition).

[9] Thus Mises emphasizes (*Wahrscheinlichkeit, Statistik und Wahrheit*, pp. 90ff), that the behavior of reality corresponding to mathematical probability is not provable. "A statement which should express something concerning reality is mathematically derivable only if one establishes something to begin the derivation and only if one places at the summit of practical knowledge isolated premises, the so-called axioms.'' Similarly with Marbe (*Gleichförmigkeit der Welt* I, p. 272).

[10] For the incomprehensibility of the causal nexus, see N. Hartmann, *Metaphysik des Erkennens*, pp. 265, 312.

[11] Also see Bertalanffy, 'Über die Bedeutung der Umwälzungen in der Physik für die Biologie', *Biol. Zentralbl.* **47**, (1927), 653ff.

[12] Cassirer in the sense of Leibniz.

SYNTHESE LIBRARY

Monographs on Epistemology, Logic, Methodology,
Philosophy of Science, Sociology of Science and of Knowledge, and on the
Mathematical Methods of Social and Behavioral Sciences

ROBERT S. COHEN and MARX W. WARTOFSKY (eds.), *Boston Studies in the Philosophy of Science*. Volume XII: *Adolf Grünbaum: Philosophical Problems of Space and Time*. Second, enlarged edition. 1973, XXIII+884 pp. (Also in paperback.)

PATRICK SUPPES (ed.), *Space, Time and Geometry*. 1973, XI + 424 pp. (Also in paperback.)

ROLAND FRAÏSSÉ, *Course of Mathematical Logic*. Volume I: *Relation and Logical Formula*. 1973, XVI + 186 pp.

I. NIINILUOTO and R. TUOMELA, *Theoretical Concepts and Hypothetico-Inductive Inference*. 1973, X + 259 pp.

RADU J. BOGDAN and ILLKA NIINILUOTO (eds.), *Logic, Language, and Probability*. 1973, X + 323 pp.

GLENN PEARCE and PATRICK MAYNARD (eds.), *Conceptual Change*. XII + 282 pp.

M. BUNGE (ed.), *Exact Philosophy – Problems, Tools, and Goals*. 1973, X + 214 pp.

ROBERT S. COHEN and MARX W. WARTOFSKY (eds.), *Boston Studies in the Philosophy of Science*. Volume IX: *A. A. Zinov'ev: Foundations of the Logical Theory of Scientific Knowledge (Complex Logic)*. Revised and Enlarged English Edition with an Appendix by G. A. Smirnov, E. A. Sidorenka, A. M. Fedina, and L. A. Bobrova. 1973, XXII + 301 pp. (Also in paperback.)

K. J. J. HINTIKKA, J. M. E. MORAVCSIK, and P. SUPPES (eds.), *Approaches to Natural Language. Proceedings of the 1970 Stanford Workshop on Grammar and Semantics*. 1973, VIII + 526 pp. (Also in papberback.)

WILLARD C. HUMPHREYS, JR. (ed.), *Norwood Russell Hanson: Constellations and Conjectures*. 1973, X + 282 pp.

MARIO BUNGE, *Method, Model and Matter*. 1973, VII + 196 pp.

MARIO BUNGE, *Philosophy of Physics*. 1973, IX + 248 pp.

LADISLAV TONDL, *Boston Studies in the Philosophy of Science*. Volume X: *Scientific Procedures*. 1973, XIII + 268 pp. (Also in paperback.)

SÖREN STENLUND, *Combinators, λ-Terms and Proof Theory*. 1972, 184 pp.

DONALD DAVIDSON and GILBERT HARMAN (eds.), *Semantics of Natural Language*. 1972, X + 769 pp. (Also in paperback.)

MARTIN STRAUSS, *Modern Physics and Its Philosophy. Selected papers in the Logic, History, and Philosophy of Science*. 1972, X + 297 pp.

‡STEPHEN TOULMIN and HARRY WOOLF (eds.), *Norwood Russell Hanson: What I Do Not Believe, and Other Essays*. 1971, XII + 390 pp.

‡ROBERT S. COHEN and MARX W. WARTOFSKY (eds.), *Boston Studies in the Philosophy of Science*. Volume VIII: *PSA 1970. In Memory of Rudolf Carnap* (ed. by Roger C. Buck and Robert S. Cohen). 1971, LXVI + 615 pp. (Also in paperback.)

‡YEHOSUA BAR-HILLEL (ed.), *Pragmatics of Natural Languages*. 1971, VII + 231 pp.

‡ROBERT S. COHEN and MARX W. WARTOFSKY (eds.), *Boston Studies in the Philosophy of Science*. Volume VII: *Milič Čapek: Bergson and Modern Physics*. 1971, XV + 414 pp.

‡CARL R. KORDIG, *The Justification of Scientific Change*. 1971, XIV + 119 pp.

‡JOSEPH D. SNEED, *The Logical Structure of Mathematical Physics*. 1971, XV + 311 pp.

‡JEAN-LOUIS KRIVINE, *Introduction to Axiomatic Set Theory*. 1971, VII + 98 pp.

‡RISTO HILPINEN (ed.), *Deontic Logic: Introductory and Systematic Readings*. 1971, VII + 182 pp.

‡EVERT W. BETH, *Aspects of Modern Logic*. 1970, XI + 176 pp.

‡PAUL WEINGARTNER and GERHARD ZECHA (eds.), *Induction, Physics, and Ethics, Proceedings and Discussions of the 1968 Salzburg Colloquium in the Philosophy of Science*. 1970, X + 382 pp.

‡ROLF A. EBERLE, *Nominalistic Systems*. 1970, IX + 217 pp.

‡JAAKKO HINTIKKA and PATRICK SUPPES, *Information and Inference*. 1970, X + 336 pp.

‡KAREL LAMBERT, *Philosophical Problems in Logic. Some Recent Developments*. 1970, VII + 176 pp.

‡P. V. TAVANEC (ed.), *Problems of the Logic of Scientific Knowledge*. 1969, XII + 429 pp.

‡ROBERT S. COHEN and RAYMOND J. SEEGER (eds.), *Boston Studies in the Philosophy of Science*. Volume VI: *Ernst Mach: Physicist and Philosopher*. 1970, VIII + 295 pp.

‡MARSHALL SWAIN (ed.), *Induction, Acceptance, and Rational Belief*. 1970, VII + 232 pp.

‡NICHOLAS RESCHNER *et al.* (eds.), *Essays in Honor of Carl G. Hempel. A Tribute on the Occasion of his Sixty-Fifth Birthday*. 1969, VII + 272 pp.

‡PATRICK SUPPES, *Studies in the Methodology and Foundations of Science. Selected Papers from 1911 to 1969*. 1969, XII + 473 pp.

‡JAAKKO HINTIKKA, *Models for Modalities. Selected Essays*. 1969, IX + 220 pp.

‡D. DAVIDSON and J. HINTIKKA (eds.), *Words and Objections: Essays on the Work of W. V. Quine*. 1969, VIII + 366 pp.

‡J. W. DAVIS, D. J. HOCKNEY and W. K. WILSON (eds.), *Philosophical Logic*. 1969, VIII + 277 pp.

‡ROBERT S. COHEN and MARX W. WARTOFSKY (eds.), *Boston Studies in the Philosophy of Science*. Volume V: *Proceedings of the Boston Colloquium for the Philosophy of Science 1966/1968*. 1969, VIII + 482 pp.

‡ROBERT S. COHEN and MARX W. WARTOFSKY (eds.), *Boston Studies in the Philosophy of Science*. Volume IV: *Proceedings of the Boston Colloquium for the Philosophy of Science 1966/1968*. 1969, VIII + 537 pp.

‡NICHOLAS RESCHER, *Topics in Philosophical Logic*. 1968, XIV + 347 pp.

‡GÜNTHER PATZIG, *Aristotle's Theory of the Syllogism. A Logical-Philosophical Study of Book A of the Prior Analytics*. 1968, XVII + 215 pp.

‡C. D. BROAD, *Induction, Probability, and Causation. Selected Papers*. 1968, XI + 296 pp.

‡ ROBERT S. COHEN and MARX W. WARTOFSKY (eds.), *Boston Studies in the Philosophy of Science*. Volume III: *Proceedings of the Boston Colloquium for the Philosophy of Science 1964/1966*. 1967, XLIX + 489 pp.

‡GUIDO KÜNG, *Ontology and the Logistic Analysis of Language. An Enquiry into the Contemporary Views on Universals*. 1967, XI + 210 pp.

*Evert W. Beth and Jean Piaget, *Mathematical Epistemology and Psychology*. 1966, XXII + 326 pp.

*Evert W. Beth, *Mathematical Thought. An Introduction to the Philosophy of Mathematics*. 1965, XII + 208 pp.

‡Paul Lorenzen, *Formal Logic*. 1965, VIII + 123 pp.

‡Georges Gurvitch, *The Spectrum of Social Time*. 1964, XXVI + 152 pp.

‡A. A. Zinov'ev, *Philosophical Problems of Many-Valued Logic*. 1963, XIV + 155 pp.

‡Marx W. Wartofsky (ed.), *Boston Studies in the Philosophy of Science*. Volume I: *Proceedings of the Boston Colloquium for the Philosophy of Science 1961/1962*. 1963, VIII + 212 pp.

‡B. H. Kazemier and D. Vuysje (eds.), *Logic and Language. Studies dedicated to Professor Rudolf Carnap on the Occasion of his Seventieth Birthday*. 1962, VI + 256 pp.

*Evert W. Beth, *Formal Methods. An Introduction to Symbolic Logic and to the Study of Effective Operations in Arithmetic and Logic*. 1962, XIV + 170 pp.

*Hans Freudenthal (ed.), *The Concept and the Role of the Model in Mathematics and Natural and Social Sciences. Proceedings of a Colloquium held at Utrecht, The Netherlands, January 1960*. 1961, VI + 194 pp.

‡P. L. Guiraud, *Problèmes et méthodes de la statistique linguistique*. 1960, VI + 146 pp.

*J. M. Bocheński, *A Precis of Mathematical Logic*. 1959, X + 100 pp.

SYNTHESE HISTORICAL LIBRARY

Texts and Studies
in the History of Logic and Philosophy

NABIL SHEHABY, *The Propositional Logic of Avicenna.* A Translation from *al-Shifā': al-Qiyās.* 1973, XIII + 296 pp.

JAN BERG (ed.), *Bolzano – Theory of Science.* 1973, XVI + 388 pp.

J. M. E. MORAVCSIK (ed.), *Patterns in Plato's Thought.* 1973, VIII + 212 pp.

LEWIS WHITE BECK (ed.), *Proceedings of the Third International Kant Congress.* 1972, XI + 718 pp.

‡KARL WOLF and PAUL WEINGARTNER (eds.), *Ernst Mally: Logische Schriften.* 1971, X + 340 pp.

‡LEROY E. LOEMKER (ed.), *Gottfried Wilhelm Leibnitz: Philosophical Papers and Letters.* A Selection Translated and Edited, with an Introduction. 1969, XII + 736 pp.

‡M. T. BEONIO-BROCCHIERI FUMAGALLI, *The Logic of Abelard.* Translated from the Italian. 1969, IX + 101 pp.

Sole Distributors in the U.S.A. and Canada:
*GORDON & BREACH, INC., 440 Park Avenue South, New York, N.Y. 10016
‡HUMANITIES PRESS, INC., 303 Park Avenue South, New York, N.Y. 10010